TOPICS IN THE THEORY OF
SCHRÖDINGER OPERATORS

editors

Huzihiro Araki
Kyoto University, Japan

Hiroshi Ezawa
Gakushuin University, Japan

TOPICS IN THE THEORY OF SCHRÖDINGER OPERATORS

World Scientific

NEW JERSEY • LONDON • SINGAPORE • BEIJING • SHANGHAI • HONG KONG • TAIPEI • CHENNAI

Published by

World Scientific Publishing Co. Pte. Ltd.
5 Toh Tuck Link, Singapore 596224
USA office: Suite 202, 1060 Main Street, River Edge, NJ 07661
UK office: 57 Shelton Street, Covent Garden, London WC2H 9HE

British Library Cataloguing-in-Publication Data
A catalogue record for this book is available from the British Library.

TOPICS IN THE THEORY OF SCHRÖDINGER OPERATORS

ISBN 981-238-797-8

Printed in Singapore by World Scientific Printers (S) Pte Ltd

PREFACE

The aim of this book is to give reviews accessible to graduate students and non-expert researchers on some of the recent topics in the theory of Schroedinger operators.

The article of Yajima is a survery on the theory of the Schrödinger equation when the potential depend on the time periodically, on which Yajima is a leading expert. This article is recommended not only for young students who want to enter into this field but also to researchers who want to know recent results on this subject.

The article of Martinez, Nakamura and Sordoni presents a good and practical introduction to the so-called FBI transformation (also called coherent state expansion in physics), explaining its application to the semi-classical limit of the S-matrix for Schrödinger operators. This article also serves as an introduction to mathematical analysis based on pseudo-differential (or micro-local) calculus, which is a recent trend.

The article of Isozaki is an overview of inverse spectral and scattering problems. It deals with inverse boundary value problems in 1-dimension, isospectral problems from differential geometry, and multi-dimensional inverse scattering problems for Schrödinger operators. In paticular it contains a detailed exposition of Faddeev's theory. It is a readable survey from a broad point of view with a rich reference to relevant literature and is recommended as a good introduction to this subject.

The article of Hiroshima treats the Pauli–Fierz model where a non-relativistic particle interacts with the radiation fields, the latter being described on a Fock space. In particular, the ground state is studied in detail, including its existence proof and information about its charge density. The mathematical methods used in the analysis are explained in detail starting from the beginning, a special feature being the method of functional integral.

A short introduction on elementary aspects of the theory of Schrödinger operators and relevant terminologies are given with a list of textbook references.

The four longer articles are based on lectures given at the Summer School "Mathematical Physics 2000 — Mathematical aspects of quantum mechanics". It is one in the series of annual summer schools organized by the editors of this book since 1987 and held at Gakushuin University, for the purpose of providing easily accessible lectures on various topics on Mathematical Physics for graduate students as well as for researchers (non-experts in the field chosen for that year) wishing to widen their scopes. The particular summer school in 2000 was organized in memory of late Professor Tosio Kato, one of the founders of the subject.

In conclusion, we express our gratitude to all four authors for their excellent lecture notes.

August, 2003

Huzihiro Araki
(Professor Emeritus of Kyoto University)
Hiroshi Ezawa
(Professor Emeritus of Gakushuin University)

CONTENTS

CHAPTER 1

INTRODUCTION

Huzihiro Araki

Research Institute for Mathematical Sciences, Kyoto University
Sakyoku, Kyoto 606-8502, Japan

1. Schrödinger Operator

The arena for quantum mechanics of an N particle system in a d-dimensional Euclidean space is the Hilbert space

$$\mathcal{H} = L^2((\mathbb{R}^d)^N), \tag{1.1}$$

whose unit vectors (i.e. vectors of length 1) $\Psi(x_1, ..., x_N)$ (called wave functions) describe states of the system, $|\Psi(x_1, ..., x_N)|^2$ being interpreted as the probability density (with respect to the Lebesgue measure) for the N particles at positions $x_1, ..., x_N$ in $\mathbb{R}^d$.

The dynamics (i.e. the change with time t) of the system is described by a continuous one-parameter group of unitaries

$$U(t) = e^{-itH}, \tag{1.2}$$

by which the state vector $\Psi \in \mathcal{H}$ change to $\Psi_t = U(t)\Psi$ after time t. A simple example of physical interest (especially at the beginning of quantum mechanics) for its generator H (called the Hamiltonian) in the case of a one-particle system ($N = 1$) is the (one-particle) Schrödinger operator

$$H = H_0 + V, \tag{1.3}$$

where

$$H_0 = -(2m)^{-1}\Delta = (2m)^{-1}\sum_{i=1}^{d} -\frac{\partial^2}{\partial x_i^2} \quad , \tag{1.4}$$

(m is a constant representing the mass of the particle) and the potential V is a real-valued function $V(x)$ of $x \in \mathbb{R}^d$ acting on $\Psi \in \mathcal{H}$ as a multiplication operator:

$$(V\Psi)(x) = V(x)\Psi(x).$$

(We always assume that $V(x)$ is Lebesgue measurable.)

A basic mathematical problem at the beginning of quantum mechanics was to show that the operator H given by (1.3) for physically relevant potential V (or the closure of H) is self-adjoint in order that (1.2) is a unitary group. The free Hamiltonian H_0 (by itself describing the free motion of the particle without any influence of external forces) is easily shown to be self-adjoint (on an appropriate domain $D(H_0)$) because it is a multiplication operator after the Fourier transform of wave functions. The basic result applicable to atomic Hamiltonians is the following Theorem of F. Rellich and T. Kato.

Theorem 1.1: If the domain of V includes $D(H_0)$ and

$$\| V\Psi \| \leqq a \| \Psi \| + b \| H_0\Psi \| \tag{1.5}$$

for some $a > 0$ and $1 > b \geq 0$, then $H_0 + V$ is selfadjoint on $D(H_0)$.

For physically interesting case of $d = 3$, the condition of this Theorem is shown to be satisfied if

$$V = V_1 + V_2,\ V_1 \in L^\infty(\mathbb{R}^3),\ V_2 \in L^2(\mathbb{R}^3). \tag{1.6}$$

For example, the Coulomb potential $V(x) = \alpha/|x|$ satisfies (1.6).

A nonzero solution $\Psi_E \in \mathcal{H}$ of the eigenvalue problem

$$H\Psi_E = E\Psi_E \tag{1.7}$$

gives a point spectrum of H, with E being the energy level of the particle which is trapped in a finite region of $\mathbb{R}^d$. Such a state is called a bound state. An instability of bound state under perturbation is treated in Section 7 of the article of Yajima.

A bound state with the lowest eigenvalue E_0 is called the ground state and all other bound states (i.e. those with eigenvalue $\neq E_0$) are called excited states. The lowest eigenvalue E_0 is normally at the lowest end of the spectrum of the Schrödinger operator H.

In actual atoms, there is an interaction between the atomic electron(s) and electromagnetic fields which is neglected in the Schrödinger operator (1.3). This causes the electron in an excited state to fall down to bound states with lower (energy) eigenvalues, emitting the energy difference as a photon (i.e. electromagnetic radiation viewed as a particle). After a few (possibly one) such steps of photon emissions, the electron will settle down to the ground state. (Mathematically, the excited states will then be no longer eigenstates.) Conversely, when the atom is radiated by an electromagnetic field (photons), an atomic electron may move up to an excited states (temporarily) by absorbing a photon, the energy difference of the states of the electron after and before the process being exactly the photon energy. Thus energy spectum of the emitted and absorbed photons are explained by the difference of eigenvalues of the atomic electron Schrödinger operator, which gave rise to a great success of quantum mechanics at its beginning.

The description of (free) electromagnetic field is made in a Fock space, which is explained in the article of Hiroshima, where the main result is the existence of the ground state for the particle under interaction with a quantum field. The ground state energy is at the lowest end of the continuous spectrum due to the massless photons, and to show the existence of the ground state is a nontrivial problem then.

2. Scattering Theory

We continue to consider the Schrödinger operator (1.3). Apart from the bound states, in which the particle is trapped in a finite region, there are states where the particle comes in from an infinite distance, is scattered by the potential V and then goes out to an infinite distance again. In such a situation, the particle will be located at a distant region at large positive and negative times and will behave approximately like a "free" particle with its dynamics described by

$$U_0(t) = e^{-itH_0} \tag{2.1}$$

if the potential $V(x)$ decreases sufficiently rapidly to 0 as $|x| \to \infty$. If this idea is correct, the unit vectors Ψ_+ and Ψ_- in $\mathcal{H}$, which behave approximately like a free particle state φ in $\mathcal{H}$ for $t \to \infty$ and for $t \to -\infty$, should satisfy

$$\lim_{t\to\infty} \|e^{-itH}\Psi_+ - e^{-itH_0}\varphi\| = \lim_{t\to\infty} \|\Psi_+ - e^{itH}e^{-itH_0}\varphi\| = 0, \tag{2.2}$$

$$\lim_{t\to\infty} \|e^{-itH}\Psi_- - e^{-itH_0}\varphi\| = \lim_{t\to-\infty} \|\Psi_- - e^{itH}e^{-itH_0}\varphi\| = 0. \tag{2.3}$$

Such vectors $\Psi_\pm$ are given by

$$\Psi_\pm = U_\pm(H, H_0)\varphi \tag{2.4}$$

if the following limits exist

$$U_{\pm}(H, H_0) = \lim_{t \to \pm\infty} e^{itH} e^{-itH_0}. \tag{2.5}$$

In fact, the following theorem is known for the Schrödinger operator.

Theorem 2.1: Let $N = 1$, $d = 3$, H be given by (1.3) and V satisfy (1.6) and

$$\int_{\mathbb{R}^3} |V(x)|^2 (1 + |x|)^{-1+\epsilon} dx < \infty \tag{2.6}$$

for some $\epsilon > 0$. Then the limits (2.5) exist and define isometric operators $U_{\pm}(H, H_0)$, which intertwine H and H_0 in the following sense:

$$e^{-itH} U_{\pm}(H, H_0) = U_{\pm}(H, H_0) e^{-itH_0}. \tag{2.7}$$

The isometric property of $U_{\pm}$ and the property (2.7) are immediate consequences of (2.5) while the completeness property was not easy to prove at the beginning stage of the scattering theory. The operators $U_{\pm}$ are called wave operators.

The vector

$$\Psi^{in}(\varphi) = U_{-}(H, H_0)\varphi \tag{2.8}$$

is a state vector whose behavior at infinite past is the same as the free motion of φ and sometimes called "in state". Likewise,

$$\Psi^{out}(\varphi) = U_{+}(H, H_0)\varphi \tag{2.9}$$

is a state vector whose behavior at infinite future is the same as the free motion of φ, sometimes called "out state". The inner product (we follow mathematics convention that (Ψ_1, Ψ_2) is linear in Ψ_1) between the in and out states given by

$$(\Psi^{in}(\varphi_1), \Psi^{out}(\varphi_2)) = (S\varphi_1, \varphi_2), \tag{2.10}$$

$$S = U_{+}(H, H_0)^* U_{-}(H, H_0), \tag{2.11}$$

is the probability amplitude ($|(S\varphi_1, \varphi_2)|^2$ being the probability) for an incoming particle which behaved like the free motion of φ_1 at infinite past, shows up after the scattering by the potential as an out-going particle describable by the free motion of φ_2. The operator S is called scattering operator.

In terms of the Fourier transform $\tilde{\varphi}(k)$ of $\varphi(x)$, the operator S can be described by a distribution $S(k_1, k_2)$ defined by

$$(S\varphi_1, \varphi_2) = \int \tilde{\varphi}_1(k_1)\tilde{\varphi}_2(k_2)^* S(k_1, k_2) d^3k_1 d^3k_2. \tag{2.12}$$

Due to (2.7), S commutes with e^{-itH_0} and hence $S(k_1, k_2)$ vanishes outside $\{(k_1, k_2); |k_1| = |k_2|\}$. (Any scattering in which the energy $(2m)^{-1}|k|^2$ is conserved is called an elastic scattering.) The observable quantities obtained by measurement on scattered particles such as the total cross section[a] and differential cross section[b] are determined by $S(k_1, k_2)$ through well-known formulae. $S(k_1, k_2)$ is sometimes called the S-matrix.

The operator S is unitary if and only if the ranges of $U_+(H, H_0)$ and $U_-(H, H_0)$ coincide. Due to (2.7), the spectrum of H on each of them is absolutely continuous (because H_0 has purely absolutely continuous spectrum). In particular, if each of these ranges contains exactly all absolutely continuous spectrum of H, then S is unitary. The unitarity of S is physically desired because all incoming particles should come out (either scattered or unscattered), unless there is some physical reason to the contrary.

If the ranges of U_+ and U_- coincide with the orthogonal complement of all bound state vectors, then all vectors in $\mathcal{H}$ obtains a physical interpretation as bound states and scattering states. The wave operators are said to be complete if this is the case. Then the self-adjoint operator H cannot have any singular continuous spectrum.

Theorem 2.2: Let $N = 1$, $d = 3$, H be given by (1.3) and V satisfies

$$|V(x)| \leq c(1 + |x|)^{-1-\epsilon} \tag{2.13}$$

for some constant $c > 0$ and $\epsilon > 0$, then the completeness follows.

If the potential satisfies the conditions (2.13) or (2.6), then it is called a short range potential. In the case of Coulomb potential $V(x) = \alpha/|x|$, these conditions are not satisfied (such potentials are called long-range potentials) and the asymptotic behavior of $e^{-itH}\Psi$ for large times cannot be approximated by a free motion $e^{-itH_0}\varphi$ for any φ. In this case of Coulomb scattering, a necessary modification of free motion for the description of asymptotic behavior at large times is known. The (modified) wave operators can be defined and scattering theory can be formulated in a similar fashion as above.

---

[a]the total number (probabilitywise) of scattered particles out of a uniform flow of incoming particles with a fixed momentum k, converted to the area of the incoming flow of that number of particles and interpreted as the effective target area perpendicular to the direction of k

[b]the portion of the scattered particles coming out into a fixed direction per unit solid angle

When we deal with N particles interacting with each other through two-body forces (but without external forces), we will typically deal with the Hamiltonian operator of the following form:

$$H = \sum_{i=1}^{N} -(2m_i)^{-1}\Delta_i + \sum_{i<j} V(x_i - x_j), \tag{2.14}$$

where each m_i is a constant (the mass of the i-th particle) and Δ_i is the Laplacian for the coordinate variables $x_i \in \mathbb{R}^3$ of the i-th particle.

For $N = 2$, we arrive at the Hamiltonian (1.3) after separating out the (free) motion of the center of mass. For a general N, the subsystem of two particles, for example, may have a bound state (i.e. a point spectrum after separating out the free motion of the center of mass). In that case, such a bound state may come out of an N-body scattering as some of the outgoing particle-like entities. Such a bound state can also be some of the incoming particle-like entities in a N-body scattering. Therefore, we are forced to include the free motion of the center of mass of a bound state in the asymptotic description of the scattering state. Thus the wave operators $U_\pm$ are to be taken as an isometric operator from the asymptotic state space $\mathcal{H}_{asymp}$ to $\mathcal{H}$, the two Hilbert spaces $\mathcal{H}_{asymp}$ and $\mathcal{H}$ being different in general.

After a suitable modification of the formulation hinted above, completeness (i.e. the interpretation of all vectors in $\mathcal{H}$ as scattering states of particles including bound states of subsystems, corresponding to the absolutely continuous spectrum of H, as well as N-body bound states, corresponding to the point spectrum of H) for a reasonable class of the potential $V(x)$ has been shown.

In the article of Martinez, Nakamura and Sordoni, the wave operators are shown to exist and to be complete. Then an exponential decay of multichannel tunneling terms of the S-matrix in the semiclassical limit is shown.

In the article of Yajima, the scattering theory is treated for a periodically time-dependent potentials and the large time asymptotics and dispersive property of scattering solutions are investigated.

In physics, it often happens that one would like to determine the potential from the observed data of scattering. In the situation discussed above, the relevant mathematical problem then is the determination of the potential $V(x)$ from the scattering operator. This is called the inverse problem of the potential scattering. Such inverse problems are reviewed in the article of Isozaki.

3. References

Standard textbooks for mathematically-minded readers are:

(1) T. Kato, Perturbation Theory for Linear Operators (Springer Verlag, 1966).
(2) S. T. Kuroda, An Introduction to Scattering Theory (Aarhus University, 1978).
(3) M. Reed and B. Simon, Methods of Modern Mathematical Physics, Vol. III *Scattering Theory in Quantum Mechanics* (Academic Press, 1979).

Topics after (3) are collected in:

(4) H. L. Cycon, R. G. Froese, W. Kirsch and B. Simon, Schrödinger Operators with Application to Quantum Mechanics and Global Geometry (Springer Verlag, 1982).

The following text books may have somewhat more physical color:

(5) R. G. Newton, Scattering Theory of Waves and Particles (McGraw-Hill, 1966)
(6) W. O. Amrein, J. M. Jauch and K. B. Sinha, Scattering Theory in Quantum Mechanics (Benjamin, 1977)

About completeness for many-body scattering theory, the following is probably the only text book available; it uses the time dependent method.

(7) J.Derezinski and C. Gérard, Scattering Tehrory of Classical and Quantum N-Particle Systems (Springer Verlag, 1997).

Some other text books are:

(8) D. B. Pearson, Quantum Spectral and Scattering Theory (Academic Press, 1988).
(9) D. R. Yafaev, *Mathematical Scattering Theory*, Translation of Mathematical Monograph Vol. **105**, AMS, 1992.

The latter uses the stationary method.

The following is a paper often cited:

(10) S. Agmon, *Spectral Properties of Schrödinger Operators and Scattering Theory*, Ann. Scoula Norm. Sup. Pisa Cl. Sci.(4) **2** (1975), 151-218.

On semiclassical limit of Schrödinger operators, there are the following text books:

(11) M. Dimassi and J. Sjöstrad, *Spectral Asymptotics in the Semiclassical Limit*, London Math. Soc. Lecture Notes Series **268** (1999)
(12) A. Martinez, An Introduction to Semiclassical and Microlocal Analysis, UTX series (Springer Verlag, 2002).

Representative text books on random Schrödinger operators are:

(13) R. Carmona and J. Lacroix, Spectral Theory of Random Schrödinger Operators (Birkhäuser, 1990).
(14) L.Pastur and A. Figotin, Spectra of Random and Almost-Periodic Operators (Springer Verlag, 1992).
(15) P. Stollmann, Caught by Disorder: Bound States in Random Media (Birkhüser, Boston, 2001).

CHAPTER 2

TIME-PERIODIC SCHRÖDINGER EQUATIONS

Kenji Yajima

Department of Mathematics, Gakushuin University
1-5-1 Mejiro, Toshima-ku, Tokyo 171-8588, Japan
E-mail: kenji.yajima@gakushuin.ac.jp

1. Introduction

In this note we review some topics on the time dependent Schrödinger equation

$$i\partial_t u = \left(-\frac{1}{2}\Delta + V(t,x)\right)u \tag{1.1}$$

with the potential $V(t,x)$ which depend periodically on time t, $V(t+2\pi,x) = V(t,x)$, and which satisfies $V(t,x) \to 0$ as $|x| \to \infty$. We always assume that $V(t,x)$ is a real valued measurable function of (t,x). Equation (1.1) describes the dynamics of quantum particles subject to the external forces given by the potential $V(t,x)$, typical examples being the dynamics of the hydrogen atom placed in a linearly polarized monochromatic electric field or a light particle in the restricted three-body problem in which two other heavy particles are set on prescribed periodic orbits. We mainly discuss the large time behavior of solutions $u(t,x)$; however, we also discuss some other topics which we think are interesting, and the following is the table of contents of the note:

(1) Existence and uniqueness of the propagator
(2) Scattering theory
(3) Large time asymptotics of scattering solutions
(4) Dispersive property of scattering solutions
(5) Instability of bound states

Each topic will be presented in a separate section with a brief introduction. As the space is limited, we omit proofs of the results which can be found in easily accessible literature; otherwise, we give an almost complete proof.

We use the following notation and conventions throughout the note:

- $\mathbf{R}$, $\mathbf{Z}$, $\mathbf{N}$ and $\mathbf{C}$ are the set of real numbers, the integers, the natural numbers and complex numbers. $\overline{\mathbf{N}} = \mathbf{N} \cup \{0\}$.
- $\Re z$ and $\Im z$ are the real and imaginary parts of $z \in \mathbf{C}$.
- We consider (1.1) on $\mathbf{R}^d$, but, we sometimes work only on $d = 3$.
- We write $\langle x \rangle = (1 + |x|^2)^{1/2}$.
- For $1 \le p \le \infty$, $L^p(\mathbf{R}^d)$ is the Lebesgues space. $\mathcal{H} = L^2(\mathbf{R}^d)$.
- For $\delta \in \mathbf{R}$, $\mathcal{H}_\delta = L^2(\mathbf{R}^d, \langle x \rangle^{2\delta} dx)$ is the weighted L^2-space.
- For $s \in \mathbf{R}$, $H^s(\mathbf{R}^d)$ is the Sobolev space of order s.
- For a Banach space $\mathcal{X}$ and $k \in \overline{\mathbf{N}}$, $C^k(I, \mathcal{X})$ is the space of $\mathcal{X}$-valued C^k functions on the interval I and $C(I, \mathcal{H}) = C^0(I, \mathcal{H})$. $L^p(I, \mathcal{X})$ is the space of $\mathcal{X}$-valued L^p functions.
- For an operator T, $D(T)$ stands for its domain and Image T is its image. For a closed operator T in a Banach space $\sigma(T)$ and $\rho(T)$ are respectively the spectrum and the resolvent set of T.
- For a self-adjoint or unitary operator T on a Hilbert space, $\sigma_p(T)$, $\sigma_c(T)$ and $\sigma_{ac}(T)$ are the point, the continuous and the absolutely continuous spectrum of T; $E_T(\Delta)$ for Borel set Δ is the spectral projection. $\mathcal{H}_p(T)$ and $\mathcal{H}_c(T)$ are respectively the point and the continuous spectral subspaces of $\mathcal{H}$ for T; and the orthogonal projections to respective subspaces are denoted by $P_p(T)$ and $P_c(T)$.
- For Banach spaces $\mathcal{X}$ and $\mathcal{Y}$, $B(\mathcal{X}, \mathcal{Y})$ (respectively $B_\infty(\mathcal{X}, \mathcal{Y})$) is the space of bounded (resp. compact) operators from $\mathcal{X}$ to $\mathcal{Y}$, $B(\mathcal{X}) = B(\mathcal{X}, \mathcal{X})$ (resp. $B_\infty(\mathcal{X}) = B_\infty(\mathcal{X}, \mathcal{X})$).

2. Existence and Uniqueness of Propagator

We start by discussing the very basic question, viz. the existence and the uniqueness of the solution to (1.1) which satisfies the initial condition

$$u(s, x) = u_0(x). \tag{2.1}$$

We consider (1.1) in the sense of distributions and, we say $u(t,x)$ *satisfies* (1.1) *when* $V(t,x)u(t,x) \in L^1_{\text{loc}}(\mathbf{R}^{d+1})$ *and it satisfies*

$$\int u(t,x) \left(i\partial_t \phi - \frac{1}{2}\Delta\phi + V(t,x)\phi(t,x) \right) dxdt = 0 \tag{2.2}$$

for all $\phi \in C_0^\infty(\mathbf{R}^1 \times \mathbf{R}^d)$. Since the initial condition (2.1) is given on the characteristic hyperplane $t = s$ for (1.1), the initial value problem (IVP for short) has in general an infinite number of solutions (see Ref. 15). We define solutions appropriate for our purpose as follows:

Definition 2.1: A function $u(t,x)$ is said to be a solution of (1.1) if $u \in C(\mathbf{R}, \mathcal{H})$, $\|u(t,\cdot)\|_{\mathcal{H}}$ is independent of $t \in \mathbf{R}$ and it satisfies (1.1).

The condition that $\|u(t,\cdot)\|_{\mathcal{H}}$ is independent of t is required in accordance with the law of the conservation of probability of quantum mechanics. The following lemma should be obvious.

Lemma 2.2: *Suppose that for any* $s \in \mathbf{R}$ *and* $u_0 \in \mathcal{H}$, *the IVP* (1.1) *with the intial condition* (2.1) *has a unique solution* u. *For* $t, s \in \mathbf{R}$, *define the operator* $U(t,s)$ *in* $\mathcal{H}$ *by* $U(t,s)u_0 = u(t,\cdot)$. *Then,* $\{U(t,s)\}$ *is a two parameter family of unitary operators in* $\mathcal{H}$ *which is strongly continuous with respect to* (t,s). *It satisfies the semigroup property: For all* $t, s, r \in \mathbf{R}$,

$$U(t,s)U(s,r) = U(t,r), \quad U(t,t) = \text{identity operator}.$$

We call $\{U(t,s) : t, s \in \mathbf{R}\}$ the **propagator** for (1.1) and we say that the IVP for (1.1) is **wellposed** if (1.1) with (2.1) has a unique solution for any $s \in \mathbf{R}$ and $u_0 \in \mathcal{H}$. In this section, we discuss the wellposedness of the IVP.

When $V = 0$, the solution of (1.1) with $u(s) = u_0 \in \mathcal{H}$ may be obtained explicitly via Fourier transform: [48]

$$U(t,s)u_0 = \frac{e^{\mp \frac{i\pi d}{4}}}{(2\pi|t-s|)^{\frac{d}{2}}} \int e^{\frac{i(x-y)^2}{2(t-s)}} u_0(y)dy, \quad \pm(t-s) > 0. \tag{2.3}$$

We denote this propagator by $U_0(t,s)$ and call the free propagator. As $U_0(t-s)$ depends only on $t-s$, we sometimes write $U_0(t,s) = U_0(t-s)$ and call $U_0(t)$ also the free propagator.

2.1. *Autonomous case, self-adjointness*

We assume $V(t,x) = V(x)$ is independent of t in this subsection. We first show that the wellposedness of the IVP for (1.1) is equivalent to the essentially self-adjointness of the differential operator $L = -(1/2)\Delta + V$.

Suppose that the IVP for (1.1) is well posed. Then, the propagator $U(t,s)$ depends on $t-s$ only and, if we define $U(t)$ by $U(t) = U(t,0)$, $\{U(t)\}$ is a strongly continuous family of unitary operators in $\mathcal{H}$ which satisfies $U(t)U(s) = U(t+s)$, viz. $\{U(t)\}$ is a strongly continuous unitary group. We recall the celebrated Stone theorem (see Ref. 47):

Theorem 2.3: The family of operators $\{U(t) : t \in \mathbf{R}\}$ in a Hilbert space $\mathcal{H}$ is a strongly continuous unitary group if and only if $U(t) = e^{-itH}$, $\forall t \in \mathbf{R}$, for a unique self-adjoint operator H in $\mathcal{H}$.

Note that the free propagator $U_0(t)$ may be written as

$$U_0(t) = e^{-itH_0}, \quad H_0 = -\frac{1}{2}\Delta \text{ with } D(H_0) = H^2(\mathbf{R}^d).$$

We assume $V \in L^2_{\text{loc}}(\mathbf{R}^d)$ in the rest of this subsection. Associated with the formal differential operator L we define the minimal operator L_{min} by

$$D(L_{\text{min}}) = C_0^\infty(\mathbf{R}^d), \quad L_{\text{min}}u = Lu \quad \text{for} \quad u \in D(L_{\text{min}})$$

and the maximal operator L_{max} by

$$D(L_{\text{max}}) = \{u \in \mathcal{H} : Lu \in \mathcal{H}\}, \quad L_{\text{max}}u = Lu \quad \text{for} \quad u \in D(L_{\text{max}}).$$

Lemma 2.4: *We have the following statements:*

(1) *The operator L_{min} is symmetric and L_{max} is its adjoint: $L_{\text{max}} = L^*_{\text{min}}$.*
(2) *Let H be a self-adjoint operator in $\mathcal{H} = L^2(\mathbf{R}^d)$. Suppose that $e^{-itH}u_0$ is a solution of (1.1) for any $u_0 \in D(H)$. Then,*

$$L_{\text{min}} \subset H \subset L_{\text{max}}. \tag{2.4}$$

Proof: We leave the proof of (1) for readers' exercise. For $u_0 \in D(H)$, set $u(t) = e^{-itH}u_0$. Then $u \in C^1(\mathbf{R}, \mathcal{H})$ and it satisfies $idu/dt = Hu$. Since $u(t, x)$ satisfies (1.1), we have

$$Hu(t, x) = -\frac{1}{2}\Delta u(t, x) + V(x)u(t, x). \tag{2.5}$$

Here $Hu \in C(\mathbf{R}, \mathcal{H})$ and $Vu \in C(\mathbf{R}, L^1_{\text{loc}}(\mathbf{R}^d))$. It follows that $-\Delta u \in C(\mathbf{R}, L^1_{\text{loc}}(\mathbf{R}^d))$ and that (2.5) holds for almost all $x \in \mathbf{R}^d$ at every $t \in \mathbf{R}$. Thus $Hu_0 = -(1/2)\Delta u_0 + Vu_0 \in \mathcal{H}$ for all $u_0 \in D(H)$ and $H \subset L_{\text{max}}$. Then, by virtue of (1) we have $L_{\text{min}} \subset [L_{\text{min}}] = L^*_{\text{max}} \subset H$, where $[L_{\text{min}}]$ is the minimal closed extension (the closure) of L_{min}. □

Lemma 2.4 implies that the IVP for (1.1) is wellposed if and only if there is a unique self-adjoint operator H such that $L_{\text{min}} \subset H \subset L^*_{\text{min}}$. Notice, however, that L_{min} maps real valued functions to real valued functions and L_{min} admits self-adjoint extensions by von Neumann's theorem (Ref. 48, Theorem X.3). We say $L = -(1/2)\Delta + V$ is **essentially self-adjoint** on $C_0^\infty(\mathbf{R}^d)$ if the self-adjoint extension of L_{min} is unique, that is, its closure $[L_{\text{min}}]$ is self-adjoint.

There is an extensive literature on the problem of essentially self-adjointness. We state here only one result among them, referring to Refs. 48 and 9 for further informations. For $1 \leq p < \infty$, weak L^p space, $L^p_w(\mathbf{R}^d)$, is the set of functions such that

$$\|f\|_{L^p_w} = (\sup_{t>0} t^p \lambda_f(t))^{\frac{1}{p}} < \infty, \tag{2.6}$$

where $\lambda_f(t) = m(\{x : |u(x)| > t\})$ is the distribution function of f and m is the Lebesgue measure. Though this $\|u\|_{L^p_w}$ is only a quasi-norm, one can define a norm on L^p_w which is equivalent to $\|u\|_{L^p_w}$ and with which L^p_w becomes a Banach space.[44] We denote the norm by the same symbol $\|u\|_{L^p_w}$. The following combines Theorems X.15, X.20 and X.21 of Ref. 48

Theorem 2.5:

(1) Let $d = 3$ and let $V \in L^2(\mathbf{R}^d) + L^\infty(\mathbf{R}^d)$. Then L is essentially self-adjoint on $C_0^\infty(\mathbf{R}^3)$ and self-adjoint on $H^2(\mathbf{R}^3)$.
(2) Let $d = 4$ and let $V \in L^p(\mathbf{R}^d) + L^\infty(\mathbf{R}^d)$ for some $p > d$. Then L is essentially self-adjoint on $C_0^\infty(\mathbf{R}^d)$ and self-adjoint on $H^2(\mathbf{R}^d)$.
(3) Let $d \geq 5$ and let $V = V_1 + V_2$ with $V_1 \in L_w^{d/2}(\mathbf{R}^d)$ and $V_2 \in L^\infty(\mathbf{R}^d)$. If $\|V_1\|_{L_w^{d/2}} < C_d$ for some constant depending only on d, then L is essentially self-adjoint on $C_0^\infty(\mathbf{R}^d)$ and self-adjoint on $H^2(\mathbf{R}^d)$.

2.2. *Nonautonomous case*

When V is t-dependent, the abstract theory of semigroups provides a simple method for proving the wellposedness of the IVP and, we first give a very brief review of this method. This method, however, tends to require unnecessarily strong conditions for $V(t, x)$ when applied to the equation of the form (1.1). In the second part, we shall show that better results may be produced by using the traditional method of converting the differential equation into an integral equation and solving the latter via the contraction mapping theorem.

2.2.1. *Semigroup theory*

The equation (1.1) may be considered as an evolution equation in the Hilbert space $\mathcal{H}$ and we may apply the abstract theory of evolution equations to prove that the IVP is wellposed. There is a vast literature on this subject and we refer readers to the textbooks (Refs. 45, 61 or 36) for the proof of following theorem. We consider an abstract evolution equation

$$\frac{du}{dt} = H(t)u \tag{2.7}$$

in a general Hilbert space $\mathcal{H}$.

Theorem 2.6: Suppose that $H(t)$ is self-adjoint in $\mathcal{H}$ with the common domain $\mathcal{D}$ and that, for any $u \in \mathcal{D}$, $\mathbf{R} \ni t \mapsto H(t)u \in \mathcal{H}$ is strongly differentiable. Then there exists a unique unitary propagator $\{U(t,s) : t, s \in \mathbf{R}\}$ such that $U(t,s)\mathcal{D} = \mathcal{D}$ and such that for $u_0 \in \mathcal{D}$, $U(t,s)u_0$ is strongly C^1 with respect (t,s) and

$$i\partial_t U(t,s)u_0 = H(t)U(t,s)u_0, \quad i\partial_s U(t,s)u_0 = -U(t,s)H(s)u_0. \tag{2.8}$$

Combining Theorem 2.6 with Theorem 2.5, we immediately obtain the following theorem:

Corollary 2.7: *If $V(t,x)$ satisfies the following conditions, the IVP for (1.1) is wellposed. Moreover, the propagator $U(t,s)$ maps $H^2(\mathbf{R}^d)$ isomorphically onto $H^2(\mathbf{R}^d)$ and, if $u_0 \in H^2(\mathbf{R}^d)$, $U(t,s)u_0$ satisfies (2.8) with $H(t) = -(1/2)\Delta + V(t,x)$ with the domain $D(H(t)) = H^2(\mathbf{R}^d)$.*

(1) *If $d = 3$, then $t \to V(t,\cdot) \in L^2(\mathbf{R}^d) + L^\infty(\mathbf{R}^d)$ is differentiable.*
(2) *If $d \geq 4$, then $t \to V(t,\cdot) \in L^p(\mathbf{R}^d) + L^\infty(\mathbf{R}^d)$ is differentiable for some $p > d/2$.*
(3) *If $d \geq 5$, then $V(t,x)$ may be written as $V(t,x) = V_1(t,x) + V_2(t,x)$ with $\|V_1(t,\cdot)\|_{L^{d/2}_w(\mathbf{R}^d)} < C_d$ and $V_2(t,\cdot) \in L^\infty(\mathbf{R}^d)$, and $t \to V(t,\cdot) \in L^{d/2}_w(\mathbf{R}^d) + L^\infty(\mathbf{R}^d)$ is differentiable. Here C_d is the constant in Theorem 2.5.*

2.2.2. *The method of integral equations*

We write $U_0(t)$ for the free propagator (2.3). Then the Duhamel formula implies that (1.1) is, at least formally, equivalent to the integral equation

$$u(t) = U_0(t)\phi - i\int_0^t U_0(t-s)V(s)u(s)ds\,. \tag{2.9}$$

We will solve (2.9) by applying the contraction mapping theorem.

By using the explicit expression (2.3) and the Hausdorff-Young inequality, we obtain the following L^p-L^q estimates for $U_0(t)$:

$$\|U_0(t)u_0\|_{L^p} \leq \left(\frac{1}{2\pi|t|}\right)^{d\left(\frac{1}{2}-\frac{1}{p}\right)} \|u_0\|_{L^q} \tag{2.10}$$

where $1/p + 1/q = 1$ and $p \geq 2$. Then the following **Strichartz inequality** follows from (2.10) via certain real variable methods (see Ref. 27). We use the notation $L^{p,q}$, $1 \leq p, q \leq \infty$ to denote Lorentz spaces, viz. the space of functions such that

$$\|f\|_{p,q} = \left(\int_0^\infty (t^{\frac{1}{p}} f^*(t))^q \frac{dt}{t}\right)^{\frac{1}{q}} < \infty \tag{2.11}$$

where $f^*(t) = \inf\{\sigma : \lambda_f(\sigma) \leq t\}$ is the non-increasing rearragement of f. We have the following:[44]

(1) $L^{p,p} = L^p$, $L^{p,\infty} = L^p_w$ and $L^{p,q} \subset L^{p,r}$ if $q \leq r$.

(2) There exists a norm on $L^{p,q}$ which is equivalent to $\|f\|_{p,q}$ and with respect to which $L^{p,q}$ becomes a Banach space. We denote the norm by $\|f\|_{L^{p,q}}$ again.

(3) Let $f \in L^{p_1,q_1}$ and $g \in L^{p_2,q_2}$. If $\dfrac{1}{p_1} + \dfrac{1}{p_2} = \dfrac{1}{r} < 1$, then the product $f \cdot g \in L^{r,s}$ for any s such that $\dfrac{1}{q_1} + \dfrac{1}{q_2} \geq \dfrac{1}{s}$. Moreover,

$$\|f \cdot g\|_{L^{r,s}} \leq r' \|f\|_{L^{p_1,q_1}} \|g\|_{L^{p_2,q_2}}. \tag{2.12}$$

If $\dfrac{1}{p_1} + \dfrac{1}{p_2} = 1$ and $\dfrac{1}{q_1} + \dfrac{1}{q_2} \geq 1$, then the product $f \cdot g \in L^1$ and

$$\|f \cdot g\|_{L^1} \leq \|f\|_{L^{p_1,q_1}} \|g\|_{L^{p_2,q_2}}. \tag{2.13}$$

Lemma 2.8: *Let $(p, \theta, d) \neq (\infty, 2, 2)$ be such that $0 \leq \dfrac{2}{\theta} = d\left(\dfrac{1}{2} - \dfrac{1}{p}\right) \leq 1$ and let θ' and p' be such that $\dfrac{1}{p} + \dfrac{1}{p'} = 1$ and $\dfrac{1}{\theta} + \dfrac{1}{\theta'} = 1$. Then there exists a constant $C > 0$ such that*

$$\left(\int_{\mathbf{R}} \|U_0(t)\phi\|^\theta_{L^p(\mathbf{R}^d_x)} dt\right)^{\frac{1}{\theta}} \leq C\|\phi\|_2. \tag{2.14}$$

For $\theta = 2$ and $d \geq 3$, we have more general estimate

$$\left(\int_{\mathbf{R}} \|U_0(t)\phi\|^2_{L^{\frac{2d}{d-2},2}(\mathbf{R}^d_x)} dt\right)^{\frac{1}{2}} \leq C_d\|\phi\|_2. \tag{2.15}$$

The integral operator defined by

$$\Gamma u(t) = \int_0^t U_0(t-s)u(s)ds \tag{2.16}$$

is bounded from $L^{\theta'}(\mathbf{R}_t, L^{p'}(\mathbf{R}^d_x))$ to $L^\theta(\mathbf{R}_t, L^p(\mathbf{R}^d_x))$ for any $d \geq 1$; when $d \geq 3$, from $L^2(\mathbf{R}_t, L^{\frac{2d}{d+2},2}(\mathbf{R}^d_x))$ to $L^2(\mathbf{R}_t, L^{\frac{2d}{d-2},2}(\mathbf{R}^d_x))$:

$$\|\Gamma u\|_{L^2(\mathbf{R}_t, L^{\frac{2d}{d-2},2}(\mathbf{R}^d_x))} \leq C^2 \|u\|_{L^2(\mathbf{R}_t, L^{\frac{2d}{d+2},2}(\mathbf{R}^d_x))}. \tag{2.17}$$

The following theorem slightly generalizes previously known results.[68]

Theorem 2.9: Let $d \geq 3$. Then there exists a constant D_d depending only on d such that the following statement is satisfied: Suppose that $V = V_1 + V_2$ with $V_1 \in L^\infty(\mathbf{R}_t, L_w^{d/2}(\mathbf{R}_x^d))$ and $V_2 \in L^1(\mathbf{R}_t, L^\infty(\mathbf{R}_x^d))$. Suppose in addition $\|V_1\|_{L^\infty(\mathbf{R}_t, L_w^{d/2}(\mathbf{R}_x^d))} < C_d$. Then, for any $u_0 \in L^2(\mathbf{R}^d)$, there exists a unique solution u of (2.9) such that

$$u \in C(\mathbf{R}, L^2(\mathbf{R}^d)) \cap L^2(\mathbf{R}, L^{\frac{2d}{d-2},2}(\mathbf{R}^d)).$$

The solution $u(t,x)$ satisfies (1.1) in the sense of Definition 2.1.

Proof: We prove the case $d \geq 3$. We first show the theorem on the interval $[0,a]$ for a sufficiently small $a > 0$. Define Banach spaces $\mathcal{X}$ and $\mathcal{Y}$ by

$$\mathcal{X} = C([0,a], L^2(\mathbf{R}^d)) \cap L^2([0,a], L^{\frac{2d}{d-2},2}(\mathbf{R}^d)).$$
$$\mathcal{Y} = L^1([0,a], L^2(\mathbf{R}^d)) + L^2([0,a], L^{\frac{2d}{d+2},2}(\mathbf{R}^d)).$$

Then, by Strichartz's inequality, we have $U_0 u_0 \in \mathcal{X}$ and, by Hölder's inequality (2.12), the multiplication operator V maps $\mathcal{X}$ into $\mathcal{Y}$:

$$\|Vu\|_{\mathcal{Y}} = \left(\frac{2d}{d-2}\|V_1\|_{L^\infty([0,a],L^{d/2,\infty})} + \|V_2\|_{L^1([0,a],L^\infty)}\right)\|u\|_{\mathcal{X}}$$

We prove that the operator Γ defined by (2.16) maps $\mathcal{Y}$ to $\mathcal{X}$. We have

$$\|\Gamma u(t)\|^2 = \int_0^t \int_0^t (U_0(r-s)u(s), u(r))dsdr \leq \|u\|^2_{L^1([0,a],L^2)}$$

which implies $\Gamma : L^1([0,a],L^2) \to L^\infty([0,a],L^2)$. But if $u \in \mathcal{S}(\mathbf{R}^{d+1})$, we clearly have $\Gamma u \in C([0,a],L^2)$. Since $\mathcal{S}(\mathbf{R}^{d+1})$ is dense in $L^1([0,a],L^2)$ and $C([0,a],L^2)$ is closed in $L^\infty([0,a],L^2)$, it follows that Γ maps $L^1([0,a],L^2)$ into $C([0,a],L^2))$ with norm ≤ 1. We may write $\|\Gamma u(t)\|^2$ in the form

$$\begin{aligned}
&\int_0^t \int_0^r (U_0(r-s)u(s), u(r))dsdr + \int_0^t \int_r^t (U_0(r-s)u(s), u(r))dsdr \\
&= \int_0^t \int_0^r (U_0(r-s)u(s), u(r))dsdr + \int_0^t \int_0^s (u(s), U_0(s-r)u(r))drds \\
&= \int_0^t (\Gamma u(r), u(r))dr + \int_0^t (u(s), \Gamma u(s))ds,
\end{aligned} \tag{2.18}$$

and by estimating the right hand side using (2.17) and (2.13), we have

$$\|\Gamma u(t)\|^2 \leq C\|u\|^2_{L_t^2([0,a],L^{\frac{2d}{d+2},2}(\mathbf{R}^d))}. \tag{2.19}$$

It follows, as previously, that

$$\Gamma u \in C([0,a],L^2) \quad \text{for} \quad u \in L_t^2([0,a],L^{\frac{2d}{d+2},2}(\mathbf{R}^d))$$

and, we have shown that that

$$\|\Gamma u\|_{C([0,a],L^2)} \le C\|u\|_{\mathcal{Y}}.$$

By Minkowski's inequality and (2.15), it follows that

$$\begin{aligned}\|\Gamma u\|_{L^2([0,a],L^{\frac{2d}{d-2},2})} &\le \int_0^a \|U_0(t)U_0(-s)u(s)\|_{L^2([0,a]_t,L^{\frac{2d}{d-2},2})} ds \\ &\le C\int_0^a \|u(s)\|_{L^2} ds = C\|u\|_{L^1([0,a],L^2)}\end{aligned}$$

Combining this with (2.17), we conclude $\Gamma : \mathcal{Y} \to L^2([0,a],L^{\frac{2d}{d-2},2})$ with norm $\le C$. The integral equation (2.9) may be written in the form

$$u = U_0(t)u_0 - i(\Gamma V u)(t) \tag{2.20}$$

We consider (2.20) as an equation in $\mathcal{X}$. Note that all constants which appear above depend only on the dimension d. Hence, if a and C_d are sufficiently small, we have $\|\Gamma V\|_{B(\mathcal{X})} < 1$ and, by the contraction mapping principle, the equation (2.20) has a unique solution u in $\mathcal{X}$ for any $u_0 \in \mathcal{H}$.

We clearly have $u \in C([0,a],\mathcal{H})$ and $Vu \in L^1_{\mathrm{loc}}(\mathbf{R}^{d+1})$. We next show $\|u(t)\|_{\mathcal{H}} = \|u_0\|_{\mathcal{H}}$. Using the unitarity of $U_0(t)$, we have

$$\begin{aligned}\|u(t)\|^2 &= \|U_0(t)u_0 - i\Gamma V u(t)\|^2 \\ &= \|u_0\|^2 - (U_0(t)u_0, i\Gamma V u(t)) \\ &\quad - (i\Gamma V u(t), U_0(t)u_0) + \|\Gamma V u(t)\|^2 .\end{aligned} \tag{2.21}$$

Using (2.18) and the equation $u(t) = U_0(t)u_0 - i\Gamma V u(t)$, we obtain

$$\begin{aligned}\|\Gamma V u(t)\|^2 &= \int_0^t (\Gamma V u(s), V u(s))ds + \int_0^t (V u(s), \Gamma V u(s))ds \\ &= \int_0^t \{(iu(s) - iU_0(s)u_0, Vu(s)) + (Vu(s), iu(s) - iU_0(s)u_0)\}ds \\ &= \int_0^t \{(U_0(t)u_0, iU_0(t-r)V(r)u(r)) + (U_0(t-s)V(s)u(s)ds, U_0(t)u_0)\}ds \\ &= (U_0(t), i\Gamma V u(t)) + (i\Gamma V u(t), U_0(t)u_0).\end{aligned}$$

Combining these two identities we have $\|u(t)\| = \|u_0\|$ for all $t \in [0,a]$. Then, we may continue the solution $u(t)$ to the whole real line by the standard method and this u satisfies the integral equation (2.9) on $\mathbf{R}$. Then differentiating (2.9) with respect to t considering the integral as an $\mathcal{S}'(\mathbf{R}^d)$-valued function of s, we see that $u(t,x)$ is a solution of (1.1) in the sense of Definition 2.1. □

Remark 2.10: Note that for $d \le 4$, $V(t,\cdot)$ is not in general in $L^2_{\mathrm{loc}}(\mathbf{R}^d)$ and it is meaningless to consider the minimal operator for

$$L(t) = -(1/2)\Delta + V(t,x).$$

This is related to the fact that the uniqueness of the solution of (2.9) is proved in the space $\mathcal{X} = C(\mathbf{R}, L^2(\mathbf{R}^d)) \cap L^2_{\mathrm{loc}}(\mathbf{R}, L^{\frac{2d}{d-2},2}(\mathbf{R}^d))$. It is not known if the solution of (2.9) is unique or not in $C(\mathbf{R}, L^2(\mathbf{R}^d))$. If the solution is unique in $C(\mathbf{R}, L^2(\mathbf{R}^d))$, the IVP is said to be unconditionally wellposed. See Ref. 25 for related topics.

Under a slightly stronger condition, we may prove the following regularity property of the solution obtained in Theorem 2.9. Note that $V(t,\cdot) \in L^2_{\mathrm{loc}}$ is assumed in the following theorem. See Ref. 68 for a proof.

Theorem 2.11: Let $1 \le p < d/2$ and suppose $V \in C(\mathbf{R}, L^{\max\{2,p\}}(\mathbf{R}^d)) + C(\mathbf{R}, L^\infty(\mathbf{R}^d))$. Suppose further $\partial_t V \in L^\alpha(\mathbf{R}, L^{p_1}(\mathbf{R}^d)) + L^1(\mathbf{R}, L^\infty(\mathbf{R}^d))$ for

$$1/p_1 = 2/d + 1/2p \text{ if } d \ge 5, \quad 1/p_1 < 1/2 + 1/2p \text{ if } d = 4,$$
$$1/p_1 = 1/2 + 1/2p \text{ if } d = 3; \quad 1/\alpha < 1 - (d/4p).$$

Then $H(t)$ is self-adjoint with the common domain $\mathcal{D} = H^2(\mathbf{R}^d)$ and the propagator $U(t,s)$ satisfies $U(t,s)\mathcal{D} = \mathcal{D}$ for all $t,s \in \mathbf{R}$. Moreover, if $u_0 \in \mathcal{D}$, then $U(t,s)u_0$ is strongly C^1 in $\mathcal{H}$ and $u(t) = U(t,s)u_0$ satisfies

$$i\frac{d}{dt}U(t,s)u_0 = H(t)U(t,s)u_0, \quad i\frac{d}{ds}U(t,s)u_0 = -U(t,s)H(s)u_0.$$

The equation (1.1) is unconditionally wellposed in this case.

3. Floquet Hamiltonian

From this section on, we assume that $V(t,x)$ is 2π periodic with respect to the variable t and that $V \in \mathcal{V}_\beta$ for some $\beta > 0$, where

$$\mathcal{V}_\beta = \left\{ V \colon \|V\|_\beta \equiv \sum_{j=0}^{2} \sup_{x\in\mathbf{R}^d} \langle x\rangle^\beta \left(\int_0^{2\pi} |\partial_t^j V(t,x)|^2 dt \right)^{\frac{1}{2}} < \infty \right\}. \tag{3.1}$$

The space $\mathcal{V}_\beta$ is a Banach space with the norm $\|V\|_\beta$. We write $\mathbf{T} = \mathbf{R}/2\pi\mathbf{Z}$.

When $V \in \mathcal{V}_\beta$, $\beta > 0$, then V satisfies the condition of Theorem 2.11 and (1.1) generates a unique unitary propagator $\{U(t,s)\}$. By virtue of the periodicity, $U(t,s)$ satisfies

$$U(t+2\pi, s+2\pi) = U(t,s), \quad \forall t, s \in \mathbf{R}.$$

The unitary operators $\mathcal{U}_t = U(2\pi + t, t)$, $t \in \mathbf{T}$ are called the **Floquet operators** for (1.1). They satisfy

$$\mathcal{U}_t = U(2\pi + t, 2\pi)U(2\pi, 0)U(0,t) = U(t,0)\mathcal{U}_0 U(0,t); \tag{3.2}$$

as $\mathcal{U}_t$ are mutually unitarily equivalent, they have the same spectral properties. The eigenvalues of $\mathcal{U}_t$ are called **Floquet multipliers.** It is well known that the Floquet operators play a central role in the analysis of time periodic systems. In this section we show that the spectral properties of $\mathcal{U}_t$ may be studied via the more explicit operator. We define **the extended phase space** by

$$\mathcal{K} = L^2(\mathbf{T}, \mathcal{H}) = \int_{\mathbf{T}}^{\oplus} \mathcal{H} dt.$$

Using the propagator $U(t,s)$ for (1.1), we define a one parameter family of operators $\{\mathcal{U}(\sigma) : \sigma \in \mathbf{R}\}$ on $\mathcal{K}$ by

$$\mathcal{U}(\sigma)u(t) = U(t, t-\sigma)u(t-\sigma), \quad t \in \mathbf{T}. \tag{3.3}$$

The proof of the following theorem is left to the readers as an exercise.

Lemma 3.1: *The family $\{\mathcal{U}(\sigma) : \sigma \in \mathbf{R}\}$ defines a strongly continuous unitary group on $\mathcal{K}$. The infinitesimal generator K of $\mathcal{U}(\sigma)$: $\mathcal{U}(\sigma) = e^{-i\sigma K}$ is self-adjoint in $\mathcal{K}$ and is defined by*

$$Ku = \left(-i\partial_t - \frac{1}{2}\Delta + V(t,x)\right) u$$

with the domain

$$D(K) = \left\{u \in \mathcal{K} : \left(-i\partial_t - \frac{1}{2}\Delta + V(t,x)\right) u \in \mathcal{K}\right\}.$$

If $\mathcal{T}_n$, $n \in \mathbf{Z}$, is a unitary operator in $\mathcal{K}$ defined by the multiplication by e^{int}, then

$$\mathcal{T}_n K \mathcal{T}_n^* = K - n \tag{3.4}$$

and $\sigma(K)$ is invariant under the translation by integers.

We call the self-adjoint operator K the **Floquet Hamiltonian** for (1.1). When $V = 0$, we write

$$K_0 = -i\partial_t - \frac{1}{2}\Delta, \quad D(K_0) = \left\{u \in \mathcal{K}\colon \left(-i\partial_t - \frac{1}{2}\Delta\right) u \in \mathcal{K}\right\} \tag{3.5}$$

and call K_0, the free Floquet Hamiltonian. If we define the Fourier coefficients with respect to t by

$$(\mathcal{F}u)(n, x) = \frac{1}{\sqrt{2\pi}} \int_{\mathbf{T}} e^{-int} u(t, x) dt, \quad n \in \mathbf{Z}, \tag{3.6}$$

then $\mathcal{F}$ is a unitary operator from $\mathcal{K}$ to $\bigoplus_{n\in\mathbf{Z}} \mathcal{H}$, the direct sum of copies of $\mathcal{H}$, and K_0 has a direct sum decomposition:

$$\mathcal{F} K_0 \mathcal{F}^* = \bigoplus_{n\in\mathbf{Z}} \left(n - \frac{1}{2}\Delta\right) \tag{3.7}$$

It follows that the spectrum of K_0 is purely absolutely continuous and it fills up the whole real line. We often use the following simple lemma. We write $R(z) = (K - z)^{-1}$ and $R_0(z) = (K_0 - z)^{-1}$ for the resolvents.

Lemma 3.2: *For $z \notin \mathbf{R}$, we have*

$$R_0(z) = \mathcal{F}^* \left(\bigoplus_{n\in\mathbf{Z}} r_0(z-n)\right) \mathcal{F}, \quad r_0(z) = \left(-\frac{1}{2}\Delta - z\right)^{-1}. \tag{3.8}$$

If F is a multiplication operator by a function $F(t, x)$ such that $F(t, x) \to 0$ as $|x| \to \infty$ uniformly with respect to t, then $FR_0(z)$ is compact in $\mathcal{K}$.

Proof: The equation (3.8) is obvious from (3.7). To prove the second statement we may assume F is real valued and independent of $t \in \mathbf{T}$ and it has a compact support. Then, as well known, $Fr_0(z)$ is compact in $\mathcal{H}$. Moreover,

$$\|F(r_0(z-n) - r_0(\bar{z}-n))F\|_{B(\mathcal{H})} = 2|\Im z| \, \|Fr_0(z-n)\|^2$$

converges to 0 as $n \to \pm\infty$ (Ref. 1). Thus, $FR_0(z)$ is compact in $\mathcal{K}$. □

The following simple equation (3.9) relates $\mathcal{U}_t$ and K and plays a basic role. In the following lemma, we regard

$$L^2(\mathbf{T}, \mathcal{H}) = \int_{\mathbf{T}}^{\oplus} \mathcal{H}(t) dt, \quad \mathcal{H}(t) = \mathcal{H} \ \ \forall t \in \mathbf{T}.$$

Lemma 3.3: *For any Borel function f on the unit circle of the complex plane, we have the identity*

$$f(e^{-2\pi i K}) = \int_{\mathbf{T}}^{\oplus} f(\mathcal{U}_t) dt. \tag{3.9}$$

Proof: By virtue of the periodicity, we have

$$(e^{-2\pi iK}u)(t) = U(t,t-2\pi)u(t-2\pi) = \mathcal{U}_t \cdot u(t), \quad u \in \mathcal{K}.$$

Then (3.9) follows by functional calculus. □

Lemma 3.4: *Let ϕ be an eigenfunction of $\mathcal{U}_s$ with eigenvalue $e^{-i2\pi\lambda}$. Then, for any $n \in \mathbf{Z}$, $u(t) = e^{i(t-s)(\lambda+n)}U(t,s)\phi$, $t \in \mathbf{T}$, is an eigenfunction of K with eigenvalue $\lambda + n$. Conversely, if $u \in \mathcal{K}$ is an eigenfunction of K with eigenvalue λ, then $u \in C(\mathbf{T}, \mathcal{H})$ and $u(t) = e^{i(t-s)\lambda}U(t,s)u(s)$ for any $t, s \in \mathbf{R}$. In particular, $u(s)$ is an eigenfunction of $\mathcal{U}_s$ with eigenvalue $e^{-i2\pi\lambda}$.*

Proof: Define $u(t) = e^{i(t-s)(\lambda+n)}U(t,s)\phi$ for $t \in \mathbf{R}$. Then, since we have $U(s+2\pi, s)\phi = e^{-i2\pi\lambda}\phi$, $u(t)$ is 2π periodic and $u \in \mathcal{K}$. Moreover, we have

$$\begin{aligned}(e^{-i\sigma K}u)(t) &= e^{i(t-\sigma-s)(\lambda+n)}U(t,t-\sigma)U(t-\sigma,s)\phi \\ &= e^{-i\sigma(\lambda+n)}u(t), \ \sigma \in \mathbf{R}\end{aligned}$$

and u is an eigenfunction of K with eigenvalue $\lambda + n$. Suppose, conversely, $Ku = \lambda u$, $u \in D(K)$. Then, for any $\sigma \in \mathbf{R}$, we have

$$e^{-i\sigma\lambda}u(t) = (e^{-i\sigma K}u)(t) = U(t,t-\sigma)u(t-\sigma), \quad \text{a.e.} \quad t \in \mathbf{T} \qquad (3.10)$$

or $u(t+\sigma) = e^{i\sigma\lambda}U(t+\sigma,t)u(t)$ for almost all $t \in \mathbf{T}$. It follows by the strong continuity of $U(t,s)$ that $u(t)$ is $\mathcal{H}$-valued continuous and (3.10) holds for all $t \in \mathbf{T}$ and $\sigma \in \mathbf{R}$. Replacing σ by $t-s$ in (3.10), we obtain $u(t) = e^{i(t-s)\lambda}U(t,s)u(s)$. Hence $u(s) = u(s+2\pi) = e^{i2\pi\lambda}U(s+2\pi,s)u(s)$, and $u(s)$ is an eigenfunction of $\mathcal{U}_s$ with eigenvalue $e^{-i2\pi\lambda}$. □

We write $P_c(t)$ and $P_p(t)$ for the orthogonal projections onto $\mathcal{H}_c(\mathcal{U}_t)$ and $\mathcal{H}_p(\mathcal{U}_t)$ respectively. By virtue of (3.2), we have

$$U(t,s)P_c(s) = P_c(t)U(t,s) \quad \text{and} \quad U(t,s)P_p(s) = P_p(t)U(t,s). \qquad (3.11)$$

Lemma 3.5: *We have the following identities:*

$$P_p(K) = \int_{\mathbf{T}}^{\oplus} P_p(t)dt, \quad P_c(K) = \int_{\mathbf{T}}^{\oplus} P_c(t)dt. \qquad (3.12)$$

Proof: Let $\{e^{-i2\pi\lambda_j}\}_{j=1,2,\dots}$, $0 \le \lambda_j < 1$, be the eigenvalues of $\mathcal{U}_0$ and let $\{\psi_{j,k}\}_{k=1,2,\dots}$ be an orthonormal basis (ONB) of the eigenspace of $\mathcal{U}_0$ with the eigenvalue $e^{-2\pi i\lambda_j}$. Define, for $n \in \mathbf{Z}$ and $t \in \mathbf{T}$,

$$\psi_{j,k;n}(t) = e^{it(\lambda+n)}U(t,0)\psi_{j,k}.$$

By virtue of (3.2), for every fixed $t \in \mathbf{T}$, $\{\psi_{j,k;0}(t)\}_{k=1,2,\dots}$ is an ONB of the eigenspace of $\mathcal{U}_t$ with the same eigenvalue $e^{-2\pi i\lambda_j}$; $\psi_{j,k;n}(t)$ is 2π periodic with respect to t, and if we define

$$\phi_{j,k;n}(t,x) \equiv \frac{1}{\sqrt{2\pi}}\psi_{j,k;n}(t,x) \in \mathcal{K}$$

then by virtue of Lemma 3.4, $\{\phi_{j,k;n}\}_{k=1,2,\dots}$ is an ONB of eigenspace of K with eigenvalue $\lambda_j + n$. Hence, $\{\phi_{j,k;n}\}_{j,k,n}$ is the ONB of $\mathcal{K}_p(K)$. Thus,

$$\begin{aligned}(P_p(K)u)(t,x) =& \sum_{j,k}\frac{1}{2\pi}\sum_{n\in\mathbf{Z}}\psi_{j,k;n}(t,x)\int_{\mathbf{T}\times\mathbf{R}^d}\overline{\psi_{j,k;n}(s,y)}u(s,y)dsdy \\ =& \sum_{j,k}\frac{1}{2\pi}\sum_{n\in\mathbf{Z}}e^{int}\psi_{j,k;0}(t,x)\int_{\mathbf{T}\times\mathbf{R}^d}e^{-ins}\overline{\psi_{j,k;0}(s,y)}u(s,y)dsdy \\ =& \sum_{j,k}\psi_{j,k;0}(t,x)\int_{\mathbf{R}^3}\overline{\psi_{j,k;0}(t,y)}u(t,y)dy = P_p(t)u(t,x).\end{aligned}$$

Here, we used the Fourier inversion formula in the third step. This proves the first identity of (3.12). The second is then obvious since

$$P_c = 1 - P_p. \qquad \square$$

Remark 3.6: The same argument implies that for any $0 \le \lambda < 1$ we have

$$E_K(\mathbf{Z}+\lambda) = \int_{\mathbf{T}}^{\oplus} E_{\mathcal{U}_t}(\{e^{-2\pi i\lambda}\})dt. \tag{3.13}$$

Note that both sides of (3.13) vanish if $\lambda \in \sigma_c(K)$.

If $V(t,x) = V(x)$ is t-independent, then, by the spectral mapping theorem

$$\mathcal{U}_t = e^{-2\pi iH}, \quad H = -\frac{1}{2}\Delta + V, \quad \text{for any} \quad t \in \mathbf{T}$$

and $\sigma_p(\mathcal{U}_t) = \{e^{-2\pi i\lambda} \colon \lambda \in \sigma_p(H)\}$. It is well known that the number of negative eigenvalues of $-\Delta + V$ is bounded by a constant times $\|V\|_{L^{d/2}}$ (Ref. 35), and that, if $V \in \mathcal{V}_\beta$ with $\beta > 1$, then positive eigenvalues are absent from H. We have the corresponding results for time periodic systems. The following theorem can be found in Ref. 13 (see also Ref. 33):

Theorem 3.7: Suppose $d = 3$ and $V \in \mathcal{V}_\beta$ with $\beta > 2$. Then $\mathcal{H}_p(\mathcal{U}_s)$ is finite dimensional. Moreover,

(1) If $\mathcal{U}_s\phi = e^{-2\pi i\lambda}\phi$, $-1 < \lambda < 0$, then $\psi(t,x) = e^{i(t-s)\lambda}U(t,s)\phi \in \mathcal{K}$ satisfies for any $N = 0, 1, \ldots$,

$$\langle x\rangle^N H(t)^\alpha \partial_t^\beta \phi(t,x) \in \mathcal{K}, \quad 0 \le \alpha + \beta \le 2. \tag{3.14}$$

(2) If $\mathcal{U}_s\phi = \phi$, then for any $\varepsilon > 0$

$$\langle x\rangle^{1-\varepsilon} H(t)^\alpha \partial_t^\beta \phi(t,x) \in \mathcal{K}, \quad 0 \le \alpha + \beta \le 2. \tag{3.15}$$

Remark 3.8: We show in the next section that the eigenvalues of K are discrete outside $\mathbf{Z}$ if $V \in \mathcal{V}_\beta$ with $\beta > 1$ via the Agmon-Kuroda method (Ref. 1). Actually, the discreteness of $\sigma_p(K)$ outside $\mathbf{Z}$ and the exponential decay of associated eigenfunctions may be proved under less stringent decay condition at infinity (but with suitable differentiability assumptions) on $V(t,x)$ by applying the Mourre method (Refs. 67 and 72). Moreover, the Mourre theory has recently been generalized and the discreteness of $\sigma_p(K)$ outside thresholds, the exponential decay of eigenfunctions and the limiting absorption principle have been proved for many particle time periodic systems. (See Ref. 38 for the details.)

As a side remark, we should mention that the extended phase space $L^2(\mathbf{R}, \mathcal{H})$ and the unitary group $\mathcal{U}(\sigma)$ defined by (3.3) were first introduced by Howland (Ref. 17) for studying the scattering theory for time dependent equations. Its adaptation for time periodic equations has been made by the author(Ref. 65, where the connection between the spectral properties of $\mathcal{U}_t$ and K has been used for proving the completeness of the scattering theory (see also Ref. 18)). Note that K in $L^2(\mathbf{R}, \mathcal{H})$ is unitarily equivalent to $-i\partial_t$ and is purely absolutely continuous.

4. Scattering Theory

When $V \in \mathcal{V}_\beta$ with $\beta > 1$, we say V is of short range type. In this section we discuss the scattering theory for (1.1) when V is short range. We show that $\sigma_p(\mathcal{U}_s)$ is discrete outside $\{1\}$, $\mathcal{H} = \mathcal{H}_p(\mathcal{U}_s) \oplus \mathcal{H}_{ac}(\mathcal{U}_s)$ and that wave operators exist and are complete. Then we discuss the structure of scattering operator and show that the potential is uniquely determined from the scattering operator. The existence and the completeness of the wave operators have been extended to time periodic three particle systems,[41,32] which, however, will not be discussed here. Very little is known for scattering theory for general time periodic multiparticle systems.[38]

4.1. *RAGE theorem*

When V is t-independent, bound and scattering states are characterized in terms of the spectrum of $H = -(1/2)\Delta + V$ (RAGE theorem, see Ref. 50). This result has a natural extension to time-periodic systems.[30,11] We assume $V \in \mathcal{V}_0$ in this subsection; thus, V may not decay at infinity.

Theorem 4.1:

(1) Suppose $P_1, P_2, \ldots \in B(\mathcal{H})$ are such that for every compact interval Δ:

$$(I \otimes P_n)E_K(\Delta) \text{ is compact in } \mathcal{K} \quad \text{and} \quad \operatorname*{s-lim}_{n\to\infty} P_n = I. \tag{4.1}$$

Then $u_0 \in \mathcal{H}_p(\mathcal{U}_s)$ if and only if

$$\lim_{n\to\infty} \sup_t \|(1-P_n)U(t,s)u_0\| = 0 \tag{4.2}$$

and $u_0 \in \mathcal{H}_c(\mathcal{U}_s)$ if and only if, for any n

$$\lim_{T\to\infty} \frac{1}{T}\int_s^{s+T} \|P_n U(t,s)u_0\| dt = 0. \tag{4.3}$$

(2) Let $P_R(x) = 1$ if $|x| < R$ and $P_R(x) = 0$ if otherwise. Then the sequence $P_1, P_2, \ldots$ satisfies the condition (4.1) of statement (1).

Proof: We write $\mathcal{H}_b$ and $\mathcal{H}_s$ for the set of functions $u_0 \in \mathcal{H}$ which satisfy (4.2) and (4.3), respectively. Then, $\mathcal{H}_b$ and $\mathcal{H}_s$ are closed linear subspaces and $\mathcal{H}_b \perp \mathcal{H}_s$. Thus it suffices for proving (1) to show that $\mathcal{H}_p(\mathcal{U}_s) \subset \mathcal{H}_b$ and $\mathcal{H}_c(\mathcal{U}_s) \subset \mathcal{H}_s$. The proof of the standard RAGE theorem[50] implies $\mathcal{H}_p(\mathcal{U}_s) \subset \mathcal{H}_b$, and we prove $\mathcal{H}_c(\mathcal{U}_s) \subset \mathcal{H}_s$. Let $u_0 \in \mathcal{H}_s(\mathcal{U}_s)$ and define $u \in \mathcal{K}$ by $u(t) = U(t,s)u_0$ for $0 \le t < 2\pi$ and by periodicity elsewhere. Then (3.11) and (3.12) imply $u \in \mathcal{K}_c(K)$ and the standard RAGE theorem implies

$$\lim_{T\to\infty} \frac{1}{T}\int_0^T \|(I \otimes P_n)e^{-i\sigma K}u\|_{\mathcal{K}} d\sigma = 0. \tag{4.4}$$

Here, by the periodicity, the integral in the LHS of (4.4) is equal to

$$\begin{aligned}
&\int_0^T \int_{\mathbf{T}} \|P_n U(t, t-\sigma)u(t-\sigma)\|_{\mathcal{H}} dt d\sigma \\
&= \int_0^T \int_0^{2\pi} \|P_n U(t+\sigma, t)u(t)\|_{\mathcal{H}} dt d\sigma \\
&= \int_0^T \int_0^{2\pi} \|P_n U(t+\sigma, s)u_0\|_{\mathcal{H}} dt d\sigma \\
&= \int_0^{2\pi} \int_t^{T+t} \|P_n U(\sigma, s)u_0\|_{\mathcal{H}} d\sigma dt \\
&= 2\pi \int_0^T \|P_n U(\sigma, s)u_0\|_{\mathcal{H}} d\sigma dt + O(1),
\end{aligned}$$

where $O(1)$ is a bounded function of T. Hence (4.4) implies $\mathcal{H}_c(\mathcal{U}_s) \subset \mathcal{H}_s$.

For proving (2), it suffices to show that $(I \otimes P_R)R(z)$ is compact in $\mathcal{K}$ for $z \notin \mathbf{R}$. But Lemma 3.2 implies that $(I \otimes P_R)R_0(z)$ is compact. The result then follows immediately from the resolvent equation

$$R(z) = R_0(z)(1 - VR(z)). \qquad \square$$

4.2. *Existence and completeness of wave operators*

We assume here $V \in \mathcal{V}_\beta$ with $\beta > 1$, and show that for $u_0 \in \mathcal{H}_c(\mathcal{U}_s)$ there exits a unique $u_\pm \in \mathcal{H}$ such that

$$\lim_{t \to \pm\infty} \|U(t, s)u_0 - e^{-i(t-s)H_0}u_\pm\| = 0 \tag{4.5}$$

and that, when u_0 runs over $\mathcal{H}_c(\mathcal{U}_s)$, $u_\pm$ ranges over the whole space $\mathcal{H}$. (For long range potentials (4.5) remains true if $e^{-i(t-s)H_0}$ is replaced by the so called modified free propagator. We refer the readers to Ref. 31 for the details.)

Theorem 4.2: Suppose $V \in \mathcal{V}_\beta$ with $\beta > 1$. Then, for any $s \in \mathbf{T}$, the operators $W_\pm(s)$ defined by the limits

$$W_\pm(s)u_0 = \lim_{t \to \pm\infty} U(t, s)^{-1} e^{-i(t-s)H_0} u_0 \tag{4.6}$$

exist. We have Image $W_\pm(s) = \mathcal{H}_c(\mathcal{U}_s)$.

Theorem 4.2 implies the aformentioned property (4.5) of the solutions. Since $U(t, s)^{-1}e^{-i(t-s)H_0}$ is unitary, $W_\pm(s)$ is isometric, if it exists, and it satisfies the periodicity $W_\pm(s+2\pi) = W_\pm(s)$ and the **intertwining property**:

$$U(t,s)W_\pm(s) = W_\pm(t)e^{-i(t-s)H_0}.$$

It follows that Image $W_\pm(s) \subset \mathcal{H}_{ac}(\mathcal{U}_s)$ and the statement Image $W_\pm(s) = \mathcal{H}_c(\mathcal{U}_s)$ implies the absence of the singular continuous spectrum for $\mathcal{U}_s$, in particular. The theorem may be proved by various methods. We prove here the completeness part, viz. the part Image $W_\pm(s) = \mathcal{H}_c(\mathcal{U}_s)$, by (a classical method of) using smooth perturbation theory due to Kato.[24]

Proof: We begin with the proof of the existence of the limit (4.6) (for the case $t \to \infty$ and $s = 0$). Let $\Omega(t) = U(t,0)^{-1}e^{-itH_0}$. Since $\Omega(t)$ is unitary it suffices to show that (4.6) exists for u_0 in a dense subset $\mathcal{D}$ of $\mathcal{H}$. We take $\mathcal{D} = \mathcal{F}^{-1}C_0^\infty(\mathbf{R}^d \setminus \{0\})$. Integrating

$$\frac{d}{dt}\Omega(t)u_0 = iU(0,t)V(t,x)e^{-itH_0}u_0$$

and estimating via Minkowski's inequality, we have

$$\|\Omega(t_2)u_0 - \Omega(t_1)u_0\| \le \int_{t_1}^{t_2} \|V(t)e^{-itH_0}u_0\|dt. \tag{4.7}$$

Recall the expression (2.3) of the free propagator $U_0(t) = e^{-itH_0}$ which may be expressed also in the form

$$e^{-itH_0}u_0 = e^{\frac{ix^2}{2t}}\frac{e^{\mp\frac{i\pi d}{4}}}{|t|^{\frac{d}{2}}}\mathcal{F}(e^{\frac{iy^2}{2t}}u_0)\left(\frac{x}{t}\right). \tag{4.8}$$

Expanding $e^{iy^2/2t}$ into the Taylor series inside the Fourier transform in (4.8), we obtain the following lemma.

Lemma 4.3: *Suppose $u \in \mathcal{S}(\mathbf{R}^d)$. Then $e^{-itH_0}u_0$ has the following asymptotic expansion as $t \to \pm\infty$*

$$e^{-itH_0}u_0(x) = e^{\frac{ix^2}{2t}}\frac{e^{\mp\frac{i\pi d}{4}}}{|t|^{\frac{d}{2}}}\left(\hat{u}_0 + (i\Delta/2t)\hat{u}_0 + \frac{(i\Delta/2t)^2}{2!}\hat{u}_0 + \cdots\right)\left(\frac{x}{t}\right) \tag{4.9}$$

in the sense that for any $\ell = 1, 2, \ldots,$

$$\left\|e^{-itH_0}u_0(x) - e^{\frac{ix^2}{2t}}\frac{e^{\mp\frac{i\pi d}{4}}}{|t|^{\frac{d}{2}}}\sum_{j=0}^{\ell-1}\left[\frac{(i\Delta/2t)^j}{j!}\hat{u}_0\right]\left(\frac{x}{t}\right)\right\| \le \frac{(2\pi|t|)^{-\ell}}{\ell!}\|x^{2\ell}u_0\|$$

It follows by virtue of Lemma 4.3 that $\|V(t,x)e^{-itH_0}u_0\|$ is bounded by

$$\|V(t,tx)\hat{u}_0(x)\| + \frac{1}{2t}\|V(t,tx)\Delta\hat{u}_0(x)\| + \frac{1}{8t^2}\|V(t,\cdot)\|_\infty\|x^4 u_0\| \le C\langle t\rangle^{-\beta},$$

which is integrable, and $\lim_{t\to\infty}\Omega(t)u_0$ exists by virtue of (4.7). □

We next prove the completeness, Image $W_\pm(s) = \mathcal{H}_c(\mathcal{U}_s)$, by applying the Floquet Hamiltonian formalism.[65,18] We define

$$\mathcal{U}_0(\sigma)u(t) = (e^{-i\sigma K_0}u)(t) = e^{-i\sigma H}u(t-\sigma), \quad u \in \mathcal{K}.$$

Then, from what we have just proven, it readily follows that the limits

$$\begin{aligned}(\mathcal{W}_\pm u)(t) &= \lim_{\sigma\to\pm\infty}\mathcal{U}(\sigma)^*\mathcal{U}_0(\sigma)u(t)\\ &= \lim_{\sigma\to\pm\infty}U(t,t+\sigma)U_0(t+\sigma,t)u(t) = W_\pm(t)u(t)\end{aligned}\tag{4.10}$$

exist. Denote by $P(t)$ the projection onto Image $W_\pm(t)$. Then, because $W_\pm(s)$ is an isometry, $P(t) = W_\pm(t)W_\pm(t)^*$. Suppose, now, that we have Image $\mathcal{W}_\pm = \mathcal{K}_c(K)$. Then $P_c(K) = \mathcal{W}_\pm\mathcal{W}_\pm^*$ and by virtue of (4.10) and (3.12),

$$P_c(t)u(t) = P_c(K)u(t) = W_\pm(t)W_\pm(t)^*u(t) = P(t)u(t).$$

Thus, for proving the completeness, it suffices to show that the wave operators $\mathcal{W}_\pm$ are complete: Image$\mathcal{W}_\pm = \mathcal{K}_c(K)$. We prove this by applying Kato's smooth perturbation theory. The following two theorems may be found in Ref. 50.

Theorem 4.4: Let H be a self-adjoint operator on the Hilbert space $\mathcal{H}$ and let A be a closed operator from $\mathcal{H}$ to another Hilbert space $\mathcal{H}_1$ such that $D(H) \subset D(A)$. Then

$$\begin{aligned}&\sup_{u\in D(H),\|u\|=1}\frac{1}{2\pi}\int_{\mathbf{R}}\|Ae^{-itH}u\|^2dt\\ &= \sup_{\|u\|=1,\varepsilon>0}\frac{1}{4\pi^2}\int_{\mathbf{R}}(\|AR(\lambda+i\varepsilon)u\|^2 + \|AR(\lambda-i\varepsilon)u\|^2)d\lambda\\ &= \sup_{\lambda\in\mathbf{R},\varepsilon>0}\frac{1}{2\pi}\|A(R(\lambda+i\varepsilon)-R(\lambda-i\varepsilon))A^*\|.\end{aligned}$$

The operator A is said to be H-smooth when any (hence all) of these quantities is finite. When $\Omega \subset \mathbf{R}$ is a Borel set, we say A is H-smooth in Ω if $AE_H(\Omega)$ is H-smooth, where $E_H(\Omega)$ is the spectral measure for H.

Theorem 4.5:

(1) Let H and A be as in the previous theorem. Let $\Omega \subset \mathbf{R}$. Suppose that

$$\sup_{\lambda \in \Omega, \varepsilon > 0} \|AR(\lambda + i\varepsilon)A^*\| < \infty.$$

Then A is H-smooth on the closure $\overline{\Omega}$ of Ω.

(2) Let H and H_0 be self-adjoint operators and $H - H_0 = A^*B$ in the sense that

$$(Hu, v) - (u, H_0 v) = (Au, Bv), \quad \forall u \in D(H),\ v \in D(H_0).$$

Suppose that A is H-smooth on a bounded interval Ω and B is H_0-smooth on Ω. Then the strong limits

$$W_\pm = \lim_{t \to \pm\infty} e^{itH} e^{-itH_0} E_{H_0}(\Omega),$$
$$Z_\pm = \lim_{t \to \pm\infty} e^{itH_0} e^{-itH} E_H(\Omega)$$

exist. Moreover, $W_\pm^* = Z_\pm$, $Z_\pm W_\pm = E_{H_0}(\Omega)$ and $W_\pm Z_\pm = E_H(\Omega)$

We apply Theorem 4.5 with $H = K$ and $H_0 = K_0$. Let A and B be the multiplication operators with $|V(t,x)|^{1/2}$ and $|V(t,x)|^{1/2}\mathrm{sign}[V(t,x)]$ respectively, so that $V(t,x) = A(t,x)B(t,x)$ and $K = K_0 + A^*B$. All we need to show is the the following lemma.

Lemma 4.6: *Suppose $V \in \mathcal{V}_\beta$ with $\beta > 1$. Then:*

(1) *$\sigma_p(K)$ is discrete in $\mathbf{R} \setminus \mathbf{Z}$.*

(2) *A and B are both K and K_0-smooth on any compact interval Ω contained in $\mathbf{R} \setminus (\sigma_p(K) \cup \mathbf{Z})$.*

Proof: We first show that the multiplication C with $\langle x \rangle^{-\beta/2}$ is K_0-smooth on compact intervals $\Omega \subset \mathbf{R} \setminus \mathbf{Z}$. By virtue of the spectral translation property (3.4), we may assume $\Omega \subset (0,1)$. We decompose $R_0(z)$ as in (3.8):

$$CR_0(z)^{-1}C^* = \mathcal{F}^*(\oplus C r_0(z-n)C^*)\mathcal{F}. \tag{4.11}$$

Recall $r_0(z) = (-(1/2)\Delta - z)^{-1}$. It is well-known (Ref. 1) that:

(a) For any $\varepsilon > 0$, $\|Cr_0(z)C^*\|_{B(\mathcal{H})} \le C|z|^{-1/2}$ when $|z| > \varepsilon$ and $\Im z \neq 0$.

(b) For every $z \in \mathbf{C}^\pm$, $Cr_0(z)C^*$ is compact in $\mathcal{H}$.

(c) The analytic function $\mathbf{C}^\pm \ni z \to Cr_0(z)C^* \in B(\mathcal{H})$ has norm continuous boundary values $Cr_0^\pm(\lambda)C^*$ on $\mathbf{R} \setminus \mathbf{Z}$ and $Cr_0^\pm(\lambda)C^*$ is Hölder continuous on $\mathbf{R} \setminus (-\varepsilon, \varepsilon)$ for any $\varepsilon > 0$.

The property (a) implies that C is K_0-smooth on Ω, (a) and (b) imply that $CR_0(z)C^*$ is compact in $\mathcal{K}$ and (c) implies $CR_0(z)C^*$ has a norm continuous boundary values $CR_0(\lambda \pm i0)^{-1}C^* = CR_0^{\pm}(\lambda)C^*$ on $\lambda \in \Omega$. This implies both A and B are K_0-smooth on Ω. For $\Im z \neq 0$, we have

$$R(z) = R_0(z) - R_0(z)B(1 + AR_0(z)B)^{-1}AR_0(z). \tag{4.12}$$

(Note that $A = A^*$, $B = B^*$ and $AB = BA = V$.) By the previous argument, $AR_0(z)B$ has boundary values $AR_0^{\pm}(\lambda)B$ for $\lambda \in \Omega$ and it is compact. Thus, if $-1 \notin \sigma(AR_0^{\pm}(\lambda_0)B)$, then $AR(z)A$ and $BR(z)B$ have boundary values $AR^{\pm}(\lambda)A$ and $BR^{\pm}(\lambda)B$ near λ_0 and, hence, A and B are K-smooth on a neighborhood of λ_0. Thus, it suffices to show for $\lambda \in \Omega$ that $-1 \in \sigma(AR_0^{\pm}(\lambda)B)$ if and only if λ is an eigenvalue of K and that eigenvalues are discrete in Ω. The following is an adaptation of Agmon's classical argument.[1] We prove the "only if" part and the discreteness of $\sigma_p(K)$, leaving the proof of the "if" part to the readers. Assume $-1 \in \sigma(AR_0^{\pm}(\lambda)B)$ so that, for some $u \neq 0$, $u + AR_0^{\pm}(\lambda)Bu = 0$. Then, $Bu + VR_0^{\pm}(\lambda)Bu = 0$. This implies

$$\lim_{\varepsilon \downarrow 0} \Im(Bu, R_0(\lambda \pm i\varepsilon)Bu) = -\lim_{\varepsilon \downarrow 0} \Im(VR_0(\lambda \pm i\varepsilon)Bu, R_0(\lambda \pm i\varepsilon)Bu) = 0.$$

We compute the LHS by using Fourier transform and (3.8). Write

$$\hat{f}(n,\xi) = \frac{1}{(2\pi)^{\frac{d+1}{2}}} \int_{\mathbf{T}} \int_{\mathbf{R}^d} e^{-i(x\xi + nt)} f(t,x)\,dt\,dx$$

Since $\widehat{Bu}(n,\xi) \in H^{\beta/2}(\mathbf{R}^d)$ and, hence, $\widehat{Bu}(n,\lambda\omega)$ is an $L^2(S_\omega^{d-1})$-valued $H^{\beta/2}$ function of λ on (ε,∞) for any $\varepsilon > 0$, we have

$$\begin{aligned} 0 = \lim_{\varepsilon \downarrow 0} \varepsilon \sum_{n \in \mathbf{Z}} \int \frac{|\widehat{Bu}(n,\xi)|^2 d\xi}{(\frac{1}{2}\xi^2 + n - \lambda)^2 + \varepsilon^2} \\ = \sum_{n<\lambda} \frac{\pi}{\sqrt{2(\lambda - n)}} \int_{|\xi| = \sqrt{2(\lambda - n)}} |\widehat{Bu}(n,\xi)|^2 d\sigma(\xi). \end{aligned}$$

It follows that $\widehat{Bu}(n, \sqrt{2(\lambda - n)}\omega) = 0$ and, by the standard division theorem in Sobolev spaces, we have $Bu = -VR_0^{\pm}(\lambda)Bu \in L^2(\mathbf{T}, L^2_{(3\beta-2)/2})$. Repeating this argument k times, we see that $Bu \in L^2(\mathbf{T}, L^2_{\beta/2+k(\beta-1)})$ for any $k = 1, 2, \dots$ and $v = R_0^{\pm}(\lambda)Bu \in L^2(\mathbf{T}, L^2_s)$ for any $s > 0$. But $(K_0 + V)v = \lambda v$ and λ is an eigenvalue of K. By elaborating the argument above by using the property (a) we can also show that the eigenfunctions

v with eigenvalues $\lambda \in \Omega$ satisfy, for $\alpha + \beta \leq 2$,

$$\|\partial_t^\alpha H(t)^\beta v\| \leq C\|v\|$$

with constant C independent of $\lambda \in \Omega$ (see Ref. 13 for details). Hence, the space of eigenfunctions with eigenvalues in Ω is finite dimensional and $\sigma_p(K) \cap \Omega$ is a finite set. □

4.3. *Scattering operator*

We suppose $V \in \mathcal{V}_\beta$ with $\beta > 1$. Then we define the scattering operator $S(s)$ for (1.1) with reference time s by

$$S(s) = W_+(s)^* W_-(s).$$

The scattering operator $S(s)$ is unitary in $\mathcal{H}$ and relates the asymptotic profiles at $t = \pm\infty$ of scattering solutions in the sense that, if $u_+ = S(s)u_-$, then for solutions $u(t)$ of (1.1) we have

$$\lim_{t\to-\infty} \|u(t) - e^{-i(t-s)H_0}u_-\| = 0 \Longleftrightarrow \lim_{t\to+\infty} \|u(t) - e^{-i(t-s)H_0}u_+\| = 0.$$

By the periodicity, $S(s)$ commutes with $e^{-2\pi i H_0}$:

$$e^{-2\pi i H_0} S(s) = S(s) e^{-2\pi i H_0}, \tag{4.13}$$

and $S(s)$ and $e^{-2\pi i H_0}$ may be simultaneously diagonalized. For $\lambda > 0$, write $\mathcal{X}(\lambda) = L^2(S^{d-1})$, and define the operator $\Gamma(\lambda) \in B(L^2_\gamma, \mathcal{X}(\lambda))$, $\gamma > 1/2$, by

$$\Gamma(\lambda)u(\omega) = (2\lambda)^{\frac{d-2}{4}} \hat{u}(\sqrt{2\lambda}\omega), \quad \omega \in S^{d-1}, \tag{4.14}$$

where $\hat{u}$ is the Fourier transform of u. It can be shown that $\lambda \to \Gamma(\lambda)$ is locally Hölder continuous, and if we define operator Γ by

$$\Gamma\colon \mathcal{H} \ni u \to \{\Gamma(\lambda)u\} \in \int_{\mathbf{R}^+}^{\oplus} \mathcal{X}(\lambda) d\lambda = L^2(\mathbf{R}^+, L^2(S^{d-1})),$$

then Γ is unitary and it satisfies $(\Gamma H_0 \Gamma^* \tilde{u})(\lambda, \omega) = \lambda \tilde{u}(\lambda, \omega)$, viz. Γ is a spectral representation for H_0. Define (omitting the variable s from $\hat{S}$)

$$\hat{S} = \Gamma S(s) \Gamma^*\colon \int_{\mathbf{R}^+}^{\oplus} \mathcal{X}(\lambda) d\lambda \to \int_{\mathbf{R}^+}^{\oplus} \mathcal{X}(\lambda) d\lambda.$$

Then, by virtue of (4.13), $\hat{S}$ satisfies $e^{-2\pi i\lambda}\hat{S} = \hat{S}e^{-2\pi i\lambda}$, and $\hat{S}$ can be decomposed into a sum of components

$$\hat{S} = \sum_{n=-\infty}^{\infty} \hat{S}_n(\lambda), \quad \hat{S}_n(\lambda)\colon \mathcal{X}(\lambda) \to \mathcal{X}(\lambda - n).$$

We call $\hat{S}_n(\lambda)$ the **scattering matrix**. The scattering matrix $\hat{S}_n(\lambda)$ may be expressed in terms of the boundary value of the resolvent $R(z) = (K-z)^{-1}$ (Ref. 42) in the form

$$\hat{S}_n(\lambda) = \delta_{n0} - 2\pi i \Gamma(\lambda - n)(V + VR(\lambda - i0)V)_{n0}\Gamma(\lambda)^* \tag{4.15}$$

where $\delta_{n0} = 0$ if $n \neq 0$ and δ_{00} is the identity operator on $\mathcal{X}(\lambda)$, and, for $G \in B(\mathcal{K})$, G_{nm} is the (n,m) component of $\mathcal{F}G\mathcal{F}^*$. Recall $\mathcal{F}\colon \mathcal{K} \to \bigoplus_{n\in\mathbf{Z}} \mathcal{H}$ is the unitary operator defined by (3.6). By using the expression (4.15) and following the argument of Isozaki-Kitada,[20] we have the following properties of the operator $\hat{S}_n(\lambda)$ (Ref. 43).

Theorem 4.7: Suppose $V \in C^\infty$ satisfies $|\partial_t^k \partial_x^\alpha V(t,x)| \leq C_{\alpha,k}\langle x\rangle^{-\beta-|\alpha|}$ for any α and k, $\beta > 1$. Then:

(1) For any $n \neq 0$, $\hat{S}_n(\lambda)$ is an integral operator on $L^2(S^{d-1})$ with smooth integral kernel $\hat{S}_n(\lambda,\xi,\eta)$. For any compact interval $\Omega \cap (\sigma_p(K) \cup \mathbf{Z}) = \emptyset$ and for any α, β and N, there exists a constant C such that

$$\sup_{\xi,\eta\in S^{d-1},\lambda\in\Omega} |\partial_\xi^\alpha \partial_\eta^\beta \hat{S}_n(\lambda,\xi,\eta)| \leq C\langle n\rangle^{-N}, \quad n \neq 0.$$

(2) The operator $\hat{S}_0(\lambda) - I$ is compact in $L^2(S^{d-1})$ and it has an integral kernel $\hat{S}_0(\lambda,\xi,\eta)$ which is C^∞ outside the diagonal set $\{(\xi,\xi) : \xi \in S^{d-1}\}$.

4.4. *Uniqueness of inverse scattering*

We show that the scattering operator determines V uniquely. For the inverse scattering theory for Schrödinger operators with time independent potentials, see Isozaki's article in this proceedings.

Theorem 4.8: Let $V_j \in \mathcal{V}_\beta$ for some $\beta > (d+2)/2$, $j = 1,2$, and let $S_j(s)$ be the corresponding scattering operators. Suppose $S_1(s) = S_2(s)$ for all $s \in \mathbf{T}$. Then $V_1(t,x) = V_2(t,x)$.

Proof: Denote $R^-(\lambda) = R(\lambda - i0)$ and write $J : \mathcal{H} \to \mathcal{K}$ for the identification operator defined by $Ju(t,x) = u(x)$. Define

$$K(\lambda,n,\omega,\theta) = \langle R^-(\lambda)VJ(e^{i\sqrt{2\lambda}x\cdot\theta}), Ve^{i\sqrt{2(\lambda-n)}x\cdot\omega}e^{int}\rangle_{\mathcal{K}}. \tag{4.16}$$

Then by virtue of (4.15) the integral kernel $S_n(\lambda,\omega,\theta)$ of $S_n(\lambda)$ is given by

$$\delta_{n,o}\delta(\omega-\theta) \\ -\frac{i(\lambda(\lambda-n))^{\frac{d-2}{4}}}{2^{\frac{d}{2}}\pi^{d-1}}\left(\int V_n(x)e^{i(\sqrt{2\lambda}\omega-\sqrt{2(\lambda-n)}\theta)\cdot x}dx+K(\lambda,n,\omega,\theta)\right).$$

Let Ω_ε be an ε-neighborhood of $\sigma_p(K)\cup\mathbf{Z}$, $0<\varepsilon<<1$. For $\xi\in\mathbf{R}^n$, we may choose sequences $\Omega_\varepsilon\not\ni\lambda_k\to\infty$ and $\omega_k,\theta_k\in S^{d-1}$, $\theta_k\neq\omega_k$ in such a way that $\sqrt{2\lambda_k}\omega_k-\sqrt{2(\lambda_k-n)}\theta_k\to\xi$ as $k\to\infty$. Thus, we have only to show

$$\sup_{\theta,\omega}|K(\lambda,n,\omega,\theta)|\to 0 \quad (\lambda\to\infty,\lambda\notin\Omega_\varepsilon), \tag{4.17}$$

as we then have $\lambda_k^{-(d-2)/2}S_n(\lambda_k,\omega_k,\theta_k)\to-\pi^{-(d-2)/2}i\hat{V}_n(\xi)$ as $k\to\infty$. Define M_γ by $M_\gamma u=\langle x\rangle^\gamma u$ for $d/2<\gamma<\beta-1$ and rewrite

$$K(\lambda,n,\omega,\theta)=\langle M_\gamma VR^-(\lambda)VM_\gamma J(M_{-\gamma}e^{i\sqrt{2\lambda}x\cdot\omega}),M_{-\gamma}e^{i\sqrt{2(\lambda-n)}x\cdot\theta}e^{int}\rangle_{\mathcal{K}}$$

Since $\|M_{-\gamma}e^{i\sqrt{2\lambda}x\cdot\omega}\|_{\mathcal{K}}$ and $\|M_{-\gamma}e^{i\sqrt{2(\lambda-n)}x\cdot\theta}e^{int}\|_{\mathcal{K}}$ are uniformly bounded with respect to λ, we have

$$\sup_{\theta,\omega}|K(\lambda,n,\omega,\theta)|\le C\|M_\gamma VR^-(\lambda)VM_\gamma J\|_{B(\mathcal{K})}. \tag{4.18}$$

Write V as a Fourier series

$$V(t,x)=\sum_n e^{int}V_n(x),\quad V_n(x)=\frac{1}{2\pi}\int_{\mathbf{T}}e^{-int}V(t,x)dt. \tag{4.19}$$

We have $|V_n(x)|\le C\langle n\rangle^{-2}\langle x\rangle^{-\beta}$. By virtue of (3.8), we have

$$\|M_\gamma VR_0^-(\lambda)VM_\gamma Ju\|_{\mathcal{K}}^2=\sum_n\|\sum_m M_\gamma V_{n-m}r_0(\lambda-m)V_mM_\gamma u\|_{\mathcal{H}}^2.$$

Using property (a) in the proof of Lemma 4.6, we estimate the RHS by

$$\begin{aligned} &C\sum_n\left(\sum_m\langle n-m\rangle^{-2}\langle m\rangle^{-2}\|M_{\gamma-\beta}r_0(\lambda-m)M_{\gamma-\beta}\|_{B(\mathcal{H})}\right)^2\|u\|_{\mathcal{H}}^2\\ &\le C\sum_n\left(\sum_m\langle n-m\rangle^{-2}\langle m\rangle^{-2}\langle\lambda-m\rangle^{-\frac12}\right)^2\|u\|_{\mathcal{H}}^2\\ &\le C\sum_n\left(\sum_m\langle n-m\rangle^{-2}\langle m\rangle^{-\frac32}\right)^2\langle\lambda\rangle^{-1}\|u\|_{\mathcal{H}}^2\le C\langle\lambda\rangle^{-1}\|u\|_{\mathcal{H}}^2. \end{aligned} \tag{4.20}$$

Thus, $\|M_\gamma V R_0^-(\lambda) V M_\gamma J\|_{B(\mathcal{K})} \le C\langle\lambda\rangle^{-1/2}$ and (4.17) follows from (4.18) because $\|M_\gamma V R^-(\lambda) V M_\gamma\|_{B(\mathcal{K})}$ is uniformly bounded outside Ω_ε and

$$\|M_\gamma V R^-(\lambda) V M_\gamma J\| \le (1 + \|M_\gamma V R^-(\lambda) M_{-\gamma}\|)\|M_\gamma V R_0^-(\lambda) V M_\gamma J\|.$$

This completes the proof. □

5. Local Decay Property

In this section, we assume $d = 3$ and $V \in \mathcal{V}_\beta$ with $\beta > 2$, and study the behavior as $t \to \infty$ of the solution $U(t,s)u_0$ for u_0 belonging to the continuous spectral subspace of $\mathcal{H}$ for the Floquet operator $\mathcal{U}_s = U(2\pi + s, s)$. We show that the solution has an asymptotic expansion in inverse powers of $\sqrt{t}$ in the topology of x-weighted spaces $L^2_{-\beta/2}$. Generically the leading term is $t^{-3/2}B_1 u_0$ with a finite rank operator B_1 which maps functions of x to functions of t and x, 2π periodic in t. If $n \in \mathbf{Z}$ is a threshold resonance of the Floquet Hamiltonian K, the leading behavior is $t^{-1/2}B_0 u_0$. As was mentioned in Theorem 3.7, $\mathcal{H}_p(\mathcal{U}_s)$ is finite dimensional and if $\mathcal{U}_s\phi = e^{-i2\pi\lambda}\phi$, then $U(t,s)\phi$ represents a quasi-periodic solution. For the details of this section, we refer to Ref. 13.

5.1. *Expansion for free Schrödinger equation*

The free propagator $U_0(t) = e^{-itH_0}$ has an explicit expression (2.3):

$$U_0(t)u_0(x) = \frac{e^{\mp i\frac{3\pi}{4}}}{|2\pi t|^{\frac{3}{2}}} \int e^{\frac{i(x-y)^2}{2t}} u_0(y)dy. \tag{5.1}$$

If u_0 decays sufficiently rapidly at infinity and $u_0 \in L^2_{\delta+k}(\mathbf{R}^3)$ for $\delta > 1/2$ and $k = 1, 2, \ldots$, then by expanding as

$$e^{\frac{i(x-y)^2}{2t}} = 1 + \frac{i(x-y)^2}{2t} + \frac{1}{2!}\left(\frac{i(x-y)^2}{2t}\right)^2 + \cdots,$$

we see from (5.1) that $U_0(t)u_0$ may be expanded up to the order $t^{-(k+3)/2}$ in $L^2_{-\delta-k}(\mathbf{R}^3)$ as

$$\|U_0(t)u_0 - (t^{-\frac{3}{2}}B_1 + \cdots + t^{-\frac{k+2}{2}}B_k)u_0\|_{L^2_{-\delta-k}} = O(t^{-\frac{k+2+\varepsilon}{2}}). \tag{5.2}$$

Here $B_{2k} = 0$ and B_{2k-1}, $k = 1, 2, \ldots$, are finite rank operators defined by

$$B_{2k-1}u_0(x) = \frac{e^{-i\frac{3\pi}{4}} i^{k-1}}{(2\pi)^{\frac{3}{2}} 2^{k-1}(k-1)!} \int (x-y)^{2(k-1)} u_0(y)dy.$$

In this section, we prove the expansion formula of this type for time periodic Schrödinger equations.

Remark 5.1: In Section 4, we obtained the expansion formula (4.9) for $U_0(t)u_0$ which is somewhat different from (5.2). Though (4.9) is not quite an asymptotic expansion, it is an expansion in $L^2(\mathbf{R}^d)$ and, therefore, is on some occasions more useful than (5.2). It seems to us considerably more difficult, however, to obtain an expansion of the form (4.9) for general time periodic Schrödinger equations.

5.2. *Threshold resonance, theorems*

The asymptotic expansion formula of the form (5.2) has been obtained[21,40] for general Schrödinger equations with time independent potentials. It is known in these works that the leading power in the expansion is different if 0 is a threshold resonance of the operator $-(1/2)\Delta + V$. For time-periodic systems we define the analogous notion of threshold resonance as follows.

Definition 5.2: We say that $n \in \mathbf{Z}$ is a threshold resonance of K, if there exists a function $u(t,x)$ such that $-i\partial_t u - (1/2)\Delta u + V(t,x)u = nu$ and such that

$$u(t,x) = \frac{Ce^{int}}{|x|} + u_1(t,x), \quad u_1 \in H^1(\mathbf{T}, H^1(\mathbf{R}^3))$$

for some constant $C \neq 0$. We call such a function an n-resonant solution. We say V is of generic type if 0 is neither an eigenvalue nor a threshold resonance of K and of exceptional type otherwise.

We remark, by virtue of the shift invariance (3.4) of the spectrum of K, n is a threshold resonance if and only if so is 0. It can be shown that 0 is threshold resonance only if $-1 \in \sigma(VR^{\pm}(\lambda))$ in $\mathcal{K}_{\beta/2} = L^2(\mathbf{T}, \mathcal{H}_\delta)$. Since $VR^{\pm}(\lambda)$ is compact in $\mathcal{K}_{\beta/2}$ and depends continuously on V, the generic potentials are dense and open in $\mathcal{V}_\beta$. We write

$$\mathcal{K}_\delta^s = H^s(\mathbf{T}, \mathcal{H}_\delta)$$

for the $\mathcal{H}_\delta$ valued Sobolev space over $\mathbf{T}$. We have the following theorem.

Theorem 5.3: Let $V \in \mathcal{V}_\beta$ for $\beta > \beta_k \equiv \max\{2k+1, 4\}$, $k \in \mathbf{N}$, and let $\{\phi_j\}$ be an orthonormal basis of eigenfunctions of K with the eigenvalues $0 \le \lambda_j < 1$. Set $\delta = \beta/2$ and $\varepsilon_0 = \min\{1, \frac{\beta-\beta_k}{2}\}$. Then:

(1) Suppose V is of generic type. Then there exist finite rank operators $B_1, \ldots, B_k$ from $\mathcal{H}_\delta$ to $\mathcal{K}^1_{-\delta}$, such that $B_j = 0$, unless j is odd, and such that, for any $u_0 \in \mathcal{H}_\delta$ and for any ε, $0 < \varepsilon < \varepsilon_0$, as $t \to \infty$,

$$\|U(t,0)u_0 - \Big(\sum_j c_j e^{-it\lambda_j}\phi_j(t,\cdot) + t^{-\frac{3}{2}}B_1u_0(t,\cdot) + \cdots \\ + t^{-\frac{k+2}{2}}B_ku_0(t,\cdot)\Big)\|_{\mathcal{H}_{-\delta}} \leq C\langle t\rangle^{-\frac{k+2+\varepsilon}{2}}\|u_0\|_{\mathcal{H}_\delta}, \quad (5.3)$$

where $c_j = 2\pi(\phi_j(0), u_0)_{\mathcal{H}}$.

(2) Suppose V is of exceptional type, $\beta > \beta_k$, $k \geq 2$, and $\{\phi_{01}, \ldots, \phi_{0m}\} \subset \{\phi_j\}$ is an ONB of eigenfunctions of K with eigenvalue 0. Then there exist a 0-resonant solution $\psi(t,x)$, finite rank operators $B_1, \ldots, B_{k-2}$ from $\mathcal{H}_\delta$ to $\mathcal{K}^1_{-\delta}$, such that $B_j = 0$, unless j is odd, and such that, for any $u_0 \in \mathcal{H}_\delta$ and for any $0 < \varepsilon < \varepsilon_0$ as $t \to \infty$,

$$\|U(t,0)u_0 - \Big[\sum_j c_j e^{-it\lambda_j}\phi_j(t,x) + t^{-\frac{1}{2}}\Big(d_0\psi(t,x) + \sum_{\ell=1}^m d_\ell\phi_{0\ell}(t,x)\Big) \\ + t^{-\frac{3}{2}}B_1u_0(t,x) + \cdots + t^{-\frac{k}{2}}B_{k-2}u_0(t,x)\Big]\|_{\mathcal{H}_{-\delta}} \leq C\langle t\rangle^{-\frac{k+\varepsilon}{2}}\|u_0\|_{\mathcal{H}_\delta}, \quad (5.4)$$

where c_j are as in (1), $d_0 = \langle u_0, \psi(0)\rangle_{\mathcal{H}}$, and d_ℓ are linear functionals of u_0 of the form

$$d_\ell(u_0) = a_{\ell 1}(u_0, \phi_{01}(0))_{\mathcal{H}} + \cdots + a_{\ell n}(u_0, \phi_{0m}(0))_{\mathcal{H}}, \quad \ell = 1, \ldots, m.$$

In particular, $d_1, \ldots, d_m$ vanish if u_0 is orthogonal to the 1-eigenspace of $U(2\pi, 0)$.

5.3. *Threshold expansion of resolvent*

We prove Theorem 5.3 by using the expression of $e^{-i\sigma K}$:

$$e^{-i\sigma K} = \lim_{\varepsilon \downarrow 0}\frac{1}{2\pi i}\int e^{-i\sigma\lambda}R(\lambda + i\varepsilon)\,d\lambda \quad (5.5)$$

and by studying $R(z)$ near and on the real line. The following expansion formulae for the resolvent $R(z)$ as $z \to n$, $n \in \mathbf{Z}$, may be useful for some other purposes. For estimating the remainders, we use the following notation. We write $L^{(j)}(z)$ for the j-th derivative of $L(z)$, and $f \otimes g$ for the integral operator on $\mathbf{T} \times \mathbf{R}$ with the kernel $f(t,x)\overline{g(s,y)}$.

Definition 5.4: Let X, Y be Banach spaces, and $L(z)$ a $B(X,Y)$-valued function defined in a punctured neighborhood $U = \{z \in \overline{\mathbf{C}^+}\colon 0 < |z| < \rho\}$

of the origin of $\overline{\mathbf{C}^+}$. For $k \in \mathbf{N}_0$ and $0 \le \varepsilon < 1$, we say $L(z) = \mathcal{O}(z^{\frac{k+\varepsilon}{2}})$, if $L(z)$ satisfies the following properties:

(i) $L(z) \in C^k(U)$ and it satisfies

$$\|(d/dz)^j L(z)\| \le C_j |z|^{\frac{k+\varepsilon}{2}-j}, \quad j = 0, 1, \ldots, k, \quad z \in U. \tag{5.6}$$

(Hence, $L(z)$ is $C^{[k/2]}(U \cup \{0\})$, if we set $L(0) = 0$.)

(ii) For $\ell = [(k+2)/2]$, there exist $\mu > 0$ and $\gamma > 0$ such that, for $0 < h < \gamma$,

$$\int_{-\mu}^{\mu} \|L^{(\ell)}(z+h) - L^{(\ell)}(z)\| dz \le \begin{cases} C|h|^{\frac{\varepsilon}{2}} & \text{if } k \text{ is even,} \\ C|h|^{\frac{1+\varepsilon}{2}} & \text{if } k \text{ is odd.} \end{cases} \tag{5.7}$$

Theorem 5.5: Let $V \in \mathcal{V}_\beta$ for $\beta > \beta_k \equiv \max\{2k+1, 4\}$, $k \in \mathbf{N}$. Let $\delta = \beta/2$ and $\varepsilon_0 = \min\{1, \frac{\beta-\beta_k}{2}\}$. Suppose that V is of generic type. Then, as a $B(\mathcal{K}_\delta^s, \mathcal{K}_{-\delta}^s)$-valued function of $z \in \overline{\mathbf{C}^+}$, $s = 0, 1$, we have

$$R(z+n) = F_0(n) + \sqrt{z} F_1(n) + \cdots + z^{\frac{k}{2}} F_k(n) + \mathcal{O}(z^{\frac{k+\varepsilon}{2}}) \tag{5.8}$$

in a neighborhood of $z = 0$ for any $0 < \varepsilon < \varepsilon_0$. Here

(1) $F_j(n) = \mathcal{T}_n F_j(0) \mathcal{T}_n^*$ for all $n \in \mathbf{Z}$ and $j = 0, 1, \ldots$, where $\mathcal{T}_n$ is the multiplication by e^{-int} (see (3.4)).

(2) If j is odd, $F_j(0) = \sum a_{j\nu} \otimes b_{j\nu}$ with $a_{j\nu}, b_{j\nu} \in \mathcal{K}_{-\delta}^1$ (finite sum).

Theorem 5.6: Let $V \in \mathcal{V}_\beta$ for $\beta > \beta_k \equiv 2k+1$, $k \ge 2$ an integer. Let $\delta = \beta/2$, and $0 < \varepsilon < \varepsilon_0 = \min\{1, \frac{\beta-\beta_k}{2}\}$. Suppose that V is of exceptional type. Then, as a $B(\mathcal{K}_\delta^s, \mathcal{K}_{-\delta}^s)$-valued function of $z \in \overline{\mathbf{C}^+}$, $s = 0, 1$,

$$R(z+n) = -\frac{F_{-2}(n)}{z} + \frac{F_{-1}(n)}{\sqrt{z}} + \cdots + z^{\frac{k-2}{2}} F_{k-2}(n) + \mathcal{O}(z^{\frac{k-2+\varepsilon}{2}}) \tag{5.9}$$

in a neighborhood of $z = 0$. Here,

(1) $F_j(n) = \mathcal{T}_n F_j(0) \mathcal{T}_n^*$ for $n \in \mathbf{Z}$ and $j = -2, -1, \ldots$.

(2) If j is odd then $F_j(n) = \sum a_{j\nu} \otimes b_{j\nu}$ with $a_{j\nu}, b_{j\nu} \in \mathcal{K}_{-\delta}^1$ (finite sum).

(3) $F_{-2}(n) = E_K(\{n\})$.

(4) $F_{-1}(n) = E_K(\{n\}) V D_3(n) V E_K(\{n\}) - 4\pi i \bar{Q}_n$, where $\bar{Q}_n = \langle \cdot, \psi^{(n)} \rangle \psi^{(n)}$, $\psi^{(n)}$ is a suitably normalized n-resonant function and $D_3(n)$ is an integral operator with the kernel $Ce^{in(t-s)}|x-y|^2$.

5.4. *Outline of proof of Theorem* 5.3

We outline the proof when non-threshold eigenvalues are absent from K as they can be handled by a similar and simpler argument (see Ref. 13). We write $J\colon \mathcal{H} \to \mathcal{K}$ for the identification operator $(Ju_0)(t,x) = u_0(x)$. We apply (5.8) or (5.9) to the expression (5.5) and obtain the asymptotic expansion of $e^{-i\sigma K}J$ as $\sigma \to \infty$. When V is of generic type, the result is

$$e^{-i\sigma K}J = \sigma^{-\frac{3}{2}}Z_1(\sigma) + \cdots + \sigma^{-\frac{k+2}{2}}Z_k(\sigma) + O(\sigma^{-\frac{k+2+\varepsilon}{2}}) \tag{5.10}$$

as a $B(\mathcal{H}_\delta, \mathcal{K}^1_{-\delta})$-valued function. Here $Z_j(\sigma)$ is 2π-periodic in σ, which vanishes if j is even, and has the form $Z_j(\sigma) = C_j \sum_n e^{-in\sigma}F_j(n)J$ with a universal constant C_j if j is odd. Because $F(n) = \mathcal{T}_n F_j(0)\mathcal{T}_n^*$ and $F_j(0) = \sum_\nu a_{j\nu} \otimes b_{j\nu}$, where $a_{j\nu}, b_{j\nu} \in \mathcal{K}^1_{-\delta}$, by Theorem 5.5, the Fourier inversion formula implies

$$\begin{aligned} Z_j(\sigma)u_0(t,x) &= C_j \sum_\nu \sum_n e^{-in(\sigma-t)} a_{j\nu}(t,x) \int_{\mathbf{T}\times\mathbf{R}^3} \overline{b_{j\nu}(s,y)} e^{-ins} u_0(y)dsdy \\ &= 2\pi C_j \sum_\nu a_{j\nu}(t,x) \int_{\mathbf{R}^3} \overline{b_{j\nu}(t-\sigma,y)} u_0(y)dy. \end{aligned} \tag{5.11}$$

Since $\mathcal{K}^1_{-\delta}$ is continuously embedded in $C(\mathbf{T}, \mathcal{H}_{-\delta})$ by the Sobolev embedding theorem, (5.10) implies that, uniformly with respect to $t \in \mathbf{T}$ (hence with respect to $t \in \mathbf{R}$ by the periodicity), as $\sigma \to \infty$,

$$\begin{aligned} &\|U(t,t-\sigma)u_0 - \sigma^{-\frac{3}{2}}Z_1(\sigma)u_0(t) - \cdots \\ &\quad \cdots - \sigma^{-\frac{k+2}{2}}Z_k(\sigma)u_0(t)\|_{\mathcal{H}_{-\delta}} = O(\sigma^{-\frac{k+2+\varepsilon}{2}})\|u_0\|_{\mathcal{H}_\delta}. \end{aligned} \tag{5.12}$$

We set $t = \sigma$ in (5.12) and replace σ by t. We then obtain (5.3) with

$$B_j(t) = 2\pi C_j \sum_\nu a_{j\nu}(t,x) \otimes b_\nu(0,y). \tag{5.13}$$

The proof of Theorem 5.3 for the exceptional case can be carried out along the same line by applying (5.9) instead of (5.8).

Corollary 5.7: *Let $V \in \mathcal{V}_\beta$ with $\beta > 4$ and let $\delta = \beta/2$. Suppose that V is of generic type. Then there exists a constant $C > 0$ such that*

$$\|U(t,0)P_c(0)\phi\|_{\mathcal{H}_{-\delta}} \le C\langle t\rangle^{-\frac{3}{2}}\|\phi\|_{\mathcal{H}_\delta}, \quad u \in \mathcal{H}_\delta. \tag{5.14}$$

6. Dispersive Estimates

In Section 2, we used Strichartz estimates for the free propagator $U_0(t)$, Lemma 2.8, for studying the wellposedness of the IVP for (1.1). The estimates simultaneously represent the time decay property and the smoothing property of solutions, and $u(t) = U_0(t)u_0$ is for almost all $t \in \mathbf{R}$ smoother than $u_0 \in L^2$ is in the sense $u(t) \in L^p$ for some $p > 2$. In fact, it is known that $U_0(t)$ improves the differentiablity property by the order 1/2 in the Sobolev sense. In this section, we show that solutions of time periodic Schrödinger equation (1.1) satisfy similar properties. We provide a proof for the results of this section, as they seem to be new. We assume $d = 3$ in this section.

6.1. L^p-L^q estimates and Strichartz estimates

By applying the Hausdorff-Young inequality to (4.8), we see that the free propagator $U_0(t) = e^{-itH_0}$ satisfies

$$\|U_0(t)u\|_p \le C_p|t|^{-d(\frac{1}{2}-\frac{1}{p})}\|u\|_{\frac{p}{p-1}}, \quad t \neq 0, \tag{6.1}$$

for any $2 \le p \le \infty$. We call estimates of the type (6.1) L^p-L^q estimates. It is well known that the Strichartz estimates automatically follow from L^p-L^q estimates via certain harmonic analysis machinery. In this subsection we prove the L^p-L^q estimates for solutions $U(t,0)u_0$ with $u_0 \in \mathcal{H}_c(\mathcal{U}_s)$. L^p-L^q estimates are known if V is t-independent [22,69,70,71] or if V is a general (not necessarily time periodic) but sufficiently small potential.[51] We write $V(t)$ for the multiplication operator with $V(t,x)$ and

$$\hat{V}(t,p) = \frac{1}{(2\pi)^3}\int_{\mathbf{R}^3} e^{-ipx}V(t,x)dx.$$

Theorem 6.1: Let $V \in \mathcal{V}_\beta$ with $\beta > 4$ be of generic type. Assume

$$M \equiv \sup \|\hat{V}(t,\cdot)\|_{L^1} < \infty, \tag{6.2}$$

in addition. Then:

(1) For any p, q such that $0 \le 1/p = 1 - 1/q \le 1/2$, we have

$$\|U(t,s)P_c(s)\phi\|_{L^p} \le C_p|t-s|^{-3(\frac{1}{2}-\frac{1}{p})}\|\phi\|_{L^q}, \quad \phi \in L^2 \cap L^p. \tag{6.3}$$

(2) Let $2 \le p \le 6$ and $2 \le \theta = \dfrac{4p}{3(p-2)} \le \infty$. Then:

$$\|U(t,s)P_c(s)\phi\|_{L^\theta(\mathbf{R}^1_t, L^p(\mathbf{R}^3_x))} \le C\|\phi\|_{L^2(\mathbf{R}^3)}, \quad \phi \in L^2.$$

Our proof here heavily uses (6.2), however, it is an unwelcome condition for various reasons. Readers are encouraged to try to prove Theorem 6.1 without the condition (6.2). We prove the first statement only as the second follows from the first by the well known argument (see Ref. 27). We begin the proof with the following lemma. We borrow the idea of using the following function spaces from Ref. 52. We write $\|u\|_p$ for $\|u\|_{L^p(\mathbf{R}^3)}$ and $\|T\|_{p\to q}$ for $\|T\|_{B(L^p,L^q)}$. We define

$$\mathcal{X} = L^1 \cap L^2, \quad \mathcal{Y} = L^2 + L^\infty.$$

They are Banach spaces with respective norms

$$\|u\|_{\mathcal{X}} = \|u\|_{L^1} + \|u\|_{L^2}, \quad \|u\|_{\mathcal{Y}} = \inf_{u=u_1+u_2}(\|u_1\|_{L^2} + \|u_2\|_{L^\infty}).$$

The free propagator $U_0(t) = e^{-itH_0}$ satisfies the obvious estimate

$$\|U_0(t)u\|_{\mathcal{Y}} \le C\langle t\rangle^{-\frac{3}{2}}\|u\|_{\mathcal{X}}. \tag{6.4}$$

Lemma 6.2: *Suppose V satisfies* (6.2). *Then:*

(1) *With the constant M of* (6.2) *we have*

$$\sup_{t,s\in\mathbf{R}} \|U_0(-t)V(s)U_0(t)u\|_{L^p} \le M\|u\|_{L^p}, \quad 1 \le p \le \infty. \tag{6.5}$$

(2) *There exists a constant $C > 0$ such that for $0 \le r \le t \le \infty$*

$$\left\| \int_0^r U_0(t-s)V(s)U_0(s)uds \right\|_{L^\infty} \le C|t|^{-\frac{3}{2}} \min(1,r)\|u\|_{L^1}. \tag{6.6}$$

(3) *For $|t-s| \le T$ and $1 \le p \le \infty$,*

$$\|U_0(s-t)U(t,s)u\|_{L^p} + \|U(t,s)U_0(s-t)u\|_{L^p} \le e^{MT}\|u\|_{L^p}. \tag{6.7}$$

Proof:
(1) Since $U_0(-t)xU_0(t) = x+tD$, $D = -i\nabla_x$, we have for every s and $t \in \mathbf{R}$

$$U_0(-t)V(s,x)U_0(t) = V(s,x+tD) = \int \hat{V}(s,\xi)e^{i(x+tD)\xi}d\xi. \tag{6.8}$$

Here $e^{i(x+tD)\xi}u(x) = e^{i\xi^2/2t}e^{-ix\xi}u(x+t\xi)$ and $\|e^{i(x+tD)\xi}u\|_p = \|u\|_p$. Integration yields the desired estimate (6.5).
(2) When $0 \le r \le 2$, we write

$$\int_0^r U_0(t-s)V(s)U_0(s)uds = U_0(t)\cdot \int_0^r U_0(-s)V(s)U_0(s)uds.$$

By using (6.5), we estimate the L^∞ norm of the right by

$$\|U_0(t)\|_{1\to\infty}\int_0^r \|U_0(-s)V(s)U_0(s)u\|_1 ds \le (2\pi t)^{-\frac{3}{2}}|r|M\|u\|_1.$$

This implies (6.6) for $0 < r < 2$. When $2 < r \le t$, we split the integral

$$\left(\int_{r-1}^r + \int_1^{r-1} + \int_0^1\right) U_0(t-s)V(s)U_0(s)uds = I_1 + I_2 + I_3.$$

For I_1 and I_3, we write $U_0(t-s)V(s)U_0(s)u = U_0(t)\cdot U_0(-s)V(s)U_0(s)u$ and estimate them as in the case $0 < r < 2$. The result is

$$\|I_1\|_\infty \le M(2\pi)^{-\frac{3}{2}}|t|^{-\frac{3}{2}}\|u\|_1, \quad \|I_3\|_\infty \le M(2\pi)^{-\frac{3}{2}}|t|^{-\frac{3}{2}}\|u\|_1. \tag{6.9}$$

For I_2 we use (6.1) and estimate as follows:

$$\begin{aligned}\|U_0(t-s)V(s)U_0(s)u\|_\infty &\le \|U_0(t-s)\|_{1\to\infty}\|V(s)\|_{\infty\to 1}\|U_0(s)u\|_\infty \\ &\le (2\pi)^{-3}C_V|t-s|^{-\frac{3}{2}}|s|^{-\frac{3}{2}}\|u\|_1.\end{aligned}$$

It follows that $\|I_2\|_\infty \le C\langle t\rangle^{-3/2}$, since $|s| \ge 1$ and $|t-s| \ge 1$ on the interval of integration. Combining this with (6.9) yields (6.6) for $r \ge 2$.
(3) We prove (6.7) for $U_0(s-t)U(t,s)$ for $s = 0$ only. Since $V(t,x)$ is bounded, the following expansion converges in the operator norm of $\mathcal{H}$:

$$U(t,0) = U_0(t) + (-i)\int_{0<r<t} U_0(t-r)V(r)U_0(r)dr$$

$$+ (-i)^2\int_{0<r_1<r_2<t} U_0(t-r_2)V(r_2)U_0(r_2-r_1)V(r_1)U_0(r_1)dr_1dr_2 + \cdots.$$

Define $\Gamma(t) = U_0(0,t)V(t)U_0(t,0)$. It then follows that

$$U_0(t)^{-1}U(t,0) = \sum_{n=0}^\infty (-i)^n\int_{0<r_1<\cdots<r_n<t} \Gamma(r_n)\cdots\Gamma(r_1)dr_1\cdots dr_n$$

By virtue of (6.5), we have $\|\Gamma(r)\|_{p\to p} \le M$ and

$$\|U_0(t)^{-1}U(t,0)\|_{p\to p} \le \sum_{n=0}^\infty \frac{M^n t^n}{n!} = e^{Mt}.$$

A similar argument applies to $U(t,0)U_0(0,t)$. □

We write $U_0(t-s) = U_0(t,s)$ in accordance with the notation $U(t,s)$. Iterate the Duhamel formulae

$$U(t,s) = U_0(t,s) - i\int_s^t U_0(t,r)V(r)U(r,s)ds \tag{6.10}$$

$$= U_0(t,s) - i\int_s^t U(t,r)V(r)U_0(r,s)dr \tag{6.11}$$

and multiply the resulting equations by $P_c(s)$ from the right or by $P_c(t)$ from the left. Using (3.11), we obtain

$$
\begin{aligned}
U(t,s)P_c(s) &= U_0(t,s)P_c(s) - i\int_s^t U_0(t,r)V(r)U_0(r,s)P_c(s)dr \\
&\quad - \int_{s<r<\theta<t} U_0(t,\theta)V(\theta)U_0(\theta,r)V(r)U(r,s)P_c(s)drd\theta
\end{aligned} \tag{6.12}
$$

$$
\begin{aligned}
&= P_c(t)U_0(t,s) - iP_c(t)\int_s^t U_0(t,r)V(r)U_0(r,s)dr \\
&\quad - \int_{s<r<\theta<t} U(t,\theta)P_c(\theta)V(\theta)U_0(\theta,r)V(r)U_0(r,s)drd\theta.
\end{aligned} \tag{6.13}
$$

Insert (6.11) into (6.10) after multiplying the former by $P_c(t)$ from the left and the latter by $P_c(s)$ from the right. Using (3.11) again, we obtain

$$
\begin{aligned}
U(t,s)P_c(s) &= U_0(t,s)P_c(s) - i\int_s^t U_0(t,r)V(r)P_c(r)U_0(r,s)ds \\
&\quad - \int_{s<r<\theta<t} U_0(t,\theta)V(\theta)U(\theta,r)P_c(r)V(r)U_0(r,s)drd\theta.
\end{aligned} \tag{6.14}
$$

Lemma 6.3: *Suppose that V is of generic type. Then:*

(1) *There exists a constant $C>0$ such that*

$$\|U_0(t,s)P_c(s)\|_{1\to\infty} + \|P_c(t)U_0(t,s)\|_{1\to\infty} \le C|t-s|^{-\frac{3}{2}}. \tag{6.15}$$

(2) *Denote by $U_1(t,s)$ and $U_2(t,s)$ the second terms on the right of (6.12) and (6.13) respectively. Then we have, for $j=1,2$,*

$$
\begin{aligned}
&\|U_j(t,s)\|_{2\to 2} \le C, \\
&\|U_j(t,s)\|_{1\to\infty} \le C\min\{|t-s|^{-\frac{1}{2}}, |t-s|^{-\frac{3}{2}}\}.
\end{aligned} \tag{6.16}
$$

(3) *Let $U_3(t,s) = \int_s^t U_0(t,r)V(r)P_c(r)U_0(r,s)dr$. Then*

$$\|U_3(t,s)\|_{B(\mathfrak{X},\mathfrak{Y})} \le C\langle t-s\rangle^{-\frac{3}{2}}. \tag{6.17}$$

Proof: From the property (3.14) of eigenfunctions of K we deduce that

$$\|P_c(t)\|_{p\to p} \le C_p, \quad 1\le p\le\infty; \quad \|P_c(t)\|_{B(\mathfrak{Y},\mathfrak{X})} \le C. \tag{6.18}$$

Thus, (6.15) follows from (6.1). Choose $\gamma > 1$ such that $|\langle x\rangle^{2\gamma}V(t,x)| \le C$ and $\|\langle x\rangle^{\gamma}V(r)P_c(r)\langle x\rangle^{\gamma}\|_{2\to 2} \le C$. Since the multiplication by $\langle x\rangle^{-\gamma}$ is H_0-smooth by virtue of Theorem 4.5, we have

$$|(U_1(t,s)u,v)| = \left|\int_s^t (U_0(t,r)V(r)U_0(r,s)P_c(s)u,v)dr\right|$$
$$\le C\int_s^t \|\langle x\rangle^{-\gamma}U_0(r,s)P_c(s)u\|_2\|\langle x\rangle^{-\gamma}U_0(r,t)v\|_2 dr \le C\|u\|_2\|v\|_2$$

and the first estimate of (6.16) follows for $U_1(t,s)$. Operator $U_2(t,s)$ likewise satisfies the first of (6.16). The second of (6.16) follows from (6.6) and (6.18). By virtue of (6.4), we have

$$\|U_3(t,s)\|_{B(\mathcal{X},\mathcal{Y})}$$
$$\le \int_s^t \|U_0(t,r)\|_{B(\mathcal{X},\mathcal{Y})}\|V(r)P_c(r)\|_{B(\mathcal{Y},\mathcal{X})}\|U_0(r,s)\|_{B(\mathcal{X},\mathcal{Y})}dr$$
$$\le C\int_s^t \langle t-r\rangle^{-\frac{3}{2}}\langle r-s\rangle^{-\frac{3}{2}}dr \le C\langle t-s\rangle^{-\frac{3}{2}}.$$

This proves (6.17). □

We first show that the weighted L^2 estimate of Corollary 5.7 implies the following $\mathcal{X}$–$\mathcal{Y}$ estimate.

Lemma 6.4: *There exists $C > 0$ such that*

$$\|U(t,s)P_c(s)u\|_{\mathcal{Y}} \le C\langle t-s\rangle^{-\frac{3}{2}}\|u\|_{\mathcal{X}} \tag{6.19}$$

Proof: We prove the case $s = 0$ only. We use (6.14). By virtue of (6.4), (6.18) and (6.17), we have only to show that the last summand on the right of (6.14) satisfies (6.19). We take $\delta > 3/2$ such that $\|V(t)\|_{B(\mathcal{H}_{-\delta},\mathcal{X})}, \|V(t)\|_{B(\mathcal{Y},\mathcal{H}_\delta)} \le C$ and, by using (6.4) and (5.14) we estimate

$$\begin{aligned}
&\|U_0(t,\theta)V(\theta)U(\theta,r)P_c(r)V(r)U_0(r,0)u\|_{\mathcal{Y}}\\
&\quad\le \|U_0(t,\theta)\|_{B(\mathcal{X},\mathcal{Y})}\cdot\|V(\theta)\|_{B(\mathcal{H}_{-\delta},\mathcal{X})}\cdot\|U(\theta,r)\|_{B(\mathcal{H}_\delta,\mathcal{H}_{-\delta})}\\
&\qquad\times\|V(r)\|_{B(\mathcal{Y},\mathcal{H}_\delta)}\cdot\|U_0(r,0)\|_{B(\mathcal{X},\mathcal{Y})}\|u\|_{\mathcal{X}}\\
&\quad\le C\langle t-\theta\rangle^{-\frac{3}{2}}\langle\theta-r\rangle^{-\frac{3}{2}}\langle r\rangle^{-\frac{3}{2}}\|u\|_{\mathcal{X}}.
\end{aligned}\tag{6.20}$$

Integrating (6.20) over $0 < r < \theta < s$, we see that the third summand also satisfies the desired estimate (6.19). □

Out of the $\mathcal{X}$–$\mathcal{Y}$ estimate (6.19), we have the following $\mathcal{X}$–L^∞ estimate.

Lemma 6.5: *There exists $C > 0$ such that*

$$\|U(t,s)P_c(s)u\|_\infty \leq C|t-s|^{-\frac{3}{2}}\|u\|_{\mathcal{X}} \tag{6.21}$$

Proof: We prove the case $s = 0$ only. The following argument is an adaptation of the one of Ref. 52. We may assume $t > 1$ by virtue of Lemma 6.2 (3). We use (6.12). By virtue of Lemma 6.3, it suffices to prove that the last integral on the right of (6.12), which we denote by $U_4(t,s)$, satisfies (6.21). We write $U_4(t,0)$ in the form

$$U_4(t,0) = \int_0^t F(t,r)V(r)U(r,0)P_c(0)dr \tag{6.22}$$

by defining

$$F(t,r) = \int_r^t U_0(t,\theta)V(\theta)U_0(\theta,r)d\theta. \tag{6.23}$$

Applying (6.6) for $F(t,r)$ and Lemma 6.4 for $U(r,0)P_c(0)$, we estimate

$$\begin{aligned} &\|F(t,r)V(r)U(r,0)P_c(0)\|_{B(\mathcal{X},L^\infty)} \\ &\quad \leq \|F(t,r)\|_{1\to\infty}\|V(r)\|_{B(\mathcal{Y},L^1)}\|U(r,0)P_c(0)\|_{B(\mathcal{X},\mathcal{Y})} \\ &\quad \leq C\min\{|t-r|^{-\frac{3}{2}}, |t-r|^{-\frac{1}{2}}\}\langle r\rangle^{-\frac{3}{2}}. \end{aligned} \tag{6.24}$$

It follows that

$$\|U_4(t,0)\|_{B(\mathcal{X},L^\infty)} \leq \int_0^{t-1} \frac{\langle r\rangle^{-\frac{3}{2}}dr}{|t-r|^{\frac{3}{2}}} + \int_{t-1}^t \frac{\langle r\rangle^{-\frac{3}{2}}dr}{|t-r|^{\frac{1}{2}}} \leq C\langle t\rangle^{-\frac{3}{2}}.$$

This completes the proof. □

Completion of the Proof of Theorem 6.1

Proof: By virtue of interpolation theory, it suffices to prove (6.3) for $(p,q) = (\infty,1)$. Thanks to Lemma 6.2 [Statement (3)], we may assume $s = 0$ and $t > 2$. We use formula (6.13). By virtue of Lemma 6.3, it suffices to deal with the last integral on the right, which we decompose into three parts and write in the following form:

$$\int_0^1 U(t,\theta)P_c(\theta)V(\theta)F(\theta,0)ud\theta + \int_1^{t-1} U(t,\theta)P_c(\theta)V(\theta)F(\theta,0)ud\theta$$
$$+ \int_{t-1}^{t} P_c(t)U(t,\theta)V(\theta)U_0(\theta,t)\left(\int_0^\theta U_0(t,r)V(r)U_0(r,0)udr\right)d\theta$$
$$= I_1(t) + I_2(t) + I_3(t). \tag{6.25}$$

Here $F(t,s)$ is defined by (6.23), we used $U(t,\theta)P_c(\theta) = P_c(t)U(t,\theta)$ and decomposed $U_0(\theta,r) = U_0(\theta,t)U_0(t,r)$ in the third summand. Estimates (6.21) for $U(t,\theta)P_c(\theta)$ and (6.6) for $F(\theta,0)$ imply

$$\begin{aligned} &\|U(t,\theta)P_c(\theta)V(\theta)F(\theta,0)u\|_\infty \\ &\quad \le \|U(t,\theta)P_c(\theta)\|_{B(\mathcal{X},L^\infty)} \cdot \|V(\theta)\|_{B(L^\infty,\mathcal{X})} \cdot \|F(\theta,0)u\|_\infty \\ &\quad \le C\begin{cases} |t-\theta|^{-\frac{3}{2}}|\theta|^{-\frac{1}{2}}\|u\|_1, & 0<\theta<1, \\ |t-\theta|^{-\frac{3}{2}}|\theta|^{-\frac{3}{2}}\|u\|_1, & 1<\theta<t-1. \end{cases} \end{aligned} \tag{6.26}$$

It follows that

$$\|I_1(t)\|_\infty \le C\int_0^1 |t-\theta|^{-\frac{3}{2}}|\theta|^{-\frac{1}{2}}\|u\|_1 d\theta \le Ct^{-\frac{3}{2}}\|u\|_1, \tag{6.27}$$

$$\|I_2(t)\|_\infty \le C\int_1^{t-1} |t-\theta|^{-\frac{3}{2}}|\theta|^{-\frac{3}{2}}\|u\|_1 d\theta \le Ct^{-\frac{3}{2}}\|u\|_1. \tag{6.28}$$

By virtue of (6.6), we have for $2 < t$

$$\left\|\int_0^\theta U_0(t,r)V(r)U_0(r,0)udr\right\|_\infty \le C|t|^{-\frac{3}{2}}\|u\|_1, \quad t-1<\theta<t. \tag{6.29}$$

Estimate (6.5), statement (3) of Lemma 6.2 and $\|P_c(t)\|_{\infty\to\infty} \le C$ imply

$$\|P_c(t)U(t,\theta)V(\theta)U_0(\theta,t)\|_{\infty\to\infty} \le C, \quad t-1<\theta<t. \tag{6.30}$$

It follows by combining (6.29) and (6.30) that

$$\|I_3(t)\|_\infty \le CM\int_{t-1}^{t} e^M|t|^{-\frac{3}{2}}\|u\|_1 d\theta = CMe^M|t|^{-\frac{3}{2}}\|u\|_1. \tag{6.31}$$

The combination of (6.27), (6.28) and (6.31) completes the proof of Theorem 6.1. □

6.2. *Local smoothing estimate*

In this subsection, we prove the local smoothing estimate for $U(t,s)$ which implies that the map $U(t,s)u_0$ improves the differentiability of property u_0 by the order 1/2 in the Sobolev sense at almost all $t \in \mathbf{R}$. There is an extensive literature on the local smoothing property of various dispersive equations with coefficients depending only on spatial variables. We mention here only Refs. 5, 7, 14, 16, 26, 28, 54 and 63, among others. For similar results for multi-particle time periodic system, see Ref. 38.

Theorem 6.6: Let $V \in \mathcal{V}_\beta$ with $\beta > 2$ be of generic type and let $\gamma > 1$.

(1) Suppose $\sigma_p(\mathcal{U}_0) = \emptyset$. Then there exists a constant $C > 0$ such that

$$\|\langle x\rangle^{-\gamma}\langle D_x\rangle^{\frac{1}{2}}U(t,0)u_0\|_{L^2(\mathbf{R}^4)} \le C\|u_0\|_{L^2(\mathbf{R}^3)}. \tag{6.32}$$

(2) Suppose $\beta > 3$. Then there exists a constant $C > 0$ such that

$$\|\langle x\rangle^{-\gamma}\langle D_x\rangle^{\frac{1}{2}}U(t,0)P_c(0)u_0\|_{L^2(\mathbf{R}^4)} \le C\|u_0\|_{L^2(\mathbf{R}^3)}. \tag{6.33}$$

We let $\gamma > 1$ in what follows and define $G_0 = \langle x\rangle^{-\gamma}\langle D\rangle^{1/2}$ with domain $D(G_0) = H^{1/2}(\mathbf{R}^3)$. Denote its closure by G. We write $\hat{G} = I \otimes G$. Recall

$$R_0(z) = (K_0 - z)^{-1}, \quad \mathcal{K}_\sigma = L^2(\mathbf{T}, \mathcal{H}_\sigma) \quad \text{and} \quad \mathcal{Y}_\sigma = B(\mathcal{K}_{-\sigma}, \mathcal{K}_{-\sigma}).$$

Lemma 6.7: *We have $D(K_0) \subset D(\hat{G})$. For $z \notin \mathbf{R}$, $\hat{G}R_0(z)\hat{G}^*$ has a bounded closure in $\mathcal{K}$, which we denote by the same symbol, and it satisfies*

$$\sup_{\varepsilon>0,\lambda\in\mathbf{R}} \|\hat{G}R_0(\lambda \pm i\varepsilon)\hat{G}^*u\|_{\mathcal{K}} \le C\|u\|_{\mathcal{K}}, \quad u \in D(\hat{G}^*)\,. \tag{6.34}$$

The same estimate holds for $\langle x\rangle^{-\gamma}R_0(z)\hat{G}^*$ and $\hat{G}R_0(z)\langle x\rangle^{-\gamma}$.

Proof: Let $r_0(z) = (-(1/2)\Delta - z)^{-1}$. Then $r_0(z)$ satisfies a stronger version of the property (a) in the proof of Lemma 4.6 (see Ref. 1):

$$\sup_{\lambda\in\mathbf{R},\varepsilon>0} \|Gr_0(\lambda \pm i\varepsilon)G^*\|_{B(\mathcal{H})} \le C. \tag{6.35}$$

Since $\hat{G}R_0(z)\hat{G}^* = \oplus Gr_0(z-n)G^*$ as in (4.11), (6.35) implies that $\hat{G}R_0(z)G^*$ has a bounded extension in $\mathcal{K}$ and that (6.34) holds. The resolvent equation implies

$$2i\Im z[R_0(z)\hat{G}^*]^*R_0(z)\hat{G}^* = \hat{G}(R_0(\overline{z}) - R_0(z))\hat{G}^*$$

and we deduce that $R_0(z)\hat{G}^*$ has a bounded extension. Thus, $\|\hat{G}R_0(i)u\| \le C\|u\|$ for all $u \in \mathcal{K}$ and $D(\hat{G}) \subset D(K_0)$. The rest of the lemma may be proved similarly. □

Lemma 6.8: *Let $V \in \mathcal{V}_\beta$ with $\beta > 2$. Suppose that $\mathcal{U}_0$ has no eigenvalues nor a threshold resonance. Then there exists a constant $C > 0$ such that*

$$\sup_{\varepsilon>0,\lambda\in\mathbf{R}} \|\hat{G}R(\lambda \pm i\varepsilon)\hat{G}^* u\|_{\mathcal{K}} \le C\|u\|_{\mathcal{K}}, \quad u \in D(\hat{G}^*). \tag{6.36}$$

The same estimate holds for $\langle x\rangle^{-\gamma}R(z)\hat{G}^*$, $\hat{G}R(z)\langle x\rangle^{-\gamma}$ and $\langle x\rangle^{-\gamma}R(z)\langle x\rangle^{-\gamma}$.

Proof: We multiply (4.12):

$$R(z) = R_0(z) - R_0(z)B^*(1 + AR_0(z)B^*)^{-1}AR_0(z), \quad \text{Im } z \neq 0, \tag{6.37}$$

by $\hat{G}$ and $\hat{G}^*$ from the left and the right. Since $-1 \in \sigma(AR_0(\lambda \pm i0)B^*)$ if and only if λ is an eigenvalue or threshold resonance of K (see the proof of Lemma 4.6), we have $\|(1+AR_0(\lambda \pm i\varepsilon)B^*)^{-1}\|_{B(\mathcal{K})} \le C$ by the assumption. Estimate (6.36) follows by applying Lemma 6.7. The proof for $\langle x\rangle^{-\gamma}R(z)\hat{G}^*$, $\hat{G}R(z)\langle x\rangle^{-\gamma}$ and $\langle x\rangle^{-c}R(z)\langle x\rangle^{-\gamma}$ is similar. □

We suppose now that K has non-integral eigenvalues λ and study the resolvent $R(z)$ near λ.

Lemma 6.9: *Let $V \in \mathcal{V}_\beta$ with $\beta > 3$. Let $\delta = \beta/2$. Suppose that $\lambda \in \mathbf{R} \setminus \mathbf{Z}$ is an eigenvalue of K. Denote the projection onto the eigenspace by E_λ. Then, as a $\mathcal{Y}_\delta$-valued function of z, $R(z+\lambda) + z^{-1}E_\lambda$ is bounded and continuous in a neighborhood of 0 in $\overline{\mathbf{C}}^\pm \setminus \{0\}$.*

Proof: We may assume $3 < \beta < 5$ without losing the generality. We write $R_0^\pm(\lambda) = R_0(\lambda \pm i0)$ and $R_0^{\pm'}(\mu) = (d/d\mu)R_0^\pm(\mu)$. Define

$$\begin{aligned} S(z) &= 1 + R_0^\pm(\lambda)V + zR_0^{\pm'}(\lambda)V, \\ R_3^\pm(z,\lambda) &= R_0^\pm(z+\lambda) - R_0^\pm(\lambda) - zR_0^{\pm'}(\lambda) \end{aligned}$$

The operator $S(z) - 1$ is compact in $\mathcal{K}_{-\delta}$ and $\mathbf{C} \ni z \to S(z) \in B(\mathcal{K}_{-\delta})$ is entire. It is known (Ref. 13) using the analytic Fredholm theory (Ref. 47) that $S(z)^{-1}$ admits a Laurent expansion in $B(\mathcal{K}_{-\delta})$ of the form:

$$\begin{aligned} S(z)^{-1} &= z^{-1}S_{-1} + S_0 + zS_1 + \cdots \equiv z^{-1}S_{-1} + \tilde{S}(z), \\ S_{-1} &= E_K(\{\lambda\})V. \end{aligned}$$

We write $\rho = (\delta - 3/2)\_$. It is also known that $\overline{\mathbf{C}}^\pm \ni z \to R_3^\pm(z,\lambda)$ is C^ρ near $z = 0$ and $\|R_3^\pm(z,\lambda)\| \le C|z|^{1+\rho}$. It follows that $L(z,\lambda) \equiv$

$R_3^{\pm}(z,\lambda)VS(z)^{-1}$ is a $B(\mathcal{K}_{-\delta})$-valued C^{ρ}-function of z near $z = 0$, $\|L(z,\lambda)\|_{B(\mathcal{K}_{-\delta})} \leq C|z|^{\rho}$ and

$$1 + R_0^{\pm}(z+\lambda)V = S(z) + R_3^{\pm}(z,\lambda)V = (1 + L(z,\lambda))S(z).$$

Thus, for small $z \in \overline{\mathbf{C}}\pm$, we have

$$\begin{aligned} R(z+\lambda) &= (1 + R_0(z+\lambda)V)^{-1}R_0(\lambda+z) \\ &= \left(\sum_{j=0}^{N} + \sum_{j=N+1}^{\infty}\right)(-1)^j S(z)^{-1}L(z,\lambda)^j R_0(\lambda+z). \end{aligned} \tag{6.38}$$

If N is sufficiently large, the sum $\sum_{N+1}^{\infty}(\cdots)$ is a $\mathcal{Y}_\delta$-valued C^{ρ} function of z. Hence, it suffices to show that

$$S(z)^{-1}L(z,\lambda)^j R_0(\lambda+z) = (S(z)^{-1}R_3^{\pm}(z,\lambda)V)^j S(z)^{-1}R_0(\lambda+z) \tag{6.39}$$

are $\mathcal{Y}_\delta$–valued bounded continuous functions ($BC(\mathcal{Y}_\delta)$ for short) of z for small z if $j \geq 1$ and of the form $-z^{-1}E_\lambda + BC(\mathcal{Y}_\delta)$ if $j = 0$. We use the following identities (6.41)∼(6.45): We have

$$(1 + R_0^{\pm}(\lambda)V)E_\lambda = E_\lambda(1 + VR_0^{\pm}(\lambda)) = 0. \tag{6.40}$$

This and the resolvent equation yield

$$E_\lambda V(R_0(\lambda+z) - R_0^{\pm}(\lambda)) = -zE_\lambda R_0(\lambda+z). \tag{6.41}$$

$$(R_0(\lambda+z) - R_0^{\pm}(\lambda))VE_\lambda = -zR_0(\lambda+z)E_\lambda. \tag{6.42}$$

Dividing both sides of (6.41) and (6.42) by z and letting $z = \pm i\varepsilon \to 0$ yield

$$E_\lambda V R_0^{\pm'}(\lambda) = -E_\lambda R_0^{\pm}(\lambda), \quad R_0^{\pm'}(\lambda)VE_\lambda = -R_0^{\pm}(\lambda)E_\lambda. \tag{6.43}$$

Recall that eigenfunctions are rapidly decaying at infinity. Adding (6.41) or (6.42) and $-z\times$(6.43) we have

$$E_\lambda V R_3^{\pm}(z,\lambda) = -zE_\lambda(R_0(z+\lambda) - R_0^{\pm}(\lambda)). \tag{6.44}$$

$$R_3^{\pm}(z,\lambda)VE_\lambda = -z(R_0(z+\lambda) - R_0^{\pm}(\lambda))E_\lambda. \tag{6.45}$$

It follows from (6.40) and (6.41) that

$$S(z)^{-1}R_0(\lambda+z) = -z^{-1}E_\lambda + M_1(z,\lambda), \tag{6.46}$$

where $M_1(z,\lambda) = -E_\lambda R_0(\lambda+z) + \tilde{S}(z)R_0(z+\lambda)$ is $BC(\mathcal{Y}_\delta)$ near $z = 0$ and (6.39) with $j = 0$ is of the desired form. From (6.44) we deduce that

$$S(z)^{-1}R_3^{\pm}(z,\lambda)V = -E_\lambda(R_0(z+\lambda) - R_0^{\pm}(\lambda))V + \tilde{S}(z)R_3^{\pm}(z,\lambda)V \tag{6.47}$$

is a bounded and continuous $B(\mathcal{K}_{-\delta})$ valued function near $z=0$. Combining (6.46) and (6.47) and using (6.45) and (6.42), we write the operator (6.39) with $j=1$ in the form

$$\begin{aligned}&S(z)^{-1}R_3^{\mp}(z,\lambda)V\cdot S(z)^{-1}R_0(\lambda+z)\\&\quad=-E_\lambda R_0(z+\lambda)E_\lambda+\tilde{S}(z)(R_0(z+\lambda)-R_0^{\pm}(\lambda))E_\lambda\\&\qquad+S(z)^{-1}R_3^{\mp}(z,\lambda)VM_1(z)\end{aligned}$$

to find that it is a $BC(\mathcal{Y}_\delta)$. Then (6.47) shows that (6.39) with $j\geq 2$ are all $BC(\mathcal{Y}_\delta)$. □

Lemma 6.10: *Let $V\in\mathcal{V}_\beta$ with $\beta>3$ be of generic type and let $\hat{G}$ be as in Lemma 6.7. Then there exists a constant $C>0$ such that*

$$\sup_{\varepsilon>0,\lambda\in\mathbf{R}}\|\hat{G}R(\lambda\pm i\varepsilon)P_c(K)\hat{G}^*u\|_{\mathcal{K}}\leq C\|u\|_{\mathcal{K}},\quad u\in D(\hat{G}^*).\tag{6.48}$$

The same estimate holds when one or both of $\hat{G}$ is replaced by $\langle x\rangle^{-\gamma}$, $\gamma>1$.

Proof: We multiply (6.37) by $\hat{G}$ and $P_c(K)\hat{G}^*$ from the left and the right:

$$\begin{aligned}\hat{G}R(z)P_c(K)\hat{G}^*&=\hat{G}R_0(z)P_c(K)\hat{G}^*\\&\quad-\hat{G}R_0(z)B^*(1+AR_0(z)B^*)^{-1}AR_0(z)P_c(K)\hat{G}^*\end{aligned}\tag{6.49}$$

Since eigenfunctions of K decay rapidly as $|x|\to\infty$, $\hat{G}R_0(z)P_c(K)\hat{G}^*$ is $B(\mathcal{K})$-valued continuous bounded function ($BC(\mathcal{K})$ for short) of $z\in\mathbf{C}^{\pm}$ as in Lemma 6.7. Likewise the proof of Theorem 6.8 shows that $GR(z)P_c(K)\hat{G}^*$ is $BC(\mathcal{K})$ outside a neighborhood of $\sigma_p(K)$. Thus, it suffices to show that the second term on the right of (6.49) is also $BC(\mathcal{K})$ near each $\lambda\in\sigma_p(K)$. Denote $\tilde{R}(z)=R(z)+(z-\lambda)^{-1}E_\lambda$. The operator $\tilde{R}(z)$ is $BC(\mathcal{Y}_\delta)$ by Lemma 6.9 and

$$(1+AR_0(z)B^*)^{-1}=1-AR(z)B^*=1+(z-\lambda)^{-1}AE_\lambda B-A\tilde{R}(z)B.$$

Thus all we need is to show that

$$Q(z)\equiv\hat{G}R_0(z)V(z-\lambda)^{-1}E_\lambda VR_0(z)P_c(K)\hat{G}^*\tag{6.50}$$

is $BC(\mathcal{K})$. By virtue of (6.40) and (6.41), we have

$$E_\lambda VR_0(z)=-E_\lambda-(z-\lambda)E_\lambda R_0(z).$$

Since $E_\lambda P_c(K)=0$, it then follows that

$$Q(z)=-\hat{G}R_0(z)VE_\lambda R_0(z)P_c(K)\hat{G}^*$$

and $Q(z)$ is a $BC(\mathcal{K})$ by vitue of Lemma 6.7 and the rapid decay property of the eigenfunctions. □

Proof of Theorem 6.6: Since the proof is similar, we prove statement (2) only. Due to Lemma 6.10 and Theorem 4.5, $\hat{G}$ is $KP_c(K)$-smooth and that

$$\int_{\mathbf{R}} \|\hat{G}e^{-i\sigma K}P_c(K)u\|_{\mathcal{K}}^2 d\sigma \le C\|u\|_{\mathcal{K}}^2, \quad u \in \mathcal{K}. \tag{6.51}$$

By virtue of (3.12) and the periodicity $U(t+2\pi, s+2\pi) = U(t,s)$, the left hand side of (6.51) is equal to

$$\begin{aligned}
&\int_{\mathbf{R}} \left(\int_{t\in\mathbf{T}} \|GU(t,t-\sigma)P_c(t-\sigma)u(t-\sigma)\|_{\mathcal{H}}^2 dt \right) d\sigma \\
&\quad = \int_{\mathbf{R}} \left(\int_{t\in\mathbf{T}} \|GU(t+\sigma,t)P_c(t)u(t)\|_{\mathcal{H}}^2 dt \right) d\sigma \\
&\quad = \int_{\mathbf{T}} \left(\int_{\mathbf{R}} \|GU(\sigma,t)P_c(t)u(t)\|_{\mathcal{H}}^2 d\sigma \right) dt.
\end{aligned} \tag{6.52}$$

For any $u_0 \in \mathcal{H}$, define $u \in \mathcal{K}$ by $u(t) = U(t,0)P_c(0)u_0$ for $0 \le t < 2\pi$ and insert this u into (6.52). Then $P_c(t)u(t) = u(t)$ and we obtain the desired bound

$$\int_{\mathbf{R}} \|GU(\sigma,0)P_c(0)u_0\|_{\mathcal{H}}^2 d\sigma \le C\|u_0\|_{\mathcal{H}}^2.$$

This completes the proof.

7. Instability of Bound States

Since wave operators $\mathcal{W}_\pm$ exist, the absolutely continuous spectrum of K fills up the whole real line. It follows that any possible eigenvalues of K are embedded in the continuum. We show in this section that these embedded eigenvalues are unstable under perturbations W which satisfy the Fermi Golden rule (FGR for short) and that W's which violate FGR are very scarce. In Section 7.1, we assume that potentials are dilation analytic and discuss the resonance theory for time periodic systems. We then show that the eigenvalues of K will disappear and turn into resonances after the perturbations which satisfy FGR are switched on. In Section 7.2, we show without assuming the analyticity, that under FGR the eigenvalues are unstable and that the life time of the bound states decays exponentially for the time $o(\mu^2 \log \mu^{-1})$ where μ is the size of the perturbation.

7.1. *Resonance theory for time periodic systems*

The theory of resonances, which is of independent mathematical and physical importance (see Refs. 73, 55 and the references therein), is one of the most powerful methods for studying the instability of embedded eigenvalues. We discuss the resonance theory for time periodic system here via the dilation analyticity, the method introduced by Aguilar-Combes[3] and Balslev-Combes[6] and further developed by Simon.[53] For $0 < a < \pi/2$, we write $\Pi_a = \{\theta \in \mathbf{C} \colon |\Im\theta| < a\}$ and $\Pi_a^{\pm} = \{\theta \in \mathbf{C} \colon 0 < \pm\Im\theta < a\}$. We assume that $V(t,x)$ satisfies the following condition:

Assumption 1: For some $0 < a < \pi/2$ and $\beta > 0$, the $\mathcal{V}_\beta$-valued function $\theta \to V(t, e^\theta x)$ defined on $\mathbf{R}$ can be extended to Π_a as an analytic function.

We define a one parameter unitary group $\{U(\theta)\colon \theta \in \mathbf{R}\}$ on $\mathcal{K}$ by

$$U(\theta)u(t,x) = e^{\frac{d\theta}{2}}u(t, e^\theta x) = e^{-itA}u(t,x),$$

where $A = (x \cdot D + D \cdot x)/2$ is the generator, $D = (D_1, \ldots, D_d)$, $D_j = -i\partial/\partial x_j$. Define, for $|\Im\theta| < \pi/2$,

$$K_0(\theta) = -i\frac{\partial}{\partial t} - \frac{e^{-2\theta}}{2}\Delta$$

with the domain $D(K_0(\theta)) = \{u \in \mathcal{K} : K_0(\theta)u \in \mathcal{K}\}$. When $\theta \in \Pi_{\pi/2}^{\pm}$, we have

$$D(K_0(\theta)) = \{u \in \mathcal{K} : \partial_t u \in \mathcal{K}, -\Delta_x u \in \mathcal{K}\}$$

and $D(K_0(\theta))$ is independent of θ. The following lemma is obvious.

Lemma 7.1:

(1) *The family* $\{K_0(\theta) : \theta \in \Pi_{\pi/2}^{\pm}\}$ *is a holomorphic family of type* (A) (Ref. 23). *Moreover, for* $\theta_0 \in \mathbf{R}$ *and* $\pm\Im z > 0$, *we have*

$$\lim_{\Pi_{\pi/2}^{\pm} \ni \theta \to \theta_0} (K_0(\theta) - z)^{-1} = (K_0(\theta_0) - z)^{-1}$$

in the strong operator topology.

(2) *We have* $K_0(\theta + \tilde\theta) = U(\theta)K_0(\tilde\theta)U(\theta)^*$ *for any* $\theta \in \mathbf{R}$ *and* $\tilde\theta \in \Pi_{\pi/2}$.

(3) *For* $\theta \in \Pi_{\pi/2}$, $K_0(\theta)$ *is a normal operator and it has the spectrum*

$$\sigma(K_0(\theta)) = \bigcup_{n \in \mathbf{Z}} \{n + e^{-2\theta}\mathbf{R}^+\}.$$

We write $\mathcal{A}(A)$ for the set of (entire) analytic vectors $u \in \mathcal{K}$ for A. For $u \in \mathcal{A}(A)$, we write $u(\theta) = e^{-i\theta A}u \in \mathcal{K}$ and $u_n(\theta) \in \mathcal{H}$ for its n-th Fourier component $u_n(\theta, x) = (\mathcal{F}u(\theta))(n, x)$, $\mathcal{F}$ being defined by (3.6). Vectors $u \in \mathcal{K}$ such that $E_A(\Omega)u = u$ for some compact $\Omega \subset \mathbf{R}$ are analytic vectors for A.

Lemma 7.2: *Let $u, v \in \mathcal{A}(A)$. Let $\theta \in \Pi^{\pm}_{\pi/2}$. Then the holomorphic function $\mathbf{C}^{\pm} \ni z \to (R_0(z)u, v)$ has an analytic extension to $\Omega(\theta) \equiv \mathbf{C} \setminus \sigma(K_0(\theta))$. For two θ and $\tilde{\theta} \in \Pi^{\pm}_{\pi/2}$ such that $\Im\theta < \Im\tilde{\theta}$, the analytic extensions coincide outside $\bigcup_{n\in\mathbf{Z}}\{z : -2\Im\tilde{\theta} \le \arg(z-n) \le -2\Im\theta\}$.*

Proof: We prove the + case. Define for $\Im z > 0$ and $0 \le \Im\theta < \pi/2$

$$\begin{aligned} F(\theta, z) &= \big((K_0(\theta) - z)^{-1}u(\theta), v(\overline{\theta})\big) \\ &= \sum_{n\in\mathbf{Z}} \left((-\frac{e^{-2\theta}}{2}\Delta + n - z)^{-1}u_n(\theta), v_n(\overline{\theta}) \right) \end{aligned}$$

The function $\theta \to F(\theta, z)$ is holomorphic and, by virtue of Lemma 7.1,(2), it is independent of $\Re\theta$. It follows that $F(\theta, z)$ is independent of θ and

$$F(\theta, z) = \lim_{\theta\to 0} F(\theta, z) = (R_0(z)u, v).$$

However, $z \to F(\theta, z)$ is holomorphic in the resolvent set $\Omega(\theta)$ of $K_0(\theta)$ and $(R_0(z)u, v)$ has an analytic extension to $\Omega(\theta)$. The rest of the lemma is obvious. □

We may apply a similar argument to K. We define for $\theta \in \Pi_a$ the operator

$$K(\theta) = -i\partial_t - \frac{e^{-2\theta}}{2}\Delta + V(t, e^{\theta}x)$$

with $D(K(\theta)) = D(K_0(\theta))$. By the assumption on V, $\{K(\theta) : \theta \in \Pi^{\pm}_a\}$ is a type (A) holomorphic family of operators on $\mathcal{K}$. We have

$$K(\theta + \tilde{\theta}) = U(\theta)K(\tilde{\theta})U(\theta)^*, \quad \theta \in \mathbf{R},\ \tilde{\theta} \in \Pi_a. \tag{7.1}$$

We denote $R(z, \theta) = (K(\theta) - z)^{-1}$ for $z \in \rho(K(\theta))$.

Theorem 7.3: Let $\theta \in \Pi^{\pm}_a$. Then,

(1) The spectrum of $K(\theta)$ consists of $\bigcup_{n\in\mathbf{Z}}\{n + e^{-2\theta}\mathbf{R}^+\}$ and possibly eigenvalues $\sigma_p(K(\theta))$ of $K(\theta)$ in the closed half plane $\overline{\mathbf{C}^{\mp}}$.

(2) The eigenvalues are discrete and they can accumulate only at the threshold set $\mathbf{Z}$. We have $\sigma_p(K(\theta)) \cap \mathbf{R} = \sigma_p(K)$.
(3) Let $u, v \in \mathcal{A}(A)$. Then the holomorphic function $\mathbf{C}^{\pm} \ni z \to (R(z)u, v)$ has a unique meromorphic extension $R_{\theta,u,v}(z)$ to $\Omega(\theta) = \mathbf{C} \setminus \sigma(K_0(\theta))$.
(4) Two meromorphic extensions $R_{\theta,u,v}(z)$ and $R_{\tilde{\theta},u,v}(z)$, $\Im\theta < \Im\tilde{\theta}$, coincide outside $\bigcup_{n\in\mathbf{Z}}\{z : -2\Im\tilde{\theta} \le \arg(z-n) \le -2\Im\theta\}$.
(5) The eigenvalues of $K(\theta)$ are precisely the poles of $R_{\theta,u,v}(z)$ for some $u, v \in \mathcal{A}(A)$. The eigenvalues of $K(\theta)$ are locally independent of θ, viz. for any compact set $D \subset \Omega(\theta_0)$, $\theta_0 \in \Pi_a^{\pm}$, there exists $\varepsilon > 0$ such that $\sigma_p(K(\theta)) \cap D = \sigma_p(K(\theta_0)) \cap D$ whenever $|\Im(\theta - \theta_0)| < \varepsilon$.

Nonreal eigenvalues of $K(\theta)$ for some $\theta \in \Pi_a^{\pm}$ are called resonances of K.

Proof: We prove the theorem for $\theta \in \Pi_a^+$. We may assume $1 < \Re\theta < 2$ as $K(\theta + \tilde{\theta})$ and $K(\theta)$ are unitarily equivalent if $\tilde{\theta} \in \mathbf{R}$. We write $V(\theta)$ for the multiplication by $V(t, e^{\theta}x)$. (Do not confuse with the notation in the previous section where $V(t)$ was the multiplication by $V(t,x)$.) Then $\Omega(\theta) \ni z \to V(\theta)R_0(z,\theta)$ is an analytic family of compact operators in $\mathcal{K}$ and $\|V(\theta)R_0(z,\theta)\| \to 0$ as $\Im z \to +\infty$. It follows by the argument in the proof of Weyl's theorem of the stability of the essential spectrum of self-adjoint operators that $\sigma(K_0(\theta)) \subset \sigma(K(\theta))$. It also follows from analytic Fredholm theory that $\Omega(\theta) \ni z \to (1 + V(\theta)R_0(z,\theta))^{-1} \in B(\mathcal{K})$ is meromorphic; the residues at the poles are of finite ranks; the poles are exactly the eigenvalues of $K(\theta)$ in $\Omega(\theta)$, and we have

$$R(z,\theta) = R_0(z,\theta)(1 + V(\theta)R_0(z,\theta))^{-1}.$$

Statement (1) follows. For $z \in \rho(K(\theta))$, $\theta \in \Pi_a^+$ and $u, v \in \mathcal{A}(A)$, we consider

$$(R(z,\theta)u(\theta), v(\overline{\theta})) = (R_0(z,\theta)(1 + V(\theta)R_0(z,\theta))^{-1}u(\theta), v(\overline{\theta})). \tag{7.2}$$

When $\Im z$ is large enough, $\Im z > M$, the right hand side is holomorphic with respect to $\theta \in \Pi_b^+$ $(1 < \Re\theta < 2)$, $0 < b < a$, and it is independent of $\Re\theta$ since

$$U(\theta)R(z,\tilde{\theta})U(\theta)^* = R(z, \theta + \tilde{\theta}), \quad \theta \in \mathbf{R}. \tag{7.3}$$

It follows for $\Im z > M$ that

$$(R(z,\theta)u(\theta), v(\overline{\theta})) = \lim_{\Im\theta \to 0} (R(z,\theta)u(\theta), v(\overline{\theta})) = (R(z)u, v). \tag{7.4}$$

Since $z \to (R(z,\theta)u(\theta), v(\overline{\theta}))$ is meromoprhic in $\Omega(\theta)$, we see from (7.4) that $(R(z)u, v)$ has a meromorphic extension $R_{\theta,u,v}(z)$ to $\Omega(\theta)$ and that statements (3) and (4) are satisfied. We deduce statement (5):

$$\bigcup_{u,v,\theta\in\Pi_a^+} \{z : \text{poles of } R_{\theta,u,v}(z)\} = \bigcup_{\theta\in\Pi_a^+} \{z : \text{poles of } R(z,\theta)\} \tag{7.5}$$

also from (7.4), since $\{u(\theta) : u \in \mathcal{A}(A)\}$ is dense in $\mathcal{K}$. In particular, $R(z,\theta)$ has no poles in $\mathbf{C}^+$, since $(R(z)u, v)$ are holomorphic in $\mathbf{C}^+$. The relation (7.4) further implies that for $\lambda \in \mathbf{R}$, we have

$$(E_K(\{\lambda\})u, v) = \lim_{\varepsilon\downarrow 0} -i\varepsilon(R(\lambda+i\varepsilon)u, v) = \lim_{\varepsilon\downarrow 0} -i\varepsilon(R(\lambda+i\varepsilon,\theta)u(\theta), v(\overline{\theta})),$$

and $\lambda \in \mathbf{R}$ is an eigenvalue of K if and only if λ is a pole of $R(z,\theta)$ which is necessarily simple. This implies statement (2). □

7.2. *Instability of bound states I, analytic case*

In this section we study the stability or instability of the embedded eigenvalues of K or, equivalently, of the Floquet operator $\mathcal{U}_t$ with respect to perturbations $\mu W(t,x)$ by using the dilation analytic method developed in the previous subsection. Thus, we assume that functions

$$\mathbf{R} \ni \theta \to V(t, e^\theta x), \quad \theta \to W(t, e^\theta x) \in \mathcal{V}_\beta, \quad \beta > 2$$

can be extended to the strip $\{\theta : |\Im\theta| < a\}$ as holomorphic functions for some $0 < a < \pi/2$, and we consider the equation

$$i\frac{\partial u}{\partial t} = (-\frac{1}{2}\Delta + V(t,x) + \mu W(t,x))u \tag{7.6}$$

for small $\mu \in \mathbf{R}$. We write $U_\mu(t,s)$ for the propagator for (7.6) and K_μ for the corresponding Floquet Hamiltonian. As in the previous section $V(\theta)$ and $W(\theta)$ denote the multiplication operators by $V(t, e^\theta x)$ and $W(t, e^\theta x)$, respectively. We define, for $\mu \neq 0$ and $\theta \in \Pi_a$

$$K_\mu(\theta) = -i\partial_t - \frac{e^{-2\theta}}{2}\Delta + V(\theta) + \mu W(\theta).$$

As previously $K(\theta) = -i\partial_t - (e^{-2\theta}/2)\Delta + V(\theta)$. We suppose $\lambda \notin \mathbf{Z}$ is an eigenvalue of K and $\{\psi_1, \ldots, \psi_n\}$ is an orthonormal basis of the corresponding eigenspace.

Lemma 7.4: *For $j = 1, \ldots, n$, $\psi_j(\theta) = e^{-i\theta A}\psi_j$, $\theta \in \mathbf{R}$, can be extended to Π_a as analytic functions. For $\theta \in \Pi_a$, $\{\psi_1(\theta), \ldots, \psi_n(\theta)\}$ is a basis of* *eigenspace of $K(\theta)$ with the same eigenvalue λ and $\{\psi_1(\bar{\theta}), \ldots, \psi_n(\bar{\theta})\}$ is its dual basis. For the residue $E(\theta, \lambda)$ of $R(z, \theta)$ at the pole λ, we have*

$$E(\theta, \lambda) = \sum_{j=1}^{n} \psi_j(\theta) \otimes \psi_j(\bar{\theta}). \tag{7.7}$$

Proof: For $u, v \in \mathcal{A}(A)$, Eq. (7.4) implies

$$(E_K(\{\lambda\})u, v) = (E(\theta, \lambda)e^{-i\theta A}u, e^{-i\bar{\theta}A}v), \quad \theta \in \Pi_a.$$

Since $e^{i\theta A}u, e^{i\bar{\theta}A}v \in \mathcal{A}(A)$, we then have

$$|(E_K(\{\lambda\})e^{i\theta A}u, e^{i\bar{\theta}A}v)| = |(E(\theta, \lambda)u, v)| \leq C\|u\|\|v\|$$

and $F(\pm b) = e^{\pm bA}E_K(\{\lambda\})e^{\mp bA}$ have bounded extensions for all $0 \leq b < a$. It follows that $\psi_j \in D(e^{\pm bA})$ for any $0 \leq b < a$ and $e^{-i\theta A}\psi_j$ has an analytic extension to Π_a and (7.7) is satisfied. Since $K(\theta)\psi_j(\theta) = \lambda\psi_j(\theta)$ for $\theta \in \mathbf{R}$, the analyticity implies that the same holds for all $\theta \in \Pi_a$. Thus, $\psi_j(\theta)$ is an eigenfunction of $K(\theta)$ when $\theta \in \Pi_a$. The rest of the lemma is obvious. □

The advantage of the dilation analytic method is that it isolates the eigenvalues of $K(\theta)$ from the rest of the spectrum of $K(\theta)$ and that the real eigenvalues remain unchanged when θ is changed. Then, since $W(\theta)$ is bounded, the standard perturbation theory for isolated eigenvalues applies (Ref. 23), which implies the following theorem.

Theorem 7.5: Suppose $\lambda \in \sigma_p(K) \setminus \mathbf{Z}$ with multiplicity m. Then, for $\theta \in \Pi_a^{\pm}$ and sufficiently small $|\mu|$, there exist exactly m (θ-independent) eigenvalues $\lambda_{1,\mu}, \ldots, \lambda_{m,\mu} \in \overline{\mathbf{C}^{\mp}}$ of $K_\mu(\theta)$, counting the multiplicities, such that $\lambda_{j,\mu} \to \lambda$ as $\mu \to 0$. The eigenvalues of K_μ in a suffciently small neighborhood of λ are precisely those $\lambda_{1,\mu}, \ldots, \lambda_{m,\mu}$ which lie on the real line.

Thus, for determining whether eigenvalues survive or turn into resonances when the perturbation is switched on, it suffices to check whether or not $\Im\lambda_{j,\mu}$ vanishes for some (fixed) $\theta \in \Pi_a$. This may be done by computing the eigenvalues $\lambda_{j,\mu}$ as a perturbation series (see Ref. 53 or 66). For the later convenience, however, we employ here the so-called Feschbach or Grushin method. We solve the equation for the resolvent of $K_\mu(\theta)$, for a fixed $\theta \in \Pi_a^+$:

$$(K(\theta) + \mu W(\theta) - z)u = f, \quad z \in U, \tag{7.8}$$

where U is a small disc around λ such that

$$U \cap \sigma(K_\mu(\theta)) = \{\lambda_{1,\mu}, \dots, \lambda_{m,\mu}\}$$

for $|\mu| < \mu_0$, $\mu_0 > 0$. In the rest of this subsection, we write

$$P(\theta) = E(\theta, \lambda), \quad Q(\theta) = 1 - P(\theta), \quad \overline{K}(\theta) = Q(\theta)K(\theta)Q(\theta)$$

and often omit the variable θ when no confusion is feared. We decompose as $\mathcal{K} = P\mathcal{K} \dot{+} Q\mathcal{K}$ and write the equation (7.8) in terms of Pu and Qu:

$$(\lambda - z + \mu PWP)Pu + \mu(PWQ)Qu = Pf, \tag{7.9}$$

$$\mu(QWP)Pu + (\overline{K} + \mu QWQ - z)Qu = Qf. \tag{7.10}$$

Since $\overline{K}$ has no spectrum in U, $\overline{K} + \mu QWQ - z$ is invertible in $Q\mathcal{K}$ when $|\mu| < \mu_0$ and $z \in U$, and we may solve (7.10) for Qu:

$$Qu = \Gamma_\mu Qf - \mu\Gamma_\mu QWP \cdot Pu. \tag{7.11}$$

Here Γ_μ is defined by the following equation and, when we change θ, it is holomorphic with respect to $z \in U$ and $\theta \in \Pi_a$ as long as $U \cap \sigma(K_0(\theta)) = \emptyset$:

$$\Gamma_\mu = \Gamma_\mu(z, \theta) = (\overline{K}(\theta) + \mu Q(\theta)W(\theta)Q(\theta) - z)^{-1}.$$

Inserting (7.11) into (7.9), we obtain

$$(\lambda - z + \mu PWP - \mu^2 PWQ\Gamma_\mu QWP)Pu = Pf - \mu PWQ\Gamma_\mu Qf. \tag{7.12}$$

It follows that the eigenvalues $\lambda_{j,\mu}$ of $K_\mu(\theta)$ near λ are those z for which

$$\begin{aligned} F_\mu(z, \theta) = {} & \lambda - z + \mu P(\theta)W(\theta)P(\theta) \\ & -\mu^2 P(\theta)W(\theta)Q(\theta)\Gamma_\mu(z,\theta)Q(\theta)W(\theta)P(\theta) \end{aligned} \tag{7.13}$$

becomes singular on $P(\theta)\mathcal{K}$. Define $\Gamma_0(z,\theta) = \Gamma_{\mu=0}(z,\theta)$ so that

$$\Gamma_\mu(z,\theta) = \Gamma_0(z,\theta) - \mu\Gamma_0(z,\theta)Q(\theta)W(\theta)Q(\theta)\Gamma_\mu(z,\theta). \tag{7.14}$$

Notice that the representation matrix $\tilde{F}_\mu(z,\theta) = \{(F_\mu(z,\theta)\psi_j(\theta), \psi_k(\overline{\theta}))_{\mathcal{K}}\}$ of $F_\mu(z,\theta)$ with respect to the basis $\{\psi_j(\theta)\}$ of $P(\theta)\mathcal{K}$ and its dual basis $\{\psi_j(\overline{\theta})\}$ is independent of $\theta \in \Pi_a$ for $z = \rho \in \mathbf{R} \cap U$ because it is holomorphic in $\theta \in \Pi_a$ and independent of $\Re\theta$, it may be computed by setting $\theta = i\delta_n$ and by taking the limit $\delta_n \downarrow 0$. The result, by virtue of (7.14), is

$$(\lambda - \rho)\delta_{jk} + \mu(W\psi_j, \psi_k) - \mu^2(WQ(\overline{K} - \rho - i0)^{-1}QW\psi_j, \psi_k) + O(\mu^3),$$

where $O(\mu^3)$ is uniform with respect to $\rho \in U \cap \mathbf{R}$. Here it is important to note that *the right hand side is expressed in terms of the original* P, Q *and* W, and not in terms of $P(\theta), Q(\theta)$ or $W(\theta)$. Thus, understanding that the right hand side is the matrix representation of the operator with respect to the orthonormal basis $\{\psi_j\}$, we obtain, with $E'(\rho) = dE_K(\rho)/d\rho$,

$$\Im \tilde{F}_\mu(\rho, \theta) = -\mu^2 \pi PWQE'(\rho)QWP + O(\mu^3), \quad \rho \in \mathbf{R} \cap U. \tag{7.15}$$

Recall that $QE'(\lambda)Q$ exists if $V \in \mathcal{V}_\beta$ with $\beta > 3$ by Lemma 6.9.

Theorem 7.6: Suppose $V \in \mathcal{V}_\beta$ with $\beta > 3$. Let $\lambda \in \sigma_p(K) \setminus \mathbf{Z}$ and let P be the corresponding eigenprojection and $Q = 1 - P$. Suppose that the Fermi Golden Rule:

$$PWQE'_K(\lambda)QWP > 0 \quad \text{on} \quad P\mathcal{K} \tag{7.16}$$

is satisfied. Then there exists μ_0 such that $\mathcal{K}_\mu$ has no eigenvalue near λ for $|\mu| < \mu_0$ and λ turns into resonances under the perturbation $\mu W(t, x)$.

Proof: Since $PWQE'(\rho)QWP$ is norm continuous with respect to $\rho \in \mathbf{R}$ near $\rho = \lambda$, (7.15) implies under the assumption (7.16) that there exists a neighborhood $U_0 \subset \mathbf{R}$ of $\lambda \in \mathbf{R}$ such that we have $\Im \tilde{F}_\mu(\rho, \theta) < 0$ for all $\rho \in U_0$ and $|\mu| < \mu_0$ and $F_\mu(\rho, \theta)$ cannot be singular for $\rho \in U_0$. □

7.3. *Fermi golden rule*

Theorem 7.6 shows that eigenvalues of K are unstable under perturbations W which satisfy the Fermi Golden Rule (7.16). We show in this section that "most" W satisfy the condition, and therefore, that eigenvalues of the Floquet operators are generically unstable under perturbation. Notice that the Fermi Golden rule (7.16) is irrelevant to the dilation analyticity of V or W and we assume in this subsection $V, W \in \mathcal{V}_\beta$ with $\beta > 3$. We also assume $d = 3$ for simplicity. We let $\lambda \in \sigma_p(K) \setminus \mathbf{Z}$ and write $P = E_K(\{\lambda\})$. We assume the following unique continuation property for the corresponding eigenfunctions:

Assumption 2: If ψ is an eigenfunction of K with eigenvalue λ, then $\psi(t, x) = 0$ almost everywhere whenever $\psi(t, x) = 0$ in an open set of $\mathbf{R}^1 \times \mathbf{R}^3$.

It is well known that Assumption 2 holds if V is analytic with respect to one of the variables of (t, x) (Ref. 62), however, we do not as yet know if it generally holds for $V \in \mathcal{V}_\beta$.

We define the family of operators $\{T_0(\rho) : \rho \in \mathbf{R}\}$ by

$$T_0(\rho) = \bigoplus_{n<\rho} \Gamma(\rho - n) \colon \mathcal{K}_\delta \to \bigoplus_{n<\rho} \mathcal{X}(n),$$

$\delta > 1$. Recall (4.14) for the definition of $\Gamma(\rho)$ and that $\mathcal{X}(n) = L^2(S^2)$. $T_0(\rho)$ is the spectral representation for K_0 and in $B(\mathcal{K}_\delta, \mathcal{K}_{-\delta})$

$$\frac{dE_{K_0}}{d\rho}(\rho) = T_0(\rho)^* T_0(\rho), \quad \rho \in \mathbf{R} \setminus \mathbf{Z}.$$

Using $T_0(\rho)$ we define $T_\pm(\rho)\colon \mathcal{K}_{\beta/2} \to \bigoplus_{n<\rho} \mathcal{X}(n)$ by

$$T_\pm(\rho) = T_0(\rho)(1 - VR^\pm(\rho))Q = T_0(\rho)(1 + VR_0^\pm(\rho))^{-1}Q.$$

Then the general theory of scattering (Ref. 34) implies that $T_\pm(\rho)$ is a spectral representation of $\overline{K}$ near λ and, for ρ in a neighborhood of λ, we have

$$Q\frac{dE_K}{d\rho}(\rho)Q = T_\pm(\rho)^* T_\pm(\rho) \tag{7.17}$$

as an operator from $\mathcal{K}_{\beta/2}$ to $\mathcal{K}_{-\beta/2}$. It follows that the Fermi Golden Rule (7.16) is equivalent to that $T_\pm(\lambda)WP : P\mathcal{K} \to \bigoplus_{n<\lambda} \mathcal{X}(n)$ is one-to-one, which in turn is equivalent to that $PWT_\pm(\lambda)^*\colon \bigoplus_{n<\lambda} \mathcal{X}(n) \to P\mathcal{K}$ is surjective since $P\mathcal{K}$ is finite dimensional. Thus, W violates (7.16) if and only if

$$\langle W, \psi g\rangle_{\mathcal{K}} = 0 \quad \text{for all } \; g \in \text{Image } T(\lambda)_\pm^* \;\text{ and some } \psi \in P\mathcal{K}. \tag{7.18}$$

Lemma 7.7: *The operator $T_\pm(\lambda)^* : \bigoplus_{n>\lambda} L^2(S^2) \to \mathcal{K}_{-\beta/2}$ is one-to-one and $\mathcal{Z}_\pm(\lambda) \equiv \text{Image } T_\pm(\lambda)^* \cap C(\mathbf{T}\times\mathbf{R}^3)$ is infinite dimensional.*

Proof: Write $N_\lambda = P\mathcal{K}$. Then, by virtue of Hörmander's argument in the proof of Lemma 14.6.7 of Ref. 15, as an operator in $\mathcal{K}_{-\beta/2}$, the null space and the image of $1 + R_0^\mp(\lambda)V$ are respectively equal to N_λ and the orthogonal space of $(K_0 - \lambda)N_\lambda = VN_\lambda \subset \mathcal{K}_{\beta/2}$. It follows that

$$Q(1 + R_0^\mp(\lambda)V)^{-1} = Q - R^\pm(\lambda)QV$$

is an isomorphism from $\{(K_0 - \lambda)N_\lambda\}^\perp \subset \mathcal{K}_{-\beta/2}$ onto $Q\mathcal{K}_{-\beta/2}$. We have

$$T_\pm(\lambda)^* = Q(1 + R_0^\mp(\lambda)V)^{-1} T_0(\lambda)^*. \tag{7.19}$$

Here, as Image $T_0(\lambda)$ is clearly dense in $\bigoplus_{n<\lambda} L^2(S^2)$, $T_0(\lambda)^*$ is one-to-one and

$$\langle (K_0 - \lambda)N_\lambda, T_0(\lambda)^*\phi\rangle = 0, \quad \forall\phi \in \bigoplus_{n>\lambda} L^2(S^2),$$

or Image $T_0(\lambda)^* \subset \{(K_0 - \lambda)N_\lambda\}^\perp$. Thus, (7.19) implies $T_\pm(\lambda)^*$ is one-to-one. We next show that $\mathcal{Z}_\pm(\lambda)$ is infinite dimensional. We set

$$\mathcal{D} = \left\{ \bigoplus_{n>\lambda} \phi_n \colon \phi_n \in C^\infty(S^2), \quad \phi_n = 0 \text{ for } n > n_0 \text{ for some } n_0 \right\}.$$

and $\tilde{\mathcal{D}} \equiv C^\infty(\mathbf{T}, \mathcal{H}^3_{-\beta/2})$, where $\mathcal{H}^s_{-\beta/2} = \{u : \langle x\rangle^{-\beta/2}u \in H^s(\mathbf{R}^3)\}$ is the weighted Sobolev space. We have

$$T_0(\lambda)^*\mathcal{D} \subset \tilde{\mathcal{D}} \quad \text{and} \quad Q\tilde{\mathcal{D}} \subset C(\mathbf{T}\times\mathbf{R}^3),$$

and, since $T_\pm(\lambda)^* = QT_0(\lambda)^* - R^\mp(\lambda)QVT_0(\lambda)^*$, we are done if we show that $u = R^\pm(\lambda)QVf \in C(\mathbf{T}\times\mathbf{R}^3)$ if $f \in \tilde{\mathcal{D}}$. Define the operator $\Delta_t(h)$ by

$$\Delta_t(h)u(t,x) = h^{-1}(u(t+h,x) - u(t,x)).$$

Then, taking the limit $h \to 0$ in

$$\begin{aligned} &Q[\Delta_t(h), R^\pm(\lambda)]Q \\ &\quad = \lim_{\varepsilon\downarrow 0} Q[\Delta_t(h), R(\lambda\pm i\varepsilon)]Q = \lim_{\varepsilon\downarrow 0} QR(\lambda\pm i\varepsilon)[V,\Delta_t(h)]R(\lambda\pm i\varepsilon)Q \\ &\quad = -QR^\pm(\lambda)[V,\Delta_t(h)]R^\pm(\lambda)Q \end{aligned}$$

we have, with D_t being the differentiation with respect to t,

$$Q[D_t, R^\pm(\lambda)]Q = QR^\pm(\lambda)Q\cdot\partial_t V\cdot QR^\pm(\lambda)Q.$$

It follows that

$$\begin{aligned} &D_t\cdot QR^\pm(\lambda)QVf \\ &\qquad = [D_t,Q]QR^\pm(\lambda)QVf - QR^\pm(\lambda)Q(\partial_t V)QR^\pm(\lambda)QVf \\ &\qquad\quad + QR^\pm(\lambda)Q[D_t,Q]Vf + QR^\pm(\lambda)Q(\partial_t V)f \\ &\qquad\quad + QR^\pm(\lambda)QVD_tf, \end{aligned} \tag{7.20}$$

and

$$u = QR^\pm(\lambda)QVf \in H^1(\mathbf{T},\mathcal{H}_{-\beta/2})$$

for any $f \in H^1(\mathbf{T},\mathcal{H}_{-\beta/2})$. By repeating a similar computation, we then see that, for $f \in \tilde{\mathcal{D}}$, the right hand side of (7.20) belongs to $H^1(\mathbf{T},\mathcal{H}_{-\beta/2})$ and $u \in H^2(\mathbf{T},\mathcal{H}_{-\beta/2})$. Thus,

$$-\Delta u = -Vu - i\partial_t u + \lambda u + QVf \in C(\mathbf{T},\mathcal{H}_{-\beta/2})$$

and we conclude that $u \in C(\mathbf{T}\times\mathbf{R}^3)$ by the Sobolev embedding theorem. □

Lemma 7.8: *Assume Assumption 2. Let ψ be an eigenfunction of K with eigenvalue λ. Then $\mathcal{Z}_\pm(\lambda) \ni g \to \psi g \in \mathcal{V}^*_\beta$ is one to one, where we consider ψg as an element of the dual space $\mathcal{V}^*_\beta$ of $\mathcal{V}_\beta$ by the standard coupling $\langle V, \psi g\rangle$.*

Proof: Because ψ is continuous and decays rapidly at infinity it is obvious that $\psi g \in \mathcal{V}_\beta^*$ for $g \in \mathcal{Z}_\pm(\lambda)$. Since the set of zeros of ψ is closed and nowhere dense by Assumption 2, $\psi(t,x)g(t,x) = 0$ will imply $g(t,x) = 0$ on a dense set. Then $g = 0$ because $g \in \mathcal{Z}_\pm(\lambda)$ is continuous. □

Lemma 7.7 and Lemma 7.8 imply that $\{\psi g : g \in \mathcal{Z}_\pm(\lambda)\}$ and hence

$$\mathcal{Y}(\psi) = \{\psi g : g \in \text{Image } T_+(\lambda)^* + \text{Image } T_-(\lambda)^*\} \subset \mathcal{V}_\beta^*$$

is infinite dimensional for any eigenfunction ψ and the set of W which violates (7.16) is given by

$$\mathcal{E}(\lambda) = \bigcup_{0 \neq \psi \in P\mathcal{K}} \mathcal{Y}(\psi)^\perp. \tag{7.21}$$

If λ is a simple eigenvalue of K, then $\mathcal{E}(\lambda) = \mathcal{Y}(\psi)^\perp \subset \mathcal{V}_\beta$ is a closed subspace of infinite codimension and $\mathcal{V}_\beta \setminus \mathcal{E}(\lambda)$, the set of W which satisfies (7.16), is open and dense in $\mathcal{V}_\beta$. In general we have the following theorem. We write $D^m = \{z \in \mathbf{C}^m : |z| = 1\}$.

Theorem 7.9: Let λ be an eigenvalue of K with multiplicity m and let $\{\psi_1, \dots, \psi_m\}$ be an orthonormal basis of the eigenspace $P\mathcal{K}$. Then, the set of W which violates (7.16) is given by

$$\mathcal{E}(\lambda) = \cup\{\mathcal{Y}(\psi_\mathbf{a})^\perp : \mathbf{a} \in D^m\}, \quad \psi_\mathbf{a} = a_1\psi_1 + \cdots + a_m\psi_m.$$

The set of W which satisfies (7.16) is open and dense in $\mathcal{V}_\beta$.

Proof: We show that $\mathcal{E}(\lambda)$ is closed and nowhere dense in $\mathcal{V}_\beta$. Suppose $W_n \in \mathcal{E}(\lambda)$ and $W_n \to W$ as $n \to \infty$. Then $W_n \in \mathcal{Y}(\psi_{\mathbf{a}_n})^\perp$ for some $\mathbf{a}_n \in D^m$. Since D^m is compact, we may assume $\mathbf{a}_n \to \mathbf{a} \in D^m$. Then

$$\langle W, \psi_\mathbf{a} g\rangle = \lim_{n\to\infty} \langle W_n, \psi_{\mathbf{a}_n} g\rangle = 0, \quad \forall g \in \text{Image } T_\pm(\lambda)^*$$

and $W \in \mathcal{Y}(\psi_\mathbf{a})^\perp \subset \mathcal{E}(\lambda)$. Thus, $\mathcal{E}(\lambda)$ is closed. To show $\mathcal{E}(\lambda)$ is nowhere dense, suppose the contrary so that $\mathcal{E}(\lambda)$ contains an open set U. Let $W \in U$ and $W \in \mathcal{Y}(\psi_{\mathbf{a}_0})^\perp$. Take $N > m$ and linearly independent $g_1, \dots, g_N \in \mathcal{Z}(\lambda)$ and let $\mathcal{X}$ be the N dimensional subspace of $\mathcal{V}_\beta^*$ spaned by $\psi_{\mathbf{a}_0} g_1, \dots, \psi_{\mathbf{a}_0} g_N$. It is clear that $\mathcal{Y}(\psi_{\mathbf{a}_0})^\perp \subset \mathcal{X}^\perp$. Choose $V_1, \dots, V_N \in \mathcal{V}_\beta$ such that

$$\langle V_j, \psi_{\mathbf{a}_0} g_k\rangle = \delta_{jk}, \quad j,k = 1, \dots, N,$$

where δ_{jk} is Kroneker's delta. Then any $V \in \mathcal{V}_\beta$ may be uniquely written as

$$V = \xi_1 V_1 + \cdots + \xi_N V_N + V_0, \quad V_0 \in \mathcal{X}^\perp,$$

where $\xi_j = \langle V, \psi_{\mathbf{a}_0} g_j \rangle$ for $j = 1, \ldots, N$, and $V \in \mathcal{Y}(\psi_{\mathbf{a}_0 + \mathbf{v}})^{\perp}$, $\mathbf{v} \in \mathbf{C}^m$ only if $\xi = (\xi_1, \ldots, \xi_N)$ satisfies

$$\sum_{j=1}^{N} (\delta_{jk} + \langle V_j, \psi_{\mathbf{v}} g_k \rangle)\xi_j + \langle V_0, \psi_{\mathbf{v}} g_k \rangle = 0, \quad k = 1, \ldots, N. \tag{7.22}$$

Since $M(\mathbf{v}) = (\delta_{jk} + \langle V_j, \psi_{\mathbf{v}} g_k \rangle)_{jk}$ is invertible for small $\mathbf{v}$, (7.22) implies

$$(\xi_1, \ldots, \xi_N) = (\langle V_0, \psi_{\mathbf{v}} g_1 \rangle, \ldots, \langle V_0, \psi_{\mathbf{v}} g_N \rangle) M(\mathbf{v})^{-1}.$$

But the right side is a holomorphic function of $\mathbf{v} = (v_1, \ldots, v_m) \in \mathbf{C}^m$ in a neighborhood of 0 and, by the implicit function theorem, its image is at most m dimensional (when V_0 is fixed). Thus, any $\mathbf{a} \in D^m$ has an open neighborhood Ω such that $\bigcup \{\mathcal{Y}(\psi_{\mathbf{b}}) : \mathbf{b} \in \Omega\}^{\perp}$ can covers at most $N - m$ codimensional subset of U. Since D^m is compact, this implies $\mathcal{E}(\lambda)$ is nowhere dense in $\mathcal{V}_{\beta}$. □

7.4. *Instability of bound states II, non-analytic case*

In Sections 7.1 and 7.2, we discussed the resonance theory for K assuming that potentials are dilation analytic, and proved, under the Fermi Golden Rule, that eigenvalues of K are unstable and turn into resonances when perturbations are switched on. In this section, we discuss the same instability problem *without assuming* the analyticity. In this case, we cannot talk about resonances, however, we can still show that the eigenvalues are unstable under perturbation if the Fermi Golden Rule is assumed. We show also that the survival probability of eigenfunctions ϕ of the unperturbed Floquet operator $\mathcal{U}_0$ defined by $|(U_{\mu}(t,0)\phi, U(t,0)\phi)|^2$ decays exponetially with t for $|t| < C\varepsilon\mu^2 \log \mu^{-1}$. A similar problem has been discussed in a series of papers by Soffer and Weinstein and their coworkers[57,29,58,59,39,60] and by Merkli and Sigal[37] by what they call "time-dependent methods". The method presented here is of more stationary type and it directly relates the properties of solutions to the the spectral property of the Floquet operators and Floquet Hamiltonians. We assume $d = 3$ again and use the same notation as in the previous subsections. We first prove the following theorem.

Theorem 7.10: Let $V, W \in \mathcal{V}_{\beta}$, $\beta > 3$ and let $\lambda \notin \mathbf{Z}$ be an eigenvalue of K. Suppose that the Fermi Golden Rule (7.16) is satisfied. Then there exists μ_0 such that $K_{\mu} = K + \mu W$ has no eigenvalues near λ whenever $|\mu| < \mu_0$.

Proof: Define $P = E_K(\{\lambda\})$ and $Q = 1 - P$ and decompose the Hilbert space $\mathcal{K} = P\mathcal{K} \oplus Q\mathcal{K}$ and the equation

$$(K + \mu W - z)u = f \in \mathcal{K}_\delta, \quad z \notin \mathbf{R}$$

as in (7.9) and (7.10) (with $\theta = 0$). We solve (7.10) for Qu:

$$Qu = \Gamma_\mu(z)(Qf - \mu(QWP)Pu), \tag{7.23}$$

where, $\Gamma_\mu(z) = (\overline{K} + \mu QWQ - z)^{-1}$, and insert (7.23) into (7.9). We obtain

$$F_\mu(z)Pu = Pf - \mu PWQ\Gamma_\mu(z)Qf, \tag{7.24}$$

where $F_\mu(z)$ is an operator in $P\mathcal{K}$ defined by

$$F_\mu(z) = \lambda - z + \mu PWP - \mu^2 PWQ\Gamma_\mu(z)QWP. \tag{7.25}$$

For $\Im z \neq 0$, $F_\mu(z)$ is invertible and in the decomposition $\mathcal{K} = P\mathcal{K} \oplus Q\mathcal{K}$, the resolvent $R_\mu(z) = (K_\mu - z)^{-1}$, $K_\mu = K + \mu W$, has the expression

$$\begin{pmatrix} F_\mu(z)^{-1} & -\mu F_\mu(z)^{-1}PWQ\Gamma_\mu(z) \\ -\mu\Gamma_\mu(z)(QWP)F_\mu(z)^{-1} & \Gamma_\mu(z)(1 + \mu^2(QWP)F_\mu(z)^{-1}PWQ\Gamma_\mu(z)) \end{pmatrix}.$$

We show that $R_\mu(z)$ as an operator form $\mathcal{K}_\delta$ to $\mathcal{K}_{-\delta}$ has a continuous boundary values on the reals near $\rho = \lambda$ for small $|\mu|$ which will then prove the absence of eigenvalues for K_μ near λ.

By virtue of Lemma 6.9, the $\mathcal{Y}_\delta \equiv B(\mathcal{K}_\delta, \mathcal{K}_{-\delta})$-valued analytic function $\Gamma_0(z) = (\overline{K} - z)^{-1}$ of $z \in \mathbf{C}^\pm$ has C^1 boundary values $\Gamma_0(\rho \pm i0) = \Gamma_0^\pm(\rho)$ on the reals near $\rho = \lambda$. Hence, for sufficiently small $|\mu| < \mu_0$,

$$\Gamma_\mu(z) = \Gamma_0(z)(1 + \mu QWQ\Gamma_0(z))^{-1}$$

also has C^1 boundary values $\Gamma_\mu^\pm(\rho)$ in $\mathcal{Y}_\delta$ near $\rho = \lambda$. By virtue of the Fermi Golden Rule (7.16) there exist an open neighborhood U of λ and a constant $C > 0$ such that

$$\pm\Im PWQ\Gamma_0^\pm(\rho)QWP \geq C \quad \text{for} \quad \rho \in U.$$

Since $\Gamma_\mu^\pm(\rho) = \Gamma_0^\pm(\rho) - \mu\Gamma_\mu^\pm(\rho)QWQ\Gamma_0^\pm(\rho)$ by the resolvent equation, it follows that $\Gamma_\mu^\pm(\rho)$ also satisfies

$$\pm\Im PWQ\Gamma_\mu^\pm(\rho)QWP \geq C > 0, \quad \text{for } \rho \in U. \tag{7.26}$$

Thus, for $|\mu| < \mu_0$, $F_\mu(z)$, as an $B(P\mathcal{K})$-valued function of z, has a C^1 boundary values $F_\mu^\pm(\rho)$ on the reals near $\rho = \lambda$ and (7.26) implies

$$\mp\Im F_\mu^\pm(\rho) = \pm\mu^2 PWQ\Gamma_\mu^\pm(\rho)QWP \geq \mu^2 C, \quad \text{for } \rho \in U$$

for some $C > 0$. Hence, the $B(P\mathcal{K})$-valued analytic function $F_\mu(z)^{-1}$ of $z \in \mathbf{C}^\pm$ has also C^1 boundary values $F_\mu^\pm(\rho)^{-1}$ for $\rho \in U$ if $|\mu| < \mu_0$. This in turn implies that $R_\mu(z)$ has C^1 boundary values $R_\mu^\pm(\rho)$ for $\rho \in U$. □

We next study the behavior of $U_\mu(t,0)\phi$ for the eigenfunction ϕ of $\mathcal{U}_0$ with eigenvalue $e^{-2\pi i\lambda} \neq 1$. The matrix expression of $R_\mu(z)$ implies, for $|\mu| < \mu_0$,

$$PR_\mu^\pm(\rho)Pf = F_\mu^\pm(\rho)^{-1}Pf, \quad \rho \in U, \tag{7.27}$$

where U is a small neighborhood of λ. The proof of the following lemma is taken from Ref. 19.

Lemma 7.11: *Let $g \in C_0^\infty(U)$ be such that $g(\rho) = 1$ near λ. Then we have $\|(1-g(K_\mu))P\|_{B(\mathcal{K})} = O(\mu)$ and $\|P(1-g(K_\mu))P\|_{B(\mathcal{K})} = O(\mu^2)$ as $\mu \to 0$.*

Proof: Let $\tilde{g}(z)$ be a compactly supported almost analytic extension of $g(\rho)$. Then, by the Helffer–Sjöstrand formula (Ref. 12), we have

$$Pg(K_\mu) = -\frac{1}{2\pi i}\int_{\mathbf{C}} (\overline{\partial_z} g)(z)PR_\mu(z)dz \wedge d\bar{z}.$$

Insert the resolvent expansion $R_\mu(z) = R(z) - \mu R(z)WR_\mu(z)$ for $R(z)$ and use $PR(z) = (\lambda - z)^{-1}P$. We have

$$Pg(K_\mu) = P + \frac{\mu}{2\pi i}\int_{\mathbf{C}} (\overline{\partial_z}\tilde{g})(z)PR_\mu(z)WR(z)dz \wedge d\bar{z}$$

and we have $\|(1-g(K_\mu))P\|_{B(\mathcal{K})} = O(\mu)$ as $\mu \to 0$. Take $g_1 \in C_0^\infty(U)$ such that $g_1(\rho) = 1$ near $\rho = \lambda$ and $g_1(\rho) = 0$ when $g(\rho) \neq 1$. Then $1 - g(r) = (1-g(r))(1-g_1(r))$ and $P(1-g(K_\mu))P = P(1-g(K_\mu))(1-g_1(K_\mu))P$. Thus, $\|P(1-g(K_\mu))P\|_{B(\mathcal{K})} = O(\mu^2)$. □

Let $g \in C_0^\infty(U)$ be as in Lemma 7.11. Then we have

$$\sup_{\sigma\in\mathbf{R}} \|Pe^{-i\sigma K_\mu}P - Pe^{-iK_\mu\sigma}g(K_\mu)P\|_{B(\mathcal{K})} \leq C\mu^2. \tag{7.28}$$

Via Fourier transform we represent $e^{-i\sigma K_\mu}g(K_\mu)$ in terms of the boundary values $R_\mu^\pm(\rho)$ and replace $PR_\mu^\pm(\rho)P$ by $F_\mu^\pm(\rho)^{-1}$ by using (7.27). This yields

$$Pe^{-i\sigma K_\mu}g(K_\mu)P = \frac{1}{2\pi i}\int_{\mathbf{R}} g(\rho)e^{-i\sigma\rho}(F_\mu^+(\rho)^{-1} - F_\mu^-(\rho)^{-1})Pd\rho \tag{7.29}$$

Replacing the variable ρ of $\Gamma_\mu^\pm(\rho)$ in the expression for $F_\mu^\pm(\rho)$ by the eigenvalue λ, define the operators $\tilde{F}_\mu^\pm(\rho;\lambda)$ and $\Xi_\mu^\pm(\lambda)$ on the m-dimensional space $P\mathcal{K}$ by

$$\tilde{F}_\mu^\pm(\rho;\lambda) = \Xi_\mu^\pm(\lambda) - \rho,$$
$$\Xi_\mu^\pm(\lambda) = \lambda + \mu PWP - \mu^2 PWQ\Gamma_\mu^\pm(\lambda)QWP.$$

We have $\Xi_\mu^-(\lambda) = \Xi_\mu^+(\lambda)^*$.

Lemma 7.12: *We have the following identity for the operators in $P\mathcal{K}$ modulo the operator valued function of σ which is uniformly bounded by $C\mu^2$:*

$$\frac{1}{2\pi i}\int_{\mathbf{R}} g(\rho)e^{-i\sigma\rho}(\tilde{F}_\mu^+(\rho;\lambda)^{-1} - \tilde{F}_\mu^-(\rho;\lambda)^{-1})d\rho \equiv \begin{cases} e^{-i\sigma\Xi_\mu^+(\lambda)} & \sigma > 0, \\ e^{-i\sigma\Xi_\mu^-(\lambda)} & \sigma < 0. \end{cases} \tag{7.30}$$

Proof: By virtue of (7.26), $\Xi_\mu^\pm(\rho)$ has eigenvalues in $\mathbf{C}^\mp$ and Cauchy's formula implies that (7.30) holds if $g(\rho)$ is replaced by 1. However,

$$\int_{\mathbf{R}} \|(1-g(\rho))(\tilde{F}_\mu^+(\rho;\lambda)^{-1} - \tilde{F}_\mu^-(\rho;\lambda)^{-1})\|d\rho \le C\mu^2\int_{\mathbf{R}}\frac{|(1-g(\rho))|}{|\rho-\lambda|^2}d\rho$$

and the right hand side is obviously bounded by $C\mu^2$. The lemma follows. □

We now want to compare the integrals (7.29) and (7.30). By virtue of the C^1 property of $\Gamma_\mu(\rho)$ for $\rho \in U$, we have

$$\|\tilde{F}_\mu^\pm(\rho;\lambda) - F_\mu^\pm(\rho)\| \le C\mu^2|\rho-\lambda| \tag{7.31}$$

Since $\|\mu PWP - \mu^2 PWQ\Gamma_\mu^\pm(\lambda)QWP\| \le C|\mu|$, we have

$$\|F_\mu^\pm(\rho)^{-1}\|, \qquad \|\tilde{F}_\mu^\pm(\rho;\lambda)^{-1}\| \le C|\rho-\lambda|^{-1}, \quad \text{if} \quad |\rho-\lambda| > L|\mu|.$$

if L is large enough. It follows from (7.31) that

$$\int_{|\rho-\lambda|>L\mu} |g(\rho)|\|\tilde{F}_\mu^\pm(\rho;\lambda)^{-1} - F_\mu^\pm(\rho)^{-1}\|d\rho \le \int_{|\rho-\lambda|>L\mu}\frac{C\mu^2|g(\rho)|d\rho}{|\rho-\lambda|} \le C\mu^2|\log\mu| \tag{7.32}$$

For estimating $\|\tilde{F}_\mu^\pm(\rho;\lambda)^{-1} - F_\mu^\pm(\rho)^{-1}\|$ for $|\rho-\lambda| \le L\mu$, we decompose the interval $[\lambda - L\mu, \lambda + L\mu]$: We denote the distinct eigenvalues of PWP

by $\alpha_1 < \ldots < \alpha_\ell$ and set $\beta_j = \lambda + \mu(\alpha_j + \alpha_{j+1})/2$, $j = 1, \ldots, \ell - 1$. Define

$$\begin{aligned} I_1 &= [\lambda - L\mu, \beta_1), \\ I_\ell &= [\beta_{\ell-1}, \lambda + L\mu], \\ I_j &= [\beta_{j-1}, \beta_j), \qquad j = 2, \ldots, \ell - 1, \end{aligned}$$

so that $[\lambda - L\mu, \lambda + L\mu] = \bigcup_{j=1}^{\ell} I_j$.

Lemma 7.13: *There exists $C > 0$ such that*

$$\|F_\mu^\pm(\rho)^{-1}\| \le C(|\rho - \lambda - \mu\alpha_j| + \mu^2)^{-1}, \quad \rho \in I_j, \quad j = 1, \ldots, \ell \tag{7.33}$$

The estimate (7.33) also holds with $\tilde{F}_\mu^\pm(\rho; \lambda)$ in place of $F_\mu^\pm(\rho)$.

Proof: By (7.16), there exist $0 < C_1, C_2$ such that for $|\rho - \lambda| \le L\mu$

$$C_1\|u\|^2 \le |\Im(PWQ\Gamma_\mu^\pm(\rho)QWPu, u)|, \quad \|PWQ\Gamma_\mu^\pm(\rho)QWP\| \le C_2. \tag{7.34}$$

Let $\rho \in I_j$. We have $\|(\lambda - \rho + \mu PWP)u\| \ge |\lambda - \rho + \mu\alpha_j|\|u\|$ by the spectral theorem. It follows, if $|\rho - (\lambda + \mu\alpha_j)| \ge M\mu^2$ with large enough $M > 0$, that

$$\|F_\mu^\pm(\rho)u\| \ge C(|\lambda - \rho + \mu\alpha_j| + \mu^2)\|u\|.$$

When $|\rho - (\lambda + \mu\alpha_j)| \le M\mu^2$, we use the first estimate of (7.34) to obtain

$$\|F_\mu^\pm(\rho)u\| \ge C(|\lambda - \rho + \mu\alpha_j| + \mu^2)\|u\|.$$

(7.33) follows. The proof for $\tilde{F}_\mu^\pm(\rho; \lambda)$ is similar. □

It follows from (7.31) and Lemma 7.13 that

$$\|F_\mu^\pm(\rho)^{-1} - \tilde{F}_\mu^\pm(\rho; \lambda)^{-1}\| \le C\mu^2|\rho - \lambda|(|\rho - \lambda - \mu\alpha_j| + \mu^2)^{-2} \tag{7.35}$$

for $\rho \in I_j$, $j = 1, \ldots, \ell$. Hence,

$$\begin{aligned} &\int_{|\rho-\lambda|<L\mu} \|F_\mu^\pm(\rho)^{-1} - \tilde{F}_\mu^\pm(\rho; \lambda)^{-1}\| d\rho \\ &\le \sum_{j=1}^{\ell} \int_{I_j} \frac{C_1\mu^2|\rho - \lambda| d\rho}{(|\rho - \lambda - \mu\alpha_j| + \mu^2)^2} d\rho \le \int_{\mathbf{R}} \frac{C_2\mu^3 d\rho}{(|\rho| + \mu^2)^2} = C_3\mu. \end{aligned} \tag{7.36}$$

Combining (7.28), (7.29), (7.30), (7.36) and (7.32), we conclude that

$$\sup_{\pm\sigma>0} \|Pe^{-i\sigma K_\mu}P - e^{-i\sigma\Xi_\mu^\pm(\lambda)}\|_{B(\mathcal{K})} \le C\mu \tag{7.37}$$

Recall that $U_\mu(t, s)$ is the propagator for (7.6).

Theorem 7.14: Suppose $V, W \in \mathcal{V}_\beta$ with $\beta > 3$ and suppose $\phi \in \mathcal{H}$ is an eigenfunction of $\mathcal{U}_0$ with the eigenvalue $e^{-2\pi i\lambda} \neq 1$. Suppose W satisfies the Fermi Golden Rule (7.16). Then, we have (7.37). In particular, if $e^{-2\pi i\lambda}$ is a simple eigenvalue, then the survival probability amplitude $\Gamma(t) = (U_\mu(t,0)\phi, U(t,0)\phi)$ satisfies

$$\Gamma(t) = e^{-is(\Xi_\mu^\pm(\lambda)-\lambda)} + O(\mu) \tag{7.38}$$

uniformly with respect to $\pm t > 0$.

Proof: We have only to prove (7.38). Define $\psi(t) = (2\pi)^{-1/2}e^{i\lambda t}U(t,0)\phi$ so that $\psi(t)$ is periodic in t and $P\psi = \lambda\psi$. We have

$$(Pe^{-i\sigma K_\mu}P\psi, \psi)_{\mathcal{K}} = \frac{e^{-i\sigma\lambda}}{2\pi}\int_{\mathbf{T}}(U_\mu(t+\sigma,t)U(t,0)\phi, U(t+\sigma,0)\phi)_{\mathcal{H}}dt. \tag{7.39}$$

However, the Duhamel formula implies that

$$\sup_{s\in\mathbf{R}, 0\le t\le 2\pi}\|U_\mu(t+s,s) - U(t+s,s)\| \le C\mu,$$

It follows by writing the inner product in the right of (7.39) as

$$(U_\mu(t+\sigma,\sigma)U_\mu(\sigma,t)U(t,0)\phi, U(t+\sigma,\sigma)U(\sigma,0)\phi)_{\mathcal{H}}$$

and by replacing $U_\mu(t+\sigma,\sigma)$ and $U(t,0)$ by $U(t+\sigma,\sigma)$ and $U_\mu(t,0)$ respectively that, uniformly with respect to $\sigma \in \mathbf{R}$,

$$(Pe^{-i\sigma K_\mu}P\psi, \psi)_{\mathcal{K}} = e^{-i\sigma\lambda}(U_\mu(\sigma,0)\phi, U(\sigma,0)\phi)_{\mathcal{H}} + O(\mu).$$

Combining this with (7.37) and replacing σ by t, we obtain (7.38). □

The analogue of (7.38) holds in general and we leave the formulation of the general result and its proof to the readers.

Acknowledgement

This work was initiated while the author was visiting Mathematisches Institut der Universität München. He gratefully acknowledges the hospitality of the institute, Professor Heinz Siedentop in particular.

Bibliography

1. S. Agmon, *Spectral properties of Schrödinger operators and scattering theory*, Ann. Scuola. Norm. Sup. Pisa, Ser. IV **2** (1975), 151–218.

2. S. Agmon, I. Herbst and E. Skibsted, *Perturbation of embedded eigenvalues in the generalized N-body problem*, Commun. Math. Phys. **122** (1989), 411-438.
3. J. Aguilar and J. M. Combes, *A class of analytic perturbations for one-body Schrödinger Hamiltonians*, Commun. Math. Phys. **22** (1971), 269–279.
4. G. Artbazar and K. Yajima, *L^p-boundedness of wave operators for one-dimensional Schrödinger operators*, J. Math. Sci. Univ. Tokyo **7** (2000), 221-240.
5. M. Ben-Artzi and A. Devinatz, *Local smoothing and convergence properties of Schrödinger type equations,* J. Funct. Anal. **101** (1991), 231-254.
6. E. Balslev and J. M. Combes, *Spectral properties of Schrödinger operators with dilatation-analytic interactions*, Commun. Math. Phys. **22** (1971), 280–294.
7. P. Constantin and J. C. Saut, *Local smoothing properties of Schrödinger equations,* Indiana Univ. Math. J. **38** (1989), 791–810.
8. O. Costin, R. D. Costin, J. L. Lebowitz, and A. Rokhlenko, *Evolution of a model quantum system under time-periodic forcing: Conditions for complete ionization*, Comm. Math. Phys. **221** (2001) 1–26.
9. H. L. Cycon, R. G. Froese, W. Kirsch and B. Simon, *Schrödinger operators with application to quantum mechanics and global geometry.* Texts and Monographs in Physics, Springer Study Edition (Springer-Verlag, Berlin, 1987).
10. N. Burq, F. Planchon, J. G. Stalker and A. Shadi Tahvildar-Zadeh, *Strichartz estimates for the wave and Schrödinger equations with inverse-square potential*, reprint, 2002.
11. V. Enss and K. Veselić, *Bound states and propagating states for time-dependent Hamiltonians.* Ann. Inst. H. Poincare Sect. A (N.S.) **39** (1983), 159–191.
12. J. Derezinski and C. Gérard, *Scattering theory of classical and quantum N-particle systems*, Texts and Monographs in Physics (Springer Verlag, Berlin-Heidelberg, 1997).
13. A. Galtbayar, A. Jensen and K. Yajima, *Local time-decay of solutions to Schrödinger equations with time-periodic potentials*, to appear in J. Stat. Phys. (2003).
14. J. Ginibre and G. Velo, *Smoothing properties and retarded estimates for some dispersive evolution equations*, Commun. Math. Phys. **144** (1992), 163-188.
15. L. Hörmander, *The analysis of linear partial differential operators I, Distribution theory and Fourier analysis.* Second edition, Springer Study Edition (Springer-Verlag, Berlin, 1990).
16. T. Hoshiro, *Mourre's method and smoothing properties of dispersive equations*, Commun. Math. Phys. **202** (1999), 255-265.
17. J. Howland, *Stationary theory for time dependent Hamiltonians*, Math. Ann. **207** (1974), 315–335.
18. J. Howland, *Scattering theory for Hamiltonians periodic in time*, Indiana Univ. Math. J. **28** (1979), 471–494.
19. W. Hunziker and I. M. Sigal, *The quantum N-body problem*, J. Math. Phys. **41** (2000), 3448–3510
20. H. Isozaki and H. Kitada, *Scattering matrices for two-body Schrodinger operators*, Sci. Papers College Arts Sci. Univ. Tokyo **35** (1986), 81–107.

21. A. Jensen and T. Kato, *Spectral properties of Schrödinger operators and time-decay of the wave functions*, Duke. Math. J. **46** (1979), 583–611.
22. J.-L. Journé, A. Soffere and C.D. Sogge. *Decay estimates for Schrödinger operators*, Comm. Pure Appl. Math. **44** (1991), 573–604.
23. T. Kato, *Perturbation theory for linear operators* (Springer Verlg. New-York, 1966).
24. T. Kato, *Wave operators and similarities for some non-selfadjoint operators*, Math. Ann. **162** (1965/1966), 258–279.
25. T. Kato, *On nonlinear Schrodinger equations. II. H^s-solutions and unconditional well-posedness.* J. d'Anal. Math. **67** (1995), 281–306.
26. T. Kato and K. Yajima, *Some examples of smooth operators and the associated smoothing effect*, Rev. Math. Phys. **1** (1989) 481-496.
27. M. Keel and T. Tao, *Endpoint Strichartz estimates*, Amer. J. Math. **120** (1998), 955-980.
28. C. E. Kenig, G. Ponce and L. Vega, *Oscillatory integrals and regularity of dispersive equations*, Indiana Univ. Math. J. **40** (1991), 33–69.
29. E. Kirr and M. I. Weinstein, *Parametric excited Hamiltonian partial differential equations*, SIAM J. Math. Anal. **33** (2001), 16–52.
30. H. Kitada and K. Yajima, *Bound states and scattering states for time periodic Hamiltonians*, Ann. Inst. H. Poincaré Sect. A (N.S.) **39** (1983), 145–157.
31. H. Kitada and K. Yajima, *A scattering theory for time-dependent long-range potentials.* Duke Math. J. **49** (1982), 341–376.
32. E. L. Korotyaev, *Scattering theory for three-particle systems with time-periodic pair interactions* (Russian), Teoret. Mat. Fiz. **62** (1985), 242–252.
33. E. L. Korotyaev, *Eigenfunctions of the monodromy operator of the Schrödinger operator with a potential that is periodic with respect to time* (Russian), Mat. Sb. (N.S.) **124(166)** (1984), 431–446.
34. S. T. Kuroda, *Scattering theory for differential operators. I. Operator theory; II. Self-adjoint elliptic operators.* J. Math. Soc. Japan **25** (1973), 75–104 and 222–234.
35. E. H. Lieb, *The stability of matter: from atoms to stars.* Bull. Amer. Math. Soc. (N.S.) **22** (1990), 1–49.
36. K. Masuda, *Theory of evolution equations* (Kinokuniya-Shoten, Tokyo, 1975), in Japanese.
37. M. Merkli and I. M. Sigal, *A time-dependent theory of quantum resonances.* Comm. Math. Phys. **201** (1999), 549–576.
38. J. Moeller and E. Skibsted, *Spectral theory for time-periodic many-body problem*, preprint, Aarhus University (2003).
39. P. D. Miller, A. Soffer, and I. M. Weinstein, *Metastability of breather modes of time dependent potentials*, Nonlinearity **13** (2000), 507–568.
40. M. Murata, *Asymptotic expansions in time for solutions of Schrödinger-type equations*, J. Funct. Anal. **49** (1982), 10–56.
41. S. Nakamura, *Asymptotic completeness for three-body Schrödinger equations with time-periodic potentials*, J. Fac. Sci. Univ. Tokyo Sect. IA Math. **33** (1986), 379–402.
42. S. Nakamura, *Structure of the scattering operator for time-periodic Schrödinger equation*, J. Math. Soc. Japan **38** (1986), 261–273.

43. S. Nakamura, *Integral kernels of the scattering operator for time-periodic Schrödinger equation*, J. Func. Anal. **76** (1988), 176–192.
44. R. O'Neil, *Convolution operators and $L(p, q)$ spaces*, Duke Math. J. **30** (1963), 129-142.
45. A. Pazy, *Semigroups of linear operators and applications to partial differential equations*, Applied Mathematical Sciences, **44** (Springer-Verlag, New York, 1983).
46. G. Ponce, *Smoothing propeties of solutions to the Benjamin-Ono equation*, Analysis and partial deifferential equations, 667-679, Lect. Notes in Pure and Appl. Math. **122** (Dekker, New York, 1990).
47. M. Reed and B. Simon, *Methods of modern mathematical physics. I: Functional analysis*, Second edition (Academic Press, New York, 1980).
48. M. Reed and B. Simon, *Methods of modern mathematical physics. II: Fourier Analysis, Selfadjointness* (Academic Press, New York–London, 1975).
49. M. Reed and B. Simon, *Methods of modern mathematical physics. III: Scattering theory* (Academic Press, New York–London, 1979).
50. M. Reed and B. Simon, *Methods of modern mathematical physics. IV: Analysis of Operators* (Academic Press, New York–London, 1978).
51. I. Rodnianski and W. Schlag, *Time decay for solutions of Schrödinger equations with rough and time dependent potentials*, preprint, 2001.
52. I. Rodnianski, Schlag and A. Soffer, *Dispersive analysis of charge transfer models*, preprint.
53. B. Simon, *Resonances in n-body quantum systems with dilatation analytic potentials and the foundations of time-dependent perturbation theory*, Ann. of Math. (2) **97** (1973), 247–274.
54. P. Sjölin, *Regularity of solutions to the Schrödinger equations,* Duke Math. J. **55** (1987), 699–715.
55. J. Sjöstrand, *Quantum resonances and trapped trajectories*, Long time behaviour of classical and quantum systems (Bologna, 1999), 33–61, Ser. Concr. Appl. Math. 1, World Sci. Publishing, River Edge, NJ (2001).
56. R. S. Strichartz, *Restrictions of Fourier transforms to a quadratic surface and decay of solutions of wave equations*, Duke Math. J. **44** (1977), 704–714.
57. A. Soffer and M. I. Weinstein, *Nonautonomous Hamiltonians*, J. Statist. Phys. **93** (1998), 359–391.
58. A. Soffer and M. I. Weinstein, *Time dependent resonance theory*, Geom. Funct. Anal. **8** (1998), 1086–1128.
59. A. Soffer and M. I. Weinstein, *Ionization and scattering for short-lived potentials*, Lett. Math. Phys. **48** (1999), 339–352.
60. A. Soffer and M. I. Weinstein, *Resonances, radiation damping and instability in Hamiltonian nonlinear wave equations*, Invent. Math. **136** (1999), 9–74.
61. H. Tanabe, *Equations of evolution*, Monographs and Studies in Mathematics, **6** (Pitman (Advanced Publishing Program), Boston, Mass.-London, 1979).
62. D. Tataru, *Unique continuation for solutions to PDE's; between Hörmander's theorem and Holmgren's theorem*, Commun. in PDE **20** (1995), 855-884.
63. L. Vega, *Schrödinger equations: Pointwise convergence to the initial data*, Proc. A. M. S. **102** (1988), 874-878.

64. R. Weder, *L^p-$L^{p'}$ estimates for the Schrödinger equation on the line and inverse scattering for the nonlinear Schrödinger equation with a potential*, J. Funct. Anal. **170** (2000), 37–68.
65. K. Yajima, *Scattering theory for Schrödinger equations with potentials periodic in time*, J. Math. Soc. Japan **29** (1977), 729–743.
66. K. Yajima, *Resonances for the AC-Stark effect*, Commun. Math. Phys. **78** (1982), 331–352.
67. K. Yajima, *Exponential decay of quasi-stationary states of time-periodic Schrodinger equation with short range potentials*, Sci. Papers College Arts Sci. Univ. Tokyo **40** (1990), 27–36.
68. K. Yajima, *Existence of solutions for Schrödinger evolution equation*, Commun. Math. Phys. **110** (1987), 415–426.
69. K. Yajima, *The $W^{k,p}$-continuity of wave operators for Schrödinger operators*, J. Math. Soc. Japan **47** (1995), 551–581.
70. K. Yajima, *The $W^{k,p}$-continuity of wave operators for Schrödinger operators, III, Even dimensional case $m \geq 4$*, J. Math. Sci. Univ. Tokyo **2** (1995), 311–346.
71. K. Yajima, *L^p-boundedness of wave operators for two-dimensional Schrödinger operators*, Commun. Math. Phys. **208** (1999), 125–152.
72. K. Yokoyama, *Mourre theory for time-periodic systems*, Nagoya Math. J. **149** (1998), 193–210.
73. M. Zworski, *Quantum resonances and partial differential equations*, Proceedings of the International Congress of Mathematicians, Vol. III (Beijing, 2002), 243–252, (Higher Ed. Press, Beijing, 2002).

CHAPTER 3

AN APPLICATION OF PHASE SPACE TUNNELING TO MULTISTATE SCATTERING THEORY

André Martinez[†], Shu Nakamura[‡], Vania Sordoni[†]

[†] *Dipartmento di Matematica, Universitá di Bologna, Piazza di Porta San Donato 4, 40127 Bologna, Italy*

[‡] *Graduate School of Mathematical Science, University of Tokyo, 3-8-1 Komaba, Meguro Tokyo, 153-8914 Japan*

In this lecture,[a] we discuss an application of microlocal methods of tunneling to physical phenomena. In particular, we consider two-state Schrödinger equation on Euclidean space $\mathbb{R}^n$. If the two energy levels defined by the matrix-valued potential do not intersect, we expect the transition probabilities between different energy levels are exponentially small in the semiclassical limit, since such transition is carried by tunneling effect. We show that it in fact holds under several assumptions.

We state our main result in Section 1. In Section 2, we review our main tools to handle tunneling phenomena in the phase space, namely, $\hbar$-pseudodifferential operator calculus, the FBI transform (or, in other words, *coherent state expansion*) and exponential weight estimates in the phase space. In Section 3, we construct our framework to estimate the transition probability using the two-space scattering theory. We prove our main theorem in Section 4 combining them.

1. Introduction – Main Theorem

We consider a two-state Schrödinger operator:

$$H = -\hbar^2\triangle + V + \hbar R \qquad \text{on } \mathcal{H} = L^2(\mathbb{R}^n) \oplus L^2(\mathbb{R}^n),$$

[a] This lecture was given by S.N.

where

$$V = \begin{pmatrix} V_1(x) & 0 \\ 0 & V_2(x) \end{pmatrix}, \qquad R = \sum_{|\alpha|\le 1} c_\alpha(x)(\hbar D_\xi)^\alpha,$$

$D_x = -i\partial_x$, and $\hbar > 0$ is the Planck constant. For $\gamma > 0$, we set

$$\Gamma = \Gamma(\gamma) = \{z \in \mathbb{C}^n \mid |\mathrm{Im}\, z| < \gamma\}.$$

Assumption A: For $j = 1, 2$ and $|\alpha| \le 1$, $V_j(x)$ and $c_\alpha(x)$ are extended to holomorphic function on $\Gamma(\gamma)$ with some $\gamma > 0$. The limits

$$V_j^\infty = \lim_{|x|\to\infty} V_j(x)$$

exists for $j = 1, 2$. Moreover, there exists $\delta > 1$ such that

$$\left|\partial_z^\beta (V_j(z) - V_j^\infty)\right| \le C_\beta \langle z\rangle^{-\delta-|\beta|}, \qquad z \in \Gamma$$

$$\left|\partial_z^\beta c_\alpha(z)\right| \le C'_\alpha \langle z\rangle^{-\delta-|\beta|}, \qquad z \in \Gamma$$

for any $\beta \in \mathbb{Z}_+^n$, where $j = 1, 2$ and $|\alpha| \le 1$.

We set

$$H_0 = -\hbar^2\Delta + V^\infty, \qquad V^\infty = \begin{pmatrix} V_1^\infty & 0 \\ 0 & V_2^\infty \end{pmatrix} \qquad \text{on} \quad \mathcal{H}$$

with $\mathcal{D}(H_0) = \mathcal{D}(H) = H^2(\mathbb{R}^n) \oplus H^2(\mathbb{R}^n)$. Then, under Assumption A, the wave operators

$$W_\pm = W_\pm(H, H_0) = \operatorname*{s-lim}_{t\to\pm\infty} e^{itH}e^{-itH_0}$$

exists and are complete, i.e. $\mathrm{Ran}\, W_\pm(H, H_0) = \mathcal{H}_c(H)$. The scattering operator is defined by

$$S = S(H, H_0) = W_+^*(H, H_0)W_-(H, H_0),$$

and is unitary. Let

$$F_0 : \mathcal{H} \longrightarrow \bigoplus_{j=1}^{2} L^2([V_j^\infty, \infty) : L^2(\Sigma^{n-1}))$$

be the spectral representation of H_0 defined by

$$(F_0\varphi)_j(\lambda, \omega) = 2^{-1/2}(\lambda - V_j^\infty)^{-1/4}\hat\varphi_j((\lambda - V_j^\infty)^{1/2}\omega) \qquad \text{for} \quad \lambda > V_j^\infty,$$

where $\hat\varphi = \mathcal{F}\varphi$ is the Fourier transform of φ. It is well-known that $F_0SF_o^*$ is decomposed as

$$F_0SF_0^* = \int^\oplus S(\lambda)d\lambda$$

where $S(\lambda) = S(\lambda; H, H_0) \in \mathfrak{B}(L^2(S^{n-1}) \oplus L^2(S^{n-1}))$ is the scattering matrix for the pair H and H_0. We denote

$$S(\lambda) = \begin{pmatrix} s_{11}(\lambda) & s_{12}(\lambda) \\ s_{21}(\lambda) & s_{22}(\lambda) \end{pmatrix}$$

with $s_{ij}(\lambda) \in \mathfrak{B}(L^2(S^{n-1}))$.

Now we fix an energy $E > \max(V_1^\infty, V_2^\infty)$, and study the asymptotic behavior of $S(E)$ as $\hbar \downarrow 0$. Let

$$q_j(x,\xi) = |\xi|^2 + V_j(x), \qquad x, \xi \in \mathbb{R}^n, j = 1, 2$$

be the classical Hamiltonian corresponding to each components of H. Let $\exp(tH_{q_j})$ be the Hamilton flow on $\mathbb{R}^{2n}$ generated by q_j.

Assumption B: $E > \max(V_1^\infty, V_2^\infty)$ is nontrapping in the sense of Robert–Tamura with respect to q_j, $j = 1, 2$, i.e. if $q_j(x,\xi) = E$ then

$$|\exp(tH_{q_j})(x,\xi)| \to \infty \qquad \text{as } t \to \pm\infty.$$

We denote the energy surface with respect to q_j by

$$\Sigma_j(E) = \{(x,\xi) \in \mathbb{R}^{2n} \mid q_j(x,\xi) = E\}, \qquad j = 1, 2.$$

We also suppose that $\Sigma_1(E)$ and $\Sigma_2(E)$ are disjoint:

Assumption C: $V_1^\infty \neq V_2^\infty$ and $\Sigma_1(E) \cap \Sigma_2(E) = \emptyset$.

Now we can state our main result.

Theorem 1.1:[15] Suppose Assumptions A–C with $E > \max(V_1^\infty, V_2^\infty)$. Then there exist a neighborhood $E \in I \subset \mathbb{R}$ and $\mathbf{d} > 0$ such that

$$\|s_{12}(\lambda)\| + \|s_{21}(\lambda)\| \leq Ce^{-\mathbf{d}/\hbar}, \qquad \lambda \in I, \hbar > 0. \tag{1.1}$$

In principle, we can compute a lower bound of the constant $\mathbf{d} > 0$, but it is not necessarily optimal. We give an estimate in the proof.

2. Preliminaries

2.1. *ħ-pseudodifferential operators*

Let $a = a(\hbar; x, \xi) \in C^\infty(\mathbb{R}^{2n})$ be a symbol with a parameter $\hbar > 0$. The Weyl-quantized $\hbar$-pseudodifferential operator $a(\hbar; x, \hbar D_x)$ is defined by

$$a(\hbar; x, \hbar D_x)\varphi(x) = \frac{1}{(2\pi\hbar)^n} \int e^{i(x-y)\cdot\xi/\hbar} a\Big(\hbar; \frac{x+y}{2}, \xi\Big)\varphi(y)dyd\xi$$

for $\varphi \in \mathcal{S}(\mathbb{R}^n)$. We mainly work on symbols in the following symbol classes.

Definition 1: Let $m = m(\hbar; x, \xi)$ be a locally bounded positive function on $\mathbb{R}^{2n}$, and let $g = dx^2/\langle x\rangle^2 + d\xi^2/\langle \xi\rangle^2$. We denote $a \in S(m, g)$ if for any multi-index α and β,

$$\left|\partial_x^\alpha \partial_\xi^\beta a(\hbar; x, \xi)\right| \leq C_{\alpha\beta} m(\hbar; x, \xi)\langle x\rangle^{-|\alpha|}\langle \xi\rangle^{-|\beta|}, \qquad x, \xi \in \mathbb{R}^n.$$

For $\gamma, \gamma' > 0$, we denote

$$\Gamma = \Gamma(\gamma, \gamma') = \{(z, \zeta) \in \mathbb{C}^{2n} \mid |\mathrm{Im}\, z| < \gamma, |\mathrm{Im}\, \zeta| < \gamma'\}.$$

We also denote by $g_0 = |dz|^2 + |d\zeta|^2$ the flat metric on $\mathbb{C}^{2n}$.

Definition 2: Let $m = m(\hbar; z, \zeta)$ be a locally bounded positive function on Γ. We write $a = a(\hbar; x, \xi) \in S_\Gamma(m, g_0)$ if a is extended to a holomorphic function on Γ, and for any multi-index α and β,

$$\left|\partial_z^\alpha \partial_\zeta^\beta a(\hbar; z, \zeta)\right| \leq C_{\alpha\beta} m(\hbar; z, \zeta)\langle z\rangle^{-|\alpha|}\langle \zeta\rangle^{-|\beta|}, \qquad (z, \zeta) \in \Gamma(\gamma, \gamma').$$

We refer Ref. 4 or Ref. 13. We will use a generalization of the Beals' characterization of pseudodifferential operators (see Ref. 1 or Ref. 4 Proposition 8.3 for the $\hbar$-pseudodifferential operator case). We write

$$\begin{aligned} U(X)\varphi(y) &= e^{i(\xi\cdot x - ix\cdot D_x)/\hbar}\varphi(y) \\ &= e^{(-ix\cdot\xi/2 + iy\cdot\xi)/\hbar}\varphi(y - x), \quad y \in \mathbb{R}^n, X = (x, \xi) \in \mathbb{R}^{2n} \end{aligned}$$

for $\varphi \in \mathcal{S}(\mathbb{R}^n)$.

Proposition 2.1: *Let $A \in \mathfrak{B}(L^2(\mathbb{R}^n))$. Then there is $a \in S_\Gamma(1, g_0)$ such that $A = a(\hbar; x, \hbar D_x)$ if and only if*

$$A(X) = U(X) A U(X)^{-1} \in \mathfrak{B}(L^2(\mathbb{R}^n)), \qquad X \in \mathbb{R}^{2n}$$

is holomorphically extended to $Z = (z, \zeta) \in \Gamma$ as an $\mathfrak{B}(L^2(\mathbb{R}^n))$-valued function, and it is uniformly bounded with respect to $\hbar$.

The proof is analogous to Proposition 8.3 of Ref. 4, and we omit it.

In the following, we use the weight functions only of the form $m(\hbar; x, \xi) = \hbar^k\langle x\rangle^a\langle\xi\rangle^b$. However, it is useful to define the following class of weights:

Definition 3: We denote $X = (x, \xi) \in \mathbb{R}^{2n}$. A locally bounded positive function $m(\hbar; X)$ is called g_0-admissible weight if there are C and $N > 0$ such that

$$m(X + Y) \le Cm(X)\langle Y\rangle^N, \qquad X, Y \in \mathbb{R}^{2n}.$$

The pseudodifferential operator calculus is well-defined in $S(m, g_0)$ if m is g_0-admissible. (See Ref. 9.)

If m is an g_0-admissible, then we can find an equivalent smooth weight $m'(X)$, i.e. $C^{-1}m(X) \le m'(X) \le Cm(X)$ with some $C > 0$, such that for any α,

$$\left|\partial_X^\alpha m'(X)\right| \le C_\alpha m'(X).$$

Namely, $m' \in S(m', g_0) = S(m, g_0)$. Thus we may suppose $m \in S(m, g_0)$ without loss of generality.

2.2. *FBI transform*

For $\varphi \in \mathcal{S}(\mathbb{R}^n)$, we set

$$(T\varphi)(x, \xi) = (2\pi^2)^{-n/2}\hbar^{-3n/4}\int_{\mathbb{R}^n} e^{-|x-y|^2/2\hbar}e^{-iy\cdot\xi/\hbar}\varphi(y)dy, \qquad (x, \xi) \in \mathbb{R}^{2n}.$$

T is called FBI (Fourier–Bros–Iagornitzer) transform. It is also called the Bargman transform or the Gaussian wave packet expansion in the physics literature, up to constants. It is well-known T is an isometry from $L^2(\mathbb{R}^n)$ to $L^2(\mathbb{R}^{2n})$. We refer Ref. 13 for basic properties of T. Here we recall several basic properties of *FBI-quantized* $\hbar$-pseudodifferential operators. For $a = a(\hbar; x, \xi) \in C^\infty(\mathbb{R}^{2n})$, we set

$$\hat{a} = T^* M_a T,$$

where M_a denotes the multiplication operator by a in $L^2(\mathbb{R}^{2n})$. $\hat{a}$ is a bounded operator in $L^2(\mathbb{R}^n)$ if a is bounded.

Proposition 2.2: *Let $a \in S(1, g)$. Then there exists $b \in S(1, g)$ such that*

$$\hat{a} = b(\hbar; x, \hbar D_x).$$

Moreover, b has an asymptotic expansion

$$b(\hbar; x, \xi) \sim \sum_{m=0}^{\infty} \frac{1}{m!}\left[\frac{\hbar}{4}(\triangle_x + \triangle_\xi)\right]^m a(\hbar; x, \xi).$$

Proof: For simplicity, we write $X = (x, \xi)$, $Y = (y, \eta) \in \mathbb{R}^{2n}$. If we set

$$b(\hbar; X) = (2\pi\hbar)^{-n} \int_{\mathbb{R}^{2n}} e^{-|X-Y|^2/\hbar} a(\hbar; Y) dY,$$

then by direct computations, we can show b satisfies the required properties. The asymptotic expansion follows also from this expression. □

Corollary 2.3: *Let $a_1 \in S(1, g)$. Then there exists $a \in S(1, g)$ such that*

$$\hat{a} = a_1(\hbar; x, \hbar D_x) + R$$

with $R \in \mathfrak{B}(L^2(\mathbb{R}^n))$ such that

$$\left\| \langle x \rangle^N \langle \hbar D_x \rangle^N R \langle \hbar D_x \rangle^N \langle x \rangle^N \right\| \leq C_N \hbar^N$$

for any N. Moreover, supp $a \subset$ supp a_1 and

$$a(\hbar; x, \xi) \sim \sum_{m=0}^{\infty} \frac{1}{m!} \Big[-\frac{\hbar}{4} (\triangle_x + \triangle_\xi) \Big]^m a_1(\hbar; x, \xi).$$

2.3. *Exponential weight estimates in phase space*

At first we recall an exponential weight estimate due to Martinez[14] and Nakamura[16] in a slightly generalized form. We write the Cauchy–Riemann operator by

$$(\partial \psi)(x, \xi) = \partial_x \psi(x, \xi) + i \partial_\xi \psi(x, \xi).$$

Let $\Gamma = \Gamma(\gamma, \gamma')$ with γ, $\gamma' > 0$.

Proposition 2.4: *Let ψ be a real-valued smooth function on $\mathbb{R}^{2n}$ such that*

$$\left| \partial_x^\alpha \partial_\xi^\beta \psi(x, \xi) \right| \leq C_{\alpha\beta}, \qquad (x, \xi) \in \mathbb{R}^{2n},$$

for $|\alpha| + |\beta| > 0$, and that

$$\sup |\partial_x \psi| < \gamma', \qquad \sup |\partial_\xi \psi| < \gamma.$$

Let $A = a(\hbar; x, \hbar D_x)$ with $a \in S_\Gamma(1, g_0)$ and let m be a g_0-admissible weight function. Then there exist $b \in S(m, g_0)$ and $R \in \bigcap_{N=0}^{\infty} S(\hbar^N m, g_0)$ such that

$$\langle e^{\psi/\hbar} T f, m e^{\psi/\hbar} T A g \rangle = \langle e^{\psi/\hbar} T f, (M_b + R) e^{\psi/\hbar} T g \rangle, \qquad f, g \in \mathcal{S}(\mathbb{R}^n),$$

where M_b is a multiplication operator by b in $L^2(\mathbb{R}^{2n})$. Moreover,

$$b(\hbar; x, \xi) - m(\hbar; x, \xi) a(\hbar; x - \partial\psi, \xi + i\partial\psi) \in S(\hbar m, g_0).$$

Similarly, there exist $b' \in S(m, g_0)$ *and* $R' \in \bigcap_{N=0}^{\infty} S(\hbar^N m, g_0)$ *such that*

$$\left\|\sqrt{m}\, e^{\psi/\hbar} T A g\right\|^2 = \langle e^{\psi/\hbar} T f, (M_{b'} + R') e^{\psi/\hbar} T g \rangle, \qquad f, g \in \mathcal{S}(\mathbb{R}^n),$$

and that

$$b'(\hbar; x, \xi) - m(\hbar; x, \xi)|a(\hbar; x - \partial\psi, \xi + i\partial\psi)|^2 \in S(\hbar m, g_0).$$

We will need exponential weight estimates for non-analytic symbols. For that purpose, we use FBI-quantized operators and the following exponential weight estimates.

Assumption D: Let $\chi \in S(1, g_0)$ and let ψ be a real-valued smooth function on $\mathbb{R}^{2n}$. Let

$$\operatorname{supp} \chi = \Omega = \sum_{j=1}^{\infty} \Omega_j$$

be the decomposition to the connected components of supp χ. ψ satisfies the following conditions:

(i) $\Omega \cap \operatorname{supp} \nabla\psi = \emptyset$.
(ii) Let $\psi = \xi_j$ on Ω_j. Then there exists $\varepsilon > 0$ such that

$$|\psi(X) - \xi_j| \le \frac{1-\varepsilon}{4} \left[\operatorname{dist}(X, \Omega_j)\right]^2, \qquad X \in \mathbb{R}^{2n}.$$

Proposition 2.5: *Suppose* χ *and* ψ *satisfies Assumption D, and let* m *be a* g_0*-admissible weight function. Then there is* $C > 0$ *such that*

$$\left\|m e^{\psi/\hbar} T \hat{\chi} f - m e^{\psi/\hbar} \chi T f\right\| \le C\sqrt{\hbar}\left\|m e^{\psi/\hbar} T f\right\|, \qquad f \in \mathcal{S}(\mathbb{R}^n).$$

We use the following simple formula in the proof of Proposition 2.5. It is proved by direct computation and we omit the proof.

Lemma 2.6: $TT^* \in \mathfrak{B}(L^2(\mathbb{R}^{2n}))$ *has an integral kernel:*

$$\Pi(X, Y) = (2\pi\hbar)^{-n} e^{i\theta(X,Y)/\hbar} e^{-|X-Y|^2/4\hbar}, \qquad X, Y \in \mathbb{R}^{2n},$$

where $\theta(X, Y) = \frac{1}{2}(x - y) \cdot (\xi + \eta)$ *for* $X = (x, \xi)$, $Y = (y, \eta) \in \mathbb{R}^{2n}$.

Proof: [Proof of Proposition 2.5] We note

$$m e^{\psi/\hbar} T \hat{\chi} f - m e^{\psi/\hbar} \chi T f = m e^{\psi/\hbar}\{TT^*\chi - \chi TT^*\} e^{-\psi/\hbar} m^{-1} \left(m e^{\psi/\hbar} T f\right).$$

Then we set

$$B = m e^{\psi/\hbar}\{TT^*\chi - \chi TT^*\} e^{-\psi/\hbar} m^{-1}.$$

The integral kernel of B is given by

$$B(X,Y) = (m(X)/m(Y))e^{(\psi(X)-\psi(Y))/\hbar}\Pi(X,Y)(\chi(X)-\chi(Y)).$$

If $X, Y \notin \Omega$, then $B(X,Y) = 0$. Let us suppose $Y \in \Omega_j$. It follows from Assumption D that

$$|\psi(X) - \psi(Y)| = |\psi(X) - \xi_j| \le \frac{1-\varepsilon}{4}\left[\mathrm{dist}(X,\Omega_j)\right]^2 \le \frac{1-\varepsilon}{4}|X-Y|^2.$$

Hence we learn from Lemma 2.6 that

$$\begin{aligned}|B(X,Y)| \le\ & 2(2\pi\hbar)^{-n}e^{-\varepsilon|X-Y|^2/4\hbar}|\chi(X)-\chi(Y)|\cdot|m(X)/m(Y)| \\ & \le C\hbar^{-n}e^{-\varepsilon|X-Y|^2/4\hbar}|X-Y|\langle X-Y\rangle^N \\ & \le \sqrt{\hbar}\cdot C\hbar^{-n}\left(\frac{|X-Y|}{\sqrt{\hbar}}\left\langle\frac{X-Y}{\sqrt{\hbar}}\right\rangle^N e^{-\varepsilon|X-Y|^2/4\hbar}\right).\end{aligned}$$

We have used the g_0-admissibility of m in the second inequality. The same estimate holds for the case $X \in \Omega$. Now it follows from Schur's lemma that $\|B\| \le C\sqrt{\hbar}$, and the claim follows immediately. □

Combining Proposition 2.4 with Proposition 2.5, we have the following exponential weighted estimate for operators with non-analytic symbols, which will be used in the proof of Theorem 1.1.

Proposition 2.7: *Let $a \in S_\Gamma(1, g_0)$ and suppose ψ satisfies the conditions of Proposition 2.4. Suppose moreover that χ and ψ satisfies Assumption D. Let m be a g_0-admissible weight, and let*

$$b_0(\hbar; x,\xi) = m(\hbar; x,\xi)\left|a(\hbar; x-\partial\psi, \xi+i\partial\psi) + \chi(\hbar; x,\xi)\right|^2.$$

Let $A = a(\hbar; x, \hbar D_x)$ and let $P = A + \hat{\chi}$. Then there exists $C > 0$ such that

$$\left|\|\sqrt{m}\,e^{\psi/\hbar}TP\varphi\|^2 - \langle e^{\psi/\hbar}T\varphi, b_0e^{\psi/\hbar}T\varphi\rangle\right| \le C\sqrt{\hbar}\|\sqrt{m}\,e^{\psi/\hbar}T\varphi\|^2,$$

for $\varphi \in \mathcal{S}(\mathbb{R}^n)$.

3. Scattering Matrix

Let M be a large number, which will be fixed later. We assume $M > -\inf V_j$ $(j = 1, 2)$, so that $-M \in \rho(H)$ for sufficiently small $\hbar$. We write

$$\mu(\lambda) = -(\lambda + M)^{-1}, \qquad \lambda \in \mathbb{R},$$
$$A = \mu(H), \quad A_0 = \mu(H_0) \qquad \text{on } \mathcal{H}$$

Then by Birman's invariance principle (cf. Ref. 19, Section 2.6), we see

$$W_\pm(H, H_0) = W_\pm(A, A_0),$$

and hence

$$S(\lambda; H, H_0) = S(\mu(\lambda); A, A_0), \qquad \lambda \in \mathbb{R}.$$

We will obtain a representation formula of $S(\mu(\lambda); A, A_0)$ in this section.

3.1. *Construction of approximation systems: A_j ($j = 1, 2$)*

Without loss of generality, we may suppose $V_1^\infty < V_2^\infty$. Then it follows from Assumption C that

$$\inf\{V_2(x) - V_1(x) \mid V_j(x) \le E, j = 1, 2\} = \delta_0 > 0,$$

i.e. $V_2 \ge V_1 + \delta_0$ in the classically accessible region. We set

$$\Omega_1(a) = \{(x, \xi) \in \mathbb{R}^{2n} \mid |\xi|^2 + V_2(x) - E \ge a\},$$
$$\Omega_2(a) = \{(x, \xi) \in \mathbb{R}^{2n} \mid |\xi|^2 + V_1(x) - E \le -a\},$$

with $a \in \mathbb{R}$. We fix a sufficiently small constant $\delta_1 > 0$, and choose smooth functions $\chi_j(x, \xi)$ ($j = 1, 2$) so that

$$\chi_j(x,\xi) = \begin{cases} 1 & \text{if} \quad (x,\xi) \in \Omega_j(2\delta_1) \\ 0 & \text{if} \quad (x,\xi) \notin \Omega_j(\delta_1) \end{cases},$$
$$0 \le \chi_j(x,\xi) \le 1, \qquad (x,\xi) \in \mathbb{R}^{2n}, \quad j = 1, 2,$$

and that $\chi_j(x, \xi)$ is independent of x if $|x|$ is sufficiently large. Let $L_1 < E < L_2$, and let $L = \begin{pmatrix} L_1 & 0 \\ 0 & L_2 \end{pmatrix}$. We construct A_j on $\mathcal{H}$ so that their principal symbol is given by

$$a_j^0(x,\xi) = \mu(h^0(x,\xi))\chi_j(x,\xi) + (1 - \chi_j(x,\xi))\mu(L), \qquad j = 1, 2,$$

where $h^0(x,\xi) = \begin{pmatrix} |\xi|^2 + V_1(x) & 0 \\ 0 & |\xi|^2 + V_2(x) \end{pmatrix}$ is the principal symbol of the Hamiltonian H. We define

$$A_j = \mu(H) + T^*\left[(1 - \chi_j)(\mu(L) - \mu(h^0))\right]T,$$

where $[\cdots]$ in the right hand side is a multiplication operator in $L^2(\mathbb{R}^{2n})$, i.e. the the second term is the FBI-quantization of $(1 - \chi_j)(\mu(L) - \mu(h^0))$. It is easy to see that A_j satisfies the required property about the principal symbol. We will construct an approximation scheme of A in terms of A_1 and A_2.

Lemma 3.1: *$A_j \in OPS(1,g)$ for $j = 1,2$. Moreover, for any $\gamma' > 0$, if M is sufficiently large, $A_j \in OPS_\Gamma(1, g_0)$ with $\Gamma = \Gamma(\gamma, \gamma')$, where $\gamma > 0$ is the constant in Assumption A.*

Proof: The first assertion follows by the functional calculus of $\hbar$-pseudodifferential operators (cf. Ref 4, Chapter 8). The latter assertion follows from Proposition 2.1. □

3.2. *Construction of identification map J*

Similarly to χ_j, we choose smooth functions $j_k(x,\xi)$ $(k = 1,2)$ on $\mathbb{R}^{2n}$ so that

$$j_k(x,\xi) = \begin{cases} 1 & \text{if} \quad (x,\xi) \in \Omega_k(3\delta_1) \\ 0 & \text{if} \quad (x,\xi) \notin \Omega_k(2\delta_1) \end{cases},$$

$$0 \le j_k(x,\xi) \le 1, \qquad (x,\xi) \in \mathbb{R}^{2n}, \quad k = 1,2,$$

and that $j_k(x,\xi)$ is independent of x if $|x|$ is sufficiently large. By the construction it is easy to see that

$$\{(x,\xi) \mid j_1(x,\xi) = 1\} \supset \text{supp}\,(1 - \chi_2),$$
$$\{(x,\xi) \mid j_2(x,\xi) = 1\} \supset \text{supp}\,(1 - \chi_1)$$

provided $0 < \delta_1 < \delta_0/5$. We write

$$J_k = \hat{j}_k = T^* j_k T \in OPS(1,g), \qquad k = 1,2,$$

and we set $J : \mathcal{H} \mapsto \tilde{\mathcal{H}} = \mathcal{H} \oplus \mathcal{H}$ by

$$Jf = J_1 f \oplus J_2 f, \qquad \text{for } f \in \mathcal{H}.$$

We set

$$\tilde{A} = A_1 \oplus A_2 \qquad \text{on } \tilde{\mathcal{H}}$$

and we consider

$$T_1 = \tilde{A}J - JA_0 \quad : \quad \mathcal{H} \mapsto \tilde{\mathcal{H}}.$$

On the support of j_k, the symbol of A_k is the same as that of A. Hence, by the symbol calculus, we have

$$T_1 = (AJ_1 - J_1 A_0) \oplus (AJ_2 - J_2 A_0) + R, \qquad R \in OPS(\hbar^N \langle x\rangle^{-N} \langle\xi\rangle^{-N}, g)$$

with any N. By the resolvent formula, we obtain

$$A - A_0 \in OPS(\langle x\rangle^{-\delta}\langle\xi\rangle^{-4}, g)$$

with $\delta > 1$. On the other hand, by the construction of J_k, J_k is decomposed as

$$J_k = J_k^0 + J_k',$$

where $J_k^0 = J_k^0(\hbar D_x)$, and the symbol of J_k' has compact support. Hence

$$A_0 J_k - J_k A_0 \in OPS(\langle x\rangle^{-N}\langle\xi\rangle^{-N}, g)$$

with any N. Combining these we have

$$T_1 \in OPS(\langle x\rangle^{-\delta}\langle\xi\rangle^{-4}, g). \tag{3.1}$$

3.3. *Wave operators* $\boldsymbol{W_\pm^1 = W_\pm(\tilde{A}, A_0; J)}$

We define the wave operators $W_\pm^1$ by

$$W_\pm^1 = W_\pm(\tilde{A}, A_o; J) = \operatorname*{s-lim}_{t\to\pm\infty} e^{it\tilde{A}} J e^{-itA_0}.$$

By the Cook–Kuroda method, we have

$$e^{it\tilde{A}} J e^{-itA_0} = i\int_0^t e^{is\tilde{A}} T_1 e^{-isA_0} ds + J.$$

Now the limits $t \to \pm\infty$ exist by the standard stationary phase argument and (3.1). Thus the wave operators $W_\pm^1$ exist.

Let $(a_j^0)_{k\ell}$ be the (k,ℓ)-component of the principal symbol of A_j. By the definition, we have

$$\left[(a_j^0)_{jj}\right]^{-1}(E) = \Sigma_j(E), \qquad j = 1, 2,$$

and

$$\operatorname{dist}\big(\operatorname{Ran}\left[(a_j^0)_{kk}\right], E\big) > 0, \qquad \text{if } j \neq k.$$

Noting this and the fact $V_j \to V_j^\infty$ as $|x| \to \infty$, we arrive at the following simple observation:

Lemma 3.2: *There exists I, a small neighborhood of $\mu(E)$ such that*

$$\begin{array}{ll} \operatorname{dist}\big(\left[(a_j^0)_{jj}\right]^{-1}(I), (\Omega_j(3\delta_1))^c\big) > 0 & j = 1, 2, \\ \operatorname{dist}\big(\operatorname{Ran}\left[(a_j^0)_{kk}\right], I\big) > 0 & \textit{if } j \neq k. \end{array}$$

Let

$$a_0^0(x,\xi) = \begin{pmatrix} \mu(|\xi|^2 + V_1^\infty) & 0 \\ 0 & \mu(|\xi|^2 + V_2^\infty) \end{pmatrix}$$

be the principal symbol of A_0. We note that $\left[(a_j^0)_{jj}\right]^{-1}(I)$ converges to $\left[(a_0^0)_{jj}\right]^{-1}(I)$ as $|x| \to \infty$. Hence, by Lemma 3.2, we obtain

$$\left[(a_0^0)_{jj}\right]^{-1}(I) \cap (\Omega_j(3\delta_1))^c \qquad \text{is compact.}$$

In particular, we have

$$\left[(a_0^0)_{kk}\right]^{-1}(I) \cap \operatorname{supp}(1 - j_k) \qquad \text{is compact,}$$

for $k = 1, 2$, and

$$\left[(a_0^0)_{kk}\right]^{-1}(I) \cap \operatorname{supp} j_\ell \qquad \text{is compact}$$

if $k \neq \ell$. Let $f \in C_0^\infty(I)$. Then these imply

$$f((a_0^0)_{kk})(1 - j_k) \quad \text{and} \quad f((a_0^0)_{kk})j_\ell \quad (k \neq \ell)$$

are compactly supported. Hence,

$$(1 - J_k)(f(A_0))_{kk} \quad \text{and} \quad J_\ell(f(A_0))_{kk} \quad (k \neq \ell)$$

are compact operators. The next lemma follows from these observations and the well-known fact that $e^{-itA_0}\varphi \to 0$ weakly as $|t| \to \infty$.

Lemma 3.3: *Let I as in Lemma 3.2, and let $f \in C_0^\infty(I)$. Then*

$$\lim_{t\to\pm\infty} \left\| J_k e^{-itA_0} f(A_0)\varphi \right\|_{\mathcal{H}} = \|(f(A_0)\varphi)_k\|_{L^2(\mathbb{R}^n)}$$

for $\varphi = (\varphi_1, \varphi_2) \in \mathcal{H}$.

This implies, in particular,

$$\lim_{t\to\pm\infty} \left\| J e^{-itA_0} f(A_0)\varphi \right\|_{\tilde{\mathcal{H}}}^2 = \|f(A_0)\varphi\|_{\mathcal{H}}^2,$$

and hence $W_\pm^1 E_I(A_0)$ is a partial isometry from $E_I(A_0)\mathcal{H}$. It also follows that off-diagonal elements of

$$S^1 = (W_+^1)^* W_-^1 = \begin{pmatrix} s_{11}^1 & s_{12}^1 \\ s_{21}^1 & s_{22}^1 \end{pmatrix}$$

is zero on $E_I(A_0)\mathcal{H}$, i.e.

$$S_1 E_I(A_0) = \begin{pmatrix} s_{11}^1 & 0 \\ 0 & s_{22}^1 \end{pmatrix} E_I(A_0).$$

Now we remark on a generalization of the celebrated semiclassical resolvent estimate of Robert–Tamura:

Proposition 3.4: *Let $P = A$, A_1 or A_2, and let $s > 1/2$. Then there exists $I \subset \mathbb{R}$, a neighborhood of $\mu(E)$ such that*

$$\langle x\rangle^{-s}(P-\lambda \pm i0)^{-1}\langle x\rangle^{-s} = \lim_{\varepsilon\downarrow 0}\langle x\rangle^{-s}(P-\lambda\pm i\varepsilon)^{-1}\langle x\rangle^{-s}, \qquad \lambda\in I$$

exist in $\mathfrak{B}(\mathcal{H})$ if $\hbar > 0$ is sufficiently small. Moreover, there is $C > 0$ such that

$$\left\|\langle x\rangle^{-s}(P-\lambda\pm i0)^{-1}\langle x\rangle^{-s}\right\| \le C\hbar^{-1}, \qquad \lambda\in I, \hbar>0.$$

This proposition is proved in the same manner as Ref. 5, and we omit it. This assertion implies the limiting absorption principle for $\tilde{A}$, and hence we can construct the inverse wave operators on $E_I(\tilde{A})\tilde{\mathcal{H}}$ by using the stationary scattering theory (cf. Ref. 19). Thus we obtain the asymptotic completeness on the energy interval I. Now we summarize our results in this subsection.

Proposition 3.5: *There exists $I \subset \mathbb{R}$, a neighborhood of $\mu(E)$, such that $W_\pm^1 E_I(A_0)$ is a partial isometry with the initial space $E_I(A_0)\mathcal{H}$, and that $W_\pm^1$ is complete on I, i.e. Ran $[W_\pm^1 E_I(A_0)] = E_I(\tilde{A})$. Moreover, off-diagonal components of $S^1 E_I(A_0)$ vanishes.*

3.4. *Spectral representation of $\tilde{A}$ on I*

Let F_0 be the spectral representation of H_0 defined in Section 1. By the completeness of $W_\pm^1$ on I, $F_0(W_\pm^1)^* E_I(\tilde{A})$ gives a spectral representation of $\tilde{A}$ on I. For each λ, we set

$$(F_0 f)_j(\lambda,\omega) = (F_0(\lambda)\varphi)_j(\omega), \qquad \lambda\in\mathbb{R}, \omega\in S^{n-1}, j=1,2,$$

for $\varphi \in (\mathcal{S}(\mathbb{R}^n))^2$. We note that $F_0(\lambda)$ is a bounded map from $(L^{2,s}(\mathbb{R}^n))^2$ to $(L^2(S^{n-1}))^2$ if $s > 1/2$, where $L^{2,s}(\mathbb{R}^n)$ denotes the weighted L^2-space with the weight $\langle x\rangle^s$. Moreover, we have

$$\|F_0(\lambda)\|_{\mathfrak{B}((L^{2,s})^2,(L^2(S^{n-1})^2)} \le C\hbar^{-1}$$

if $\lambda > V_2^\infty$. We set

$$F_1(\lambda) = F_0(\lambda)(W_+^1)^*, \qquad \lambda\in I.$$

Lemma 3.6: $\{F_1(\lambda) \mid \lambda \in I\}$ *is a spectral representation of $\tilde{A}$ on $E_I(\tilde{A})\tilde{\mathcal{H}}$, and for each $\lambda\in I$, $F_1(\lambda)$ is a bounded map from $(L^{2,\delta/2}(\mathbb{R}^n))^2$*

to $(L^2(S^{n-1}))^2$, *and the operator norm is bounded by* $C\hbar^{-2}$ *(uniformly in* $\lambda \in I$*).*

Proof: By the stationary representation of wave operators, we have

$$W_{\pm}^{1} F_0(\lambda)^* = (J + (\tilde{A} - \lambda + i0)^{-1} T_1) F_0(\lambda)^*,$$

where $(\tilde{A} - \lambda + i0)^{-1} = \lim_{\varepsilon \downarrow 0} (\tilde{A} - \lambda + i\varepsilon)^{-1}$ is defined as an operator from $(L^{2,\delta/2})^2$ to $(L^{2,-\delta/2})^2$ by virtue of the semiclassical resolvent formula (Proposition 3.4). Hence we have

$$F_1(\lambda) = F_0(\lambda)(J^* + T_1^*(\tilde{A} - \lambda - i0)^{-1}).$$

The assertion follows easily by recalling (3.1). □

3.5. *Wave operators* $W_{\pm}^{2} = W_{\pm}(A, \tilde{A}; J^*)$

We set

$$W_{\pm}^{2} = W_{\pm}(A, \tilde{A}; J^*) = \operatorname*{s-lim}_{t \to \pm\infty} e^{itA} J^* e^{-it\tilde{A}}.$$

In order to show the existence of $W_{\pm}^{2}$, we compute

$$T_2 = AJ^* - J^*\tilde{A} \in \mathfrak{B}(\tilde{\mathcal{H}}, \mathcal{H}).$$

As well as T_1, we have

$$T_2\varphi = (AJ_1 - J_1 A)\varphi_1 + (AJ_2 - J_2 A)\varphi_2 + R_2\varphi \tag{3.2}$$

for $\varphi = (\varphi_1, \varphi_2) \in \tilde{\mathcal{H}}$, where

$$R_2\varphi = J_1(A - A_1)\varphi_1 + J_2(A - A_2)\varphi_2.$$

We have $R_2 \in OPS(\hbar^N \langle x\rangle^{-N} \langle \xi\rangle^{-N}, g)$ for any N since the symbol of $A - A_k$ vanishes on the support of j_k $(k = 1, 2)$.

Lemma 3.7: $T_2 \in OPS(\hbar\langle x\rangle^{-\delta-1}\langle\xi\rangle^{-4}, g)$.

Proof: Let $J_k = J_k^0 + J_k'$ be the decomposition of J_k as in Section 3.2. Then we have

$$\begin{aligned} AJ_k - J_k A &= [A, J_k^0] + [A, J_k'] \\ &= A[J_k^0, (V + \hbar R)]A + [A, J_k'] \in OPS(\hbar\langle x\rangle^{-\delta-1}\langle\xi\rangle^{-4}, g) \end{aligned}$$

since $J_k' \in OPS(\hbar^N \langle x\rangle^{-N} \langle\xi\rangle^{-N}, g)$ for any N. □

By virtue of the semiclassical resolvent estimate for A_j, we have the local decay property on $E_I(A_j)\mathcal{H}$. Hence, combining Lemma 3.7 with the

Cook–Kuroda method as in the proof of Lemma 3.3, we have the existence of

$$W^2_\pm E_I(\tilde{A}) = \operatorname*{s-lim}_{t\to\pm\infty} e^{itA} J^* e^{-it\tilde{A}} E_I(\tilde{A}).$$

We next consider

$$\begin{aligned} W^2_\pm W^1_\pm E_I(A_0) &= W^2_\pm E_I(\tilde{A}) \cdot W^1_\pm E_I(A_0) \\ &= \operatorname*{s-lim}_{t\to\pm\infty} e^{itA} J^* J e^{-itA_0} E_I(A_0). \end{aligned}$$

Note that the symbol of J^*J is given by $j_1(x,\xi)^2 + j_2(x,\xi)^2$. By the construction of $j_k(x,\xi)$, we easily see

$$\text{supp}\,(1 - (j_1(x,\xi)^2 + j_2(x,\xi)^2)) \subset \Omega_1(2\delta_1) \cap \Omega_2(2\delta_1),$$

and hence

$$\text{supp}\left(1 - \sum_{k=1}^{2} j_k^2\right) \cap \bigcup_{k=1}^{2} (a_0^0)_{kk}^{-1}(I) \qquad \text{is compact,}$$

if I is chosen as in Lemma 3.2. Now, by the same argument as in Lemma 3.3, we have

Lemma 3.8:

$$\operatorname*{s-lim}_{t\to\pm\infty} (1 - J^*J) e^{-itA_0} E_I(A_0) = 0.$$

In particular,

$$W^2_\pm W^1_\pm E_I(A_0) = W_\pm E_I(A_0), \tag{3.3}$$

where $W_\pm = W_\pm(H, H_0) = W_\pm(A, A_0)$.

We summarize our finding of this subsection:

Proposition 3.9: *There exists* $I \subset \mathbb{R}$, *a neighborhood of* $\mu(E)$, *such that* $W^2_\pm E_I(A_0)$ *is a unitary map from* $E_I(\tilde{A})\tilde{\mathcal{H}}$ *to* $E_I(A)\mathcal{H}$, *i.e.* $W^2_\pm$ *is complete on* I. *Moreover, (3.3) holds.*

Note that the completeness follows from those of $W_\pm$ and $W^1_\pm$ and the chain rule.

3.6. *Representation formula of* $S^2(\lambda)$

Now we let

$$S^2 E_I(\tilde{A}) = S(A, \tilde{A}; J^*)E_I(\tilde{A}) = (W_+^2)^* W_-^2 E_I(\tilde{A})$$

be the scattering operator for the pair of operators A and $\tilde{A}$ on the energy interval I. Using the spectral representation constructed in Section 3.4, we can define the scattering matrix by

$$S^2(\lambda) = S(\lambda; A, \tilde{A}; J^*) \in \mathfrak{B}([L^2(S^{n-1}) \oplus L^2(S^{n-1})]^2), \qquad \lambda \in I,$$

in the same way as in Section 1. By the abstract stationary theory of scattering, we have

$$S^2(\lambda) = 1 - 2\pi i F_1(\lambda)\big[T_2^* J^* - T_2^*(A - \lambda - \chi 0)^{-1} T_2\big] F_1(\lambda)^*, \quad \lambda \in I. \quad (3.4)$$

Note that the right hand side is well-defined since

- $F_1(\lambda)$ maps $[L^2(S^{n-1}) \oplus L^2(S^{n-1})]^2$ into $(L^{2,-\delta/2}(\mathbb{R}^n))^2$ by Lemma 3.6;
- T_2 maps $(L^{2,-\delta/2}(\mathbb{R}^n))^2$ to $(L^{2,-\delta/2}(\mathbb{R}^n))^2$ by Lemma 3.7;
- $(A-\lambda-i0)^{-1}$ maps $(L^{2,\delta/2}(\mathbb{R}^n))^2$ to $(L^{2,-\delta/2}(\mathbb{R}^n))^2$ by the semiclassical resolvent formula.

Given the above construction, we can now easily show the following estimates on $S^2(\lambda)$, which is valid even if the potential is not necessarily analytic.

Proposition 3.10: *Suppose Assumptions* A–C *with* $\gamma = 0$, *i.e. without the analyticity condition. Then there exists* $I \subset \mathbb{R}$, *a neighborhood of* $\mu(E)$, *such that*

$$\left\|S^2(\lambda) - 1\right\|_{\mathfrak{B}(L^2(S^{n-1})^2)} \le C_N \hbar^N, \quad \lambda \in I, \hbar > 0,$$

with any N.

Proof: Let $f \in C_0^\infty(I)$, and we consider

$$S^2(\lambda)f(\lambda)^2 = f(\lambda)^2 - 2\pi i F_1(\lambda)\Big[f(\tilde{A})T_2^* J^* f(\tilde{A}) \\ - f(\tilde{A})T_2^*(A - \lambda - i0)^{-1} T_2 f(\tilde{A})\Big] F_1(\lambda)^*, \quad \lambda \in I.$$

Let $k = 1$ or 2. By the asymptotic expansion, we learn that the symbol of $AJ_k - J_k A$ is supported in $\Omega_k(2\delta_1)\backslash\Omega_k(3\delta_1)$ modulo $S(\hbar^\infty \langle x\rangle^{-\infty}\langle\xi\rangle^{-\infty})$. On the other hand, by the functional calculus of $\hbar$-pseudodifferential operators, the symbol of $f(A_k)$ is supported in $(a_k^0)_{kk}^{-1}(I)$, a small neighborhood of $\Sigma_k(E)$. Hence, by using Lemma 3.2 again, we have

$$(AJ_k - J_kA)f(A_k) \in S(\hbar^\infty \langle x\rangle^{-\infty}\langle \xi\rangle^{-\infty}).$$

Then, by (3.2), we learn $T_2 f(\tilde{A}) \in S(\hbar^\infty \langle x\rangle^{-\infty}\langle \xi\rangle^{-\infty})$.

Now, combining this observation with Lemma 3.6 and the semiclassical resolvent estimate, we have $\|f(\lambda)^2 S^2(\lambda)\| = O(\hbar^\infty)$. The assertion follows immediately since $f \in C_0^\infty(I)$ is arbitrary. □

Corollary 3.11: *Under the same assumption as Proposition* 3.10, *there exists* $I' \in \mathbb{R}$, *a neighborhood of* E *such that*

$$\|s_{12}(\lambda)\| + \|s_{21}(\lambda)\| \le C_N \hbar^N, \qquad \lambda \in I', \hbar > 0,$$

with any $N > 0$.

Proof: It suffices to consider $S(A, A_0)$ by Birman's invariance principle. Let I be the interval in Proposition 3.10. By the chain rule (3.3), we have

$$\begin{aligned} SE_I(A_0) &= (W_+^2 W_+^1 E_I(A_0))^*(W_-^2 W_-^1 E_I(A_0)) \\ &= (W_+^1 E_I(A_0))^* S^2 E_I(\tilde{A})(W_-^1 E_I(A_0)) \\ &= S^1 E_I(A_0) + (W_+^1 E_I(A_0))^* \big[(S^2 - 1)E_I(\tilde{A})\big](W_-^1 E_I(A_0)). \end{aligned}$$

The off-diagonal components of $S^1 E_I(A_0)$ vanishes by Proposition 3.5, and the second term is $O(\hbar^\infty)$ by Proposition 3.10. Thus we have

$$\|(SE_I(A_0))_{k\ell}\|_{\mathfrak{B}(L^2(\mathbb{R}^n))} = O(\hbar^\infty) \qquad \text{if } k \ne \ell.$$

The statement on the scattering matrix follows from this estimate, since the scattering matrix is a diagonalization of the scattering operator. □

4. Exponential Estimates – Proof of Main Theorem

In view of the results in Section 3, it remains only to improve Proposition 3.10 to prove Theorem 1.1. We will obtain exponential decay estimates on, e.g., $T_2 F_1(\lambda)^*\varphi$ for $\varphi \in (L^2(S^{n-1}))^2$. We write

$$\begin{aligned} \Lambda_1(a) &= \big\{(x,\xi) \in \mathbb{R}^{2n} \;\big|\; |\xi|^2 + V_1(x) - E \ge -a\big\} = \overline{\Omega_2(a)^c} \\ \Lambda_2(a) &= \big\{(x,\xi) \in \mathbb{R}^{2n} \;\big|\; |\xi|^2 + V_2(x) - E \le a\big\} = \overline{\Omega_1(a)^c} \end{aligned}$$

for $a \in \mathbb{R}$.

Assumption E: Let $k = 1$ or 2. A real valued function $\rho_k(x,\xi) \in C^\infty(\mathbb{R}^{2n})$ satisfies the following conditions.

(i) $\rho_k(x,\xi) = 0$ if $(x,\xi) \in \Lambda_k(2\delta_2)$ with some $\delta_2 > 0$.
(ii) $\sup|\partial_\xi \rho_k| < \gamma$.
(iii) There is $\delta_3 > 0$ such that

$$|(\xi + i\partial\rho_k)^2 + V_k(x - \partial\rho_k) - E| \geq \delta_3 \quad \text{if } (x,\xi) \notin \Lambda_k(\delta_2).$$

(iv) $\rho_k(x,\xi) = \rho_k^0 > 0$ if $(x,\xi) \in \Omega_k(3\delta_1)^c \supset \text{supp}\,(1 - j_k)$.
(v) There is $\delta_4 > 0$ such that

$$|\rho_k(x,\xi) - \rho_k^0| \leq \frac{1-\delta_4}{4} \text{dist}((x,\xi), \Omega_k(2\delta_1)) \quad \text{for } (x,\xi) \in \mathbb{R}^{2n}.$$

Proposition 4.1: *Let $k = 1$ or 2. Suppose ρ_k satisfies Assumption E. Then there exist I, a neighborhood of $\mu(E)$ and $M > 0$ such that if $\psi \in (L^{2,-\delta/2}(\mathbb{R}^n))^2$ be a generalized eigenfunction of A_k with an eigenvalue $\lambda \in I$, then*

$$\|\chi_{\Lambda_k(\delta_2)}\langle x\rangle^{-\delta/2} e^{\rho_k/\hbar} T\psi\| \leq C\|\langle x\rangle^{-\delta/2}\psi\|.$$

Proof: By the assumption, if M is large enough, and if λ is in a small neighborhood of $\mu(E)$, then

$$|(a_k^0)_{jj}(x - \partial\rho_k, \xi + i\partial\rho_k) - \lambda| \geq \delta_5 > 0, \tag{4.1}$$

for $(x,\xi) \notin \Lambda_k(\delta_2), j = 1,2$. Then by Proposition 2.7, we have

$$|\langle e^{\rho_k/\hbar} T\psi, b_0 e^{\rho_k/\hbar} T\psi\rangle| \leq C\sqrt{\hbar}\|\langle x\rangle^{-\delta/2} e^{\rho_k/\hbar} T\psi\|^2,$$

where $b_0(x,\xi) = \langle x\rangle^{-\delta}|a_k^0(x - \partial\rho_k, \xi + i\partial\rho_k) - \lambda|^2$. By (4.1), we have

$$b_0(x,\xi) \geq \delta_5\langle x\rangle^{-\delta}\chi_{\Lambda_k(\delta_2)}(x,\xi),$$

and hence

$$\begin{aligned}
\delta_5\|\chi_{\Lambda_k(\delta_2)}\langle x\rangle^{-\delta/2} e^{\rho_k/\hbar} T\psi\|^2 &\leq C\sqrt{\hbar}\|\langle x\rangle^{-\delta/2} e^{\rho_k/\hbar} T\psi\|^2 \\
&\leq C\sqrt{\hbar}\|\chi_{\Lambda_k(\delta_2)}\langle x\rangle^{-\delta/2} e^{\rho_k/\hbar} T\psi\|^2 \\
&\quad + C\sqrt{\hbar}\|(1 - \chi_{\Lambda_k(\delta_2)})\langle x\rangle^{-\delta/2} e^{\rho_k/\hbar} T\psi\|^2.
\end{aligned}$$

Since $\rho_k = 0$ on $\text{supp}\,(1 - \chi_{\Lambda_k(\delta_2)})$, the last term is bounded by $C\sqrt{\hbar}\|\langle x\rangle^{-\delta/2}\psi\|^2$. Thus the assertion follows immediately if $\hbar$ is sufficiently small. □

By the symbol calculus of $\hbar$-pseudodifferential operators as in Section 3, we also have

$$\|\chi_{\Lambda_k(\delta_2)}\langle x\rangle^N T\psi\| \leq C_N \hbar^N \|\langle x\rangle^{-\delta/2}\psi\|$$

with any N. By interpolation, we obtain the following corollary:

Corollary 4.2: *Let ρ_k and ψ as in Proposition 4.1, and let $\varepsilon_1 > 0$ be arbitrary. Then for any N, there is C_N such that*

$$\left\| \langle x \rangle^N \chi_{\Lambda_k(\delta_2)} e^{(1-\varepsilon_1)\rho_k/\hbar} T\psi \right\| \leq C\hbar^N \| \langle x \rangle^{-\delta/2} \psi \|.$$

We set

$$\mathbf{c}_k = \sup\{ \rho_k^0 \mid \rho_k \text{ satisfies Assumption E} \}$$

for $k = 1, 2$, where ρ_k^0 denotes the value of ρ_k on $\Omega_k(3\delta_1)^c$ as in Assumption E-(iv). We also set

$$\mathbf{c} = \min(\mathbf{c}_1, \mathbf{c}_2).$$

Note that $\mathbf{c} > 0$, since if ρ_k satisfies the conditions (i), (ii), (iv) and (v) of Assumption E, then $a\rho_k$ satisfies the other condition (iii), provided $0 < a$ is sufficiently small.

We are now ready to estimate $T_2 F_1(\lambda)^*$.

Lemma 4.3: *Let $0 < \mathbf{d} < \mathbf{c}$, and let $N \in \mathbb{R}$. Then there exist I, a neighborhood of $\mu(E)$ and $C > 0$ such that*

$$\left\| \langle x \rangle^{\delta/2} T_2 F_1(\lambda)^* \varphi \right\| \leq C\hbar^N e^{-\mathbf{d}/\hbar} \|\varphi\|$$

for $\lambda \in I$ and $\varphi \in (L^2(S^{n-1}))^2$.

Proof: Let I as in Proposition 4.1. By the definition of T_2, we have

$$T_2\psi = (AJ_1 - J_1 A_1)\psi_1 + (AJ_2 - J_2 A_2)\psi_2,$$

for $\psi = (\psi_1, \psi_2) \in \tilde{\mathcal{H}}$. Hence it suffices to show

$$\left\| \langle x \rangle^{\delta/2} (AJ_k - J_k A_k)\psi \right\| \leq C\hbar^N e^{-\mathbf{d}/\hbar} \left\| \langle x \rangle^{-\delta/2} \psi \right\|, \qquad k = 1, 2, \tag{4.2}$$

for $\psi \in F_1(\lambda)^*\varphi \in (L^{2,-\delta/2}(\mathbb{R}^n))^2$ by Lemma 3.6. Recall that ψ is a generalized eigenfunction of A_k: $A_k\psi = \lambda\psi$, $\lambda \in I$. We write

$$AJ_k - J_k A_k = (1 - J_k)A_k - A(1 - J_k) + (A - A_k) = T(1) + T(2) + T(3).$$

We choose $\varepsilon_1 > 0$ and ρ_k so that

$$\mathbf{d} \leq (1 - \varepsilon_1)\rho_k^0 < \mathbf{c}.$$

By Corollary 4.2, we have

$$\begin{aligned}
&\|\langle x\rangle^{-\delta/2}T(1)\psi\| \\
&\quad = \|\langle x\rangle^{-\delta/2}(1-J_k)\lambda\psi\| \\
&\quad = \Big\|\lambda\big(\langle x\rangle^{\delta/2}T^*\langle x\rangle^{-\delta/2}\big)(1-j_k)e^{-(1-\varepsilon_1)\rho_k/\hbar}\big(\langle x\rangle^{\delta/2}\chi_{\Lambda_k(\delta_2)}e^{(1-\varepsilon_1)\rho_k/\hbar}T\psi\big)\Big\| \\
&\quad \le Ce^{-(1-\varepsilon_1)\rho_k^0/\hbar}\|\langle x\rangle^{\delta/2}\chi_{\Lambda_k(\delta_2)}e^{(1-\varepsilon_1)\rho_k/\hbar}T\psi\| \\
&\quad \le C\hbar^N e^{-\mathbf{d}/\hbar}\|\langle x\rangle^{-\delta/2}\psi\|.
\end{aligned}$$

Similarly, we have

$$\begin{aligned}
&\|\langle x\rangle^{-\delta/2}T(2)\psi\| \\
&\quad = \|\langle x\rangle^{-\delta/2}A(1-J_k)\psi\| \\
&\quad = \Big\|\big(\langle x\rangle^{\delta/2}AT^*\langle x\rangle^{-\delta/2}\big)(1-j_k)e^{-(1-\varepsilon_1)\rho_k/\hbar}\big(\langle x\rangle^{\delta/2}\chi_{\Lambda_k(\delta_2)}e^{(1-\varepsilon_1)\rho_k/\hbar}T\psi\big)\Big\| \\
&\quad \le C\hbar^N e^{-\mathbf{d}/\hbar}\|\langle x\rangle^{-\delta/2}\psi\|.
\end{aligned}$$

Finally, by the definition of A_k, we similarly have

$$\begin{aligned}
\|\langle x\rangle^{-\delta/2}T(3)\psi\| &= \|\langle x\rangle^{-\delta/2}T^*(\mu(L)-\mu(h^0))(1-\chi_k)T\psi\| \\
&= \Big\|\lambda\big(\langle x\rangle^{\delta/2}T^*\langle x\rangle^{-\delta/2}\big)(\mu(L)-\mu(h^0))(1-\chi_k)T\psi\Big\| \\
&\le C\hbar^N e^{-\mathbf{d}/\hbar}\|\langle x\rangle^{-\delta/2}\psi\|,
\end{aligned}$$

and (4.2) is proved. □

Proof: [Proof of Theorem 1.1] We recall the representation formula of $S^2(\lambda)$, i.e. (3.4). Then by Lemmas 4.3, 3.6, and the semiclassical resolvent estimate, we have

$$\|S^2(\lambda)-1\| \le C\hbar^{-5}\big(\hbar^N e^{-\mathbf{d}/\hbar}\big), \qquad \lambda\in I,$$

where $\mathbf{d}$ is chosen as in Lemma 4.3. We set $N=5$. Then, as in the proof of Corollary 3.11, we arrive at

$$\|s_{12}(\lambda)\| + \|s_{21}(\lambda)\| \le Ce^{-\mathbf{d}/\hbar}, \qquad \lambda\in I,$$

and this completes the proof of Theorem 1.1. □

Bibliography

1. R. BEALS, *Characterization of pseudodifferential operators and applications*, Duke Math. J. **44**, 45–57 (1977).
2. M. BENCHAOU, *Estimations de diffusion pour un opérateur de Klein-Gordon dépendant du temps*, Bull. Soc. Math. Fr. **126**, 273–294 (1998).

3. M. BENCHAOU, A.MARTINEZ, *Estimations exponentielles en théorie de la diffusion pour des opérateurs de Schrödinger matriciels*, Ann. Inst. H. Poincaré **71**, 561–594 (1999).
4. M.DIMASSI, J.SJÖSTRAND, *Spectral Asymptotics in the Semi-Classical Limit*, London Math. Soc. Lecture Notes Series **268**, 1999.
5. C. GÉRARD, A. MARTINEZ, *Principe d'absorption limite pour des opérateurs de Schrdinger à longue portée* , C. R. Acad. Sci. Paris **306**, Sér I, 121–123 (1988).
6. M.KLEIN, A.MARTINEZ, R.SEILER, X.P.WANG, *On the Born-Oppenheimer expansion for polyatomic molecules*, Commun. Math. Phys. **143**, 607–639 (1992).
7. M.KLEIN, A.MARTINEZ, X.P.WANG, *On the Born-Oppenheimer approximation of wave operators in molecular scattering theory*, Commun. Math. Phys. **152**, 73–95 (1993).
8. M.KLEIN, A.MARTINEZ, X.P.WANG, *On the Born-Oppenheimer approximation of diatomic wave operators. II: Singular potentials*, J. Math. Phys. **38**, 1373–1396 (1997).
9. L. HÖRMANDER, The Analysis of Linear Partial Differential Operators, Vol. 3 (Springer Verlag, 1985).
10. H. ISOZAKI, *Decay Rates of Scattering States for Schrödinger Operators*, J. Math. Kyoto Univ. **26**, 595–603 (1986).
11. T.JECKO, *Classical limit of elastic scattering operator of a diatomic molecule in the Born-Oppenheimer approximation*, Ann. Inst. Henri Poincaré, Phys. Theor. **69**, 83–131 (1998).
12. A. JENSEN, S. NAKAMURA, *Mapping Properties for Wave and Scattering Operators for Two-Body Schrödinger Operators*, Lett. Math. Phys. **24**, 295–305 (1992).
13. A. MARTINEZ, An Introduction to Semiclassical and Microlocal Analysis, UTX Series(Springer Verlag, 2002).
14. A. MARTINEZ, *Estimates on complex interactions in phase space*, Math. Nachr. **167**, 203–254 (1994).
15. A. MARTINEZ, S. NAKAMURA, V. SORDONI, *Phase space tunneling in multistate scattering*, J. Funct. Anal. **191**, 297–317 (2002).
16. S. NAKAMURA, *On Martinez' method of phase space tunneling*, Rev. Math. Phys. **7**, 431–441 (1995).
17. S. NAKAMURA, *Tunneling effects in momentum space and scattering*, in Spectral and scattering theory, Ikawa, Mitsuru (ed.), Notes Pure Appl. Math. **161**, 131–151 (1994).
18. G.NENCIU, V.SORDONI, *Semiclassical limit for multistate systems: Almost invariant subspaces and scattering theory*, Preprint (2001).
19. D. YAFAEV, Mathematical Scattering Theory (American Math. Soc., 1992).

CHAPTER 4

INVERSE SPECTRAL THEORY

Hiroshi Isozaki*

Department of Mathematics, Tokyo Metropolitan University, Hachioji, Minami-Osawa, 192-0397, Japan

E-mail: isozakih@comp.metro-u.ac.jp

Many self-adjoint operators appearing in mathematical physics and geometry have their spectral data: eigenvalues, informations of eigenvectors, scattering matrices. A natural attempt is the reconstruction of the original operator in terms of its spectral data. The precursor of this inverse spectral problem goes back at least to the Sturm–Liouville theory of differential operators. The systematic study of inverse spectral problems has become active from early 20th century, and the interest on this subject is unceasingly growing up since then.

The aim of this article is to give a brief survey of the inverse spectral problem for self-adjoint differential operators: boundary value problems and scattering problems for Schrödinger opeartors, Laplace–Beltrami operators on Riemannian manifolds. Both of the 1-dimensional and the multi-dimensional problems are discussed. There is so extensive literature on the inverse problem that our arguments must be restricted to limited aspects of the subject. The basic feature I would like to stress is:

One-dimensional spectral problems are smoothly deformable like C^∞-functions, while multi-dimensional problems are rigid like analytic functions (at least in Euclidean spaces).

Contents

*Current address: Department of Mathematics, Tsukuba University, Tsukubashi, Tennoudai 1-1-1, 305-8571, Japan. Email:isozakih@math.tsukuba.ac.jp

Part I One-Dimensional Problem

1. Inverse Eigenvalue Problem

Let us begin with this lecture by the 1-dimensional inverse eigenvalue problem. In Sections 1 and 2, we follow Pöschel-Trubowitz[66] and Deift.[13]

1.1. *Theorem of Borg–Levinson*

Let us consider the simplest case of the Dirichlet boundary value problem on $(0,1)$:

$$-y'' + V(x)y = \lambda y, \quad 0 < x < 1, \tag{1.1}$$

$$y(0) = y(1) = 0. \tag{1.2}$$

If $V(x)$ is real-valued, this problem has a set of eigenvalues

$$\lambda_1(V) < \lambda_2(V) < \cdots < \lambda_n(V) < \cdots .$$

The first question of the inverse eigenvalue problem is

Question: *If $\lambda_n(V_1) = \lambda_n(V_2)$ for all $n \geq 1$, does V_1 coincide with V_2?*

The answer is easily seen to be *negative*. You have only to take $V_2(x) = V_1(1-x) \neq V_1(x)$ as a counter example. A potential $V(x)$ is said to be *even* if $V(x) = V(1-x)$. This parity is the only symmetry that the above Dirichlet problem has. The following theorem due to Borg and Levinson is the first uniqueness result up to symmetry.

Theorem 1.1: Suppose V_1 and V_2 are even and $\lambda_n(V_1) = \lambda_n(V_2)\ \forall n \geq 1$. Then $V_1 = V_2$.

If the potential is not even, one needs some auxiliary condition to prove the uniqueness. The second theorem, also due to Borg and Levinson, shows that the values of the derivatives at the boundary of eigenfunctions serve for this purpose.

Let $y(x,\lambda)$ satisfy (1.1) and

$$y(0,\lambda) = 0, \quad y'(0,\lambda) = 1. \tag{1.3}$$

We put

$$k_n(V) = y'(1, \lambda_n(V)). \tag{1.4}$$

Theorem 1.2: Suppose $\lambda_n(V_1) = \lambda_n(V_2), k_n(V_1) = k_n(V_2), \forall n \geq 1$. Then $V_1 = V_2$.

1.2. *Global structure of isospectral potentials*

The set of potentials having the same eigenvalues is parametrized by $\{k_n(V)\}_{n=1}^{\infty}$, which makes it possible to deform continuously the potential keeping the eigenvalues fixed. Let us formulate it rigorously.

Let $L^2_{\mathbf{R}}(0,1)$ be the set of all real-valued L^2-functions on $(0,1)$. For $V \in L^2_{\mathbf{R}}(0,1)$, $\lambda_n(V)$ has an asymptotic expansion

$$\lambda_n(V) = n^2\pi^2 + \int_0^1 V(x)dx + \mu_n(V), \quad \sum_{n=1}^{\infty}(\mu_n(V))^2 < \infty.$$

With this in mind, we put

$$\mu_0(V) = \int_0^1 V(x)dx, \tag{1.5}$$

and let $\mu(V) = (\mu_0(V), \mu_1(V), \mu_2(V), \ldots)$. Then μ is a map

$$\mu : L^2_{\mathbf{R}}(0,1) \to \mathbf{R} \times l^2 =: S. \tag{1.6}$$

We also put $\kappa(V) = (\kappa_1(V), \kappa_2(V), \ldots)$, where

$$\kappa_n(V) = \log(-1)^n k_n(V) \tag{1.7}$$

and let l_1^2 be defined by

$$l_1^2 \ni \alpha = (\alpha_1, \alpha_2, \ldots) \Longleftrightarrow \sum_{n=1}^{\infty} (n\alpha_n)^2 < \infty. \tag{1.8}$$

Theorem 1.3: $\kappa \times \mu : L^2_{\mathbf{R}}(0,1) \to l_1^2 \times S$ is a real analytic isomorphism.

Here a map between real Hilbert spaces is said to be real analytic if it is continuously Fréchet differentiable on the complexification of the real Hilbert spaces.

Theorem 1.3 characterizes the Dirichlet spectral data, the eigenvalues and the derivatives at the boundary of the eigenfunctions.

For $q \in L^2_{\mathbf{R}}(0,1)$, let

$$M(q) = \{V \in L^2_{\mathbf{R}}(0,1) : \lambda_n(q) = \lambda_n(V), \forall n \geq 1\}. \tag{1.9}$$

Theorem 1.4: $M(q)$ is a real analytic submanifold of $L^2_{\mathbf{R}}(0,1)$ and κ is a global coordinate system on $M(q)$.

Therefore one can deform $V(x)$ continuously keeping eigenvalues fixed, by varying $k_1(V)$ for instance.

2. Isospectral Deformation

It is well-known that for two bounded operators A and B

$$\sigma(AB) \setminus \{0\} = \sigma(BA) \setminus \{0\}, \tag{2.1}$$

where $\sigma(A)$ denotes the spectrum of A. The proof of (2.1) is elementary for the finite dimensional case, and the infinite dimensional case uses the equation

$$\lambda(AB + \lambda)^{-1} + A(BA + \lambda)^{-1}B = 1. \tag{2.2}$$

For unbounded operators, (2.1) is also true if A is a densely defined closed operator and $B = A^*$. This is particularly useful in the application to one-dimensional problems. In fact, Crum[11] used this method, by a straightforward computation, to deform and remove eigenvalues of Strum–Liouville operators.

The commutation method has a long history, precursors of which are seen in the works of Jacobi[41] and Darboux.[12] We elucidate it here formally.

Let $V \in L^2_{\mathbf{R}}(0,1)$ and φ_n be a Dirichlet eigenfunction of $-d^2/dx^2 + V(x)$ with eigenvalue λ_n. Ignoring the question of domain of all relevant operators, we put formally

$$A = \varphi_n \frac{d}{dx}\left(\frac{1}{\varphi_n}\cdot\right), \quad A^* = -\frac{1}{\varphi_n}\frac{d}{dx}\left(\varphi_n\cdot\right). \tag{2.3}$$

Using $-\varphi_n'' + V\varphi_n = \lambda_n\varphi_n$, we have

$$A^*A = -\frac{d^2}{dx^2} + V(x) - \lambda_n, \tag{2.4}$$

$$AA^* = -\frac{d^2}{dx^2} + V(x) - 2\frac{d^2}{dx^2}\log\varphi_n(x) - \lambda_n. \tag{2.5}$$

Therefore we will get $\sigma(A^*A)\setminus\{0\} = \sigma(AA^*)\setminus\{0\}$. However, the potential of AA^* is not in L^2 in neighbourhoods of the zeros of $\varphi_n(x)$. This causes troubles.

The remedy comes from the *double commutation*. Namely by taking

$$u = \frac{1}{\varphi_n}\left(a + b\int_0^x \varphi_n(t)^2 dt\right) \tag{2.6}$$

and putting

$$B = u\frac{d}{dx}\left(\frac{1}{u}\cdot\right), \quad B^* = -\frac{1}{u}\frac{d}{dx}\left(u\,\cdot\right), \tag{2.7}$$

we have

$$AA^* = B^*B, \tag{2.8}$$

$$BB^* = -\frac{d^2}{dx^2} + V(x) - 2\frac{d^2}{dx^2}\log(u\varphi_n) - \lambda_n. \tag{2.9}$$

Furthermore, we have

$$BB^*\frac{1}{u} = 0. \tag{2.10}$$

This shows that $V(x)$ and $V(x) - 2\frac{d^2}{dx^2}\log(u\varphi_n)$ have the same Dirichlet eigenvalues $\{\lambda_n\}_{n=1}^{\infty}$.

Let us note that this commutation method does not work in multi-dimension. We show in Section 4 that there is no analogue of isospectral deformation of $-\Delta + V$ like Theorem 1.4.

3. Inverse Scattering

3.1. *Scattering problem*

The concept of scattering experiment is as follows. One puts a target and projects a beam of particles. By observing the scattered particles, one tries to investigate the target. In the case of potential scattering in quantum mechanics, this process is described by the following Schrödinger equation in $\mathbf{R}^3$:

$$(-\Delta + V(x))\varphi = E\varphi, \tag{3.1}$$

where $E > 0$ denotes the energy of scattering particles and $V(x)$ is a real function, which is assumed to be rapidly decreasing.

The equation (3.1) has an infinite number of solutions. However, the solution corresponding to the above scattering process can be chosen uniquely under a suitable boundary condition at infinity, and has the following asymptotic expansion

$$\varphi \simeq e^{i\sqrt{E}\omega\cdot x} + \frac{e^{i\sqrt{E}r}}{r} f(E;\theta,\omega) \tag{3.2}$$

as $r = |x| \to \infty, \theta = x/r$. Here the first term of the right-hand side of (3.2) represents the incident plane wave having direction $\omega \in S^2$, and the second term represents the scattered spherical wave. $f(E;\theta,\omega)$ is called the *scattering amplitude*. $|f(E;\theta,\omega)|^2$ is called the *differential cross section* and denotes the ratio of number of particles reflected to the θ direction to number of incident particles with direction ω. This is the physically observed quantity.

Time-dependent picture visualizes the scattering process more clearly. Let $H = -\Delta + V(x)$. Then the behavior of the particle is described by a solution of the time-dependent Schrödinger eqiuation $i\partial_t u = Hu$, namely $u(t) = e^{-itH}u$. In the remote past and the remote future, this particle runs very far from the scattering center. Since the potential $V(x)$ decays rapidly, the behavior of the particle is approximately governed by $H_0 = -\Delta$ as $t \to \pm\infty$. More precisely, there exist $u_\pm$ such that $\|e^{-itH}u - e^{-itH_0}u_\pm\| \to 0$ as $t \to \pm\infty$. This means that the incoming free particle $e^{-itH_0}u_-$ is scattered into $e^{-itH_0}u_+$ after the collision. The operator

$$S : u_- \to u_+ \tag{3.3}$$

is called the *scattering operator*. The structure of S is seen more explicitly by passing to the Fourier transformation

$$\mathcal{F}_0 u(\xi) = \hat{u}(\xi) = (2\pi)^{-3/2} \int_{\mathbf{R}^3} e^{-ix\cdot\xi} u(x)dx. \tag{3.4}$$

In fact, we have

$$\hat{u}_+(\sqrt{E}\theta) = \hat{u}_-(\sqrt{E}\theta) - \frac{\sqrt{E}}{2\pi i} \int_{S^2} f(E;\theta,\omega)\hat{u}_-(\sqrt{E}\omega)d\omega, \tag{3.5}$$

where $f(E;\theta,\omega)$ is the scattering amplitude which appeared in (3.2). This means that the operator

$$\hat{S} = \mathcal{F}_0 S \mathcal{F}_0^* \tag{3.6}$$

acts only on the variable $\omega \in S^2$. This is physically natural, since $\mathcal{F}_0 H_0 \mathcal{F}_0^* = |\xi|^2$ and the energy is conserved during the scattering process.

Let $\hat{S}(E)$ be the integral operator on $L^2(S^2)$:

$$\hat{S}(E)\psi(\theta) = \psi(\theta) - \frac{\sqrt{E}}{2\pi i} \int_{S^2} f(E;\theta,\omega)\psi(\omega)d\omega. \tag{3.7}$$

Then $\hat{S}$ is written as

$$(\hat{S}u)(\sqrt{E}\theta) = (\hat{S}(E)u(\sqrt{E}\cdot)(\theta). \tag{3.8}$$

$\hat{S}(E)$ is a unitary operator on $L^2(S^2)$ and is called the *S-matrix*.

The inverse problem of scattering is now formulated as follows:

> *Given the scattering amplitude $f(E;\theta,\omega)$,*
> *reconstruct the potential $V(x)$.*

Physically, the differential cross section is the only observable quantity. However, since $\hat{S}(E)$ is unitary, $f(E;\theta,\omega)$ satisfies an integral equation. Using this equation one can construct $f(E;\theta,\omega)$ from the square of its modulus when $|f(E;\theta,\omega)|$ is sufficiently small. [See Martin (Ref. 52).]

However, when the differential cross section is not small, this is no longer true. [See Newton (Ref. 60).]

3.2. *Spherically symmetric potentials*

When the potential $V(x)$ is spherically symmetric, $V(x) = V(|x|)$, the above problem is reduced to the one on the interval $(0,\infty)$ by the well-known procedure of partial wave expansion.

Let $k = \sqrt{E}$ and $r = |x|$. Then the solution φ of (3.1) depends only on r and the angle between ω and x. Letting θ denote this angle, we expand φ as

$$\varphi = \frac{1}{kr} \sum_{l=0}^{\infty} i^l (2l+1) u_l(r,k) P_l(\cos\theta), \tag{3.9}$$

where P_l is the Legendre polynomial. Then by (3.1) and (3.2), $u_l(r,k)$ satisfies

$$-\frac{d^2}{dr^2} u_l(r,k) + \left[V(r) + \frac{l(l+1)}{r^2} \right] u_l(r,k) = k^2 u_l(r,k), \tag{3.10}$$

$$u_l(0,k) = 0, \tag{3.11}$$

$$u_l(r,k) \simeq \sin(kr - \frac{l\pi}{2} + \delta_l) \quad \text{as} \quad r \to \infty \tag{3.12}$$

for some $\delta_l \in \mathbf{R}$. This is called the *phase shift* and the scattering amplitude is written as

$$f(k,\theta) = \frac{1}{k} \sum_{l=0}^{\infty} (2l+1) e^{i\delta_l} \sin \delta_l P_l(\cos\theta). \tag{3.13}$$

In the following we pick up the case $l = 0$.

3.3. *Gel'fand–Levitan theory*

Let us consider the scattering theory on the half line described by the following equation

$$-\frac{d^2}{dx^2}\psi(x,k) + V(x)\psi(x,k) = k^2\psi(x,k), \qquad 0 < x < \infty, \tag{3.14}$$

$$\psi(0,k) = 0, \qquad k \in \mathbf{R}. \tag{3.15}$$

If $V(x)$ decays sufficiently rapidly at infinity, the solution $\psi(x,k)$ has the following asymptotics

$$\begin{aligned} \psi(x,k) &\simeq C(k)\sin(kx - \delta(k)) \\ &= \frac{C(k)}{2i}(e^{-i\delta(k)}e^{ikx} - e^{i\delta(k)}e^{-ikx}) \end{aligned}$$

as $x \to \infty$. The term e^{ikx} represents the outgoing wave and the term e^{-ikx} represents the incoming wave. The correspondence $e^{i\delta(k)} \to e^{-i\delta(k)}$ then defines the S-matrix. Hence we put

$$S(k) = e^{-2i\delta(k)}. \tag{3.16}$$

The inverse scattering problem is rephrased as

Given $\delta(k)$, reconstruct $V(x)$.

The first complete solution of the one-dimensional inverse problem was given by Gel'fand–Levitan.[26] They reconstructed $V(x)$ from the spectral function, which is the density function in the generalized eigenfunction expansion theory for Sturm–Liouville operators. Krein and Marchenko completed the inverse scattering by showing the passage from $S(k)$ to the spectral function. The completed theory contains a characterization of the scattering matrix and the reconstruction procedure. It is summarized in the following theorem.

Theorem 3.1: In order that a **C**-valued function $S(k)$ defined on **R** be the scattering matrix of a Schrödinger operator $H = -d^2/dx^2 + V(x)$ on $(0, \infty)$ with Dirichlet boundary condition at 0 and a real-valued potential $V(x)$ satisfying

$$\int_0^\infty x|V(x)|dx < \infty,$$

it is necessary and sufficient that

(1) $\quad S(k) \in C(\mathbf{R}), \quad |S(k)| = 1, \quad \overline{S(k)} = S(-k),$

(2) $\quad S(k) - 1 = \int_{-\infty}^{\infty} e^{-ikt}(F_1(t) + F_2(t))dt,$

where

$$F_1(t) \in L^1(\mathbf{R}),\ F_2(t) \in L^2(\mathbf{R}) \cap L^\infty(\mathbf{R})$$

and

$$\int_0^\infty t|F_1'(t) + F_2'(t)|dt < \infty,$$

(3) $\quad \arg S(+0) - \arg S(+\infty) + \frac{\pi}{2}(S(0) - 1) = 2\pi m,$

where m is a non-negative integer.

If $m = 0$, the potential $V(x)$ is uniquely determined. If $m > 0$, there exists an m-parameter family of potentials having the same $S(k)$ as the S-matrix.

In the above theorem, m is the number of discrete eigenvalues of $-d^2/dx^2 + V(x)$. The deformation of the potential keeping the S-matrix fixed is carried out by the method of commutation in Section 2.

For the proof, see e.g. Faddeev (Ref. 20), or Marchenko (Ref. 51) .

3.4. *Generalized sine transformation*

To study the spectral structure of H and the properties of S-matrix, the generalized sine transformation plays a key role. Let us recall that for a self-adjoint operator A, the absolutely continuous subspace for A, $\mathcal{H}_{ac}(A)$, is the set of all u such that $(E(\lambda)u, u)$ is absolutely continuous with respect to $d\lambda$, where $A = \int_{-\infty}^{\infty} \lambda dE(\lambda)$.

Now let $\varphi(x, k)$ be the solution of the equation

$$-\varphi''(x,k) + V(x)\varphi(x,k) = k^2\varphi(x,k), \quad x > 0, \tag{3.17}$$

$$\varphi(0,k) = 0, \quad \varphi'(0,k) = 1, \tag{3.18}$$

where $' = d/dx$. Then $\varphi(x, k)$ behaves like

$$\varphi(x,k) = \frac{A(k)}{k}\sin(kx - \delta(k)) + o(1) \tag{3.19}$$

as $x \to \infty$. We define

$$\mathcal{F}u(k) = \int_0^\infty u(x)\psi^{(+)}(x,k)dx, \tag{3.20}$$

$$\psi^{(+)}(x,k) = \varphi(x,k)/M(k), \quad M(k) = A(k)e^{i\delta(k)}. \tag{3.21}$$

Then $\mathcal{F}$ is a unitary from $\mathcal{H}_{ac}(H)$ to $L^2\big((0,\infty); \frac{2k^2}{\pi}dk\big)$, and we have the inversion formula for $u \in \mathcal{H}_{ac}(H)$

$$u = \frac{2}{\pi}\int_0^\infty \overline{\psi^{(+)}(x,k)}\mathcal{F}u(k)k^2dk. \tag{3.22}$$

Moreover $\mathcal{F}$ diagonalizes H:

$$(\mathcal{F}Hu)(k) = k^2(\mathcal{F}u)(k). \tag{3.23}$$

When $V(x) = 0$, $\mathcal{F}$ reduces to the sine transformation

$$\mathcal{F}_0u(k) = \int_0^\infty u(x)\frac{\sin kx}{k}dx. \tag{3.24}$$

3.5. *The core of Gel'fand–Levitan theory*

Let us explain the essence of Gel'fand–Levitan theory. Let $\varphi(x,k)$ be as in (3.17) and (3.18). Then $\varphi(x,k)$ is an even and entire function of $k \in \mathbf{C}$ satisfying

$$\varphi(x,k) = \frac{1}{k}\sin kx + o\left(\frac{e^{|\mathrm{Im}\ k|x}}{|k|}\right), \quad |k| \to \infty. \tag{3.25}$$

Here we recall the Paley–Wiener theorem. An entire function $F(z)$ is said to be of exponential type σ if for any $\epsilon > 0$, there exists $C_\epsilon > 0$ such that $|F(z)| \leq C_\epsilon e^{(\sigma+\epsilon)|z|}$, $z \in \mathbf{C}$.

Theorem 3.2: *(Paley–Wiener)* $F(x) \in L^2(\mathbf{R})$ is extended to an entire function of exponential type σ if and only if there exists $h \in L^2(-\sigma,\sigma)$ such that

$$F(z) = \int_{-\sigma}^{\sigma} h(\xi)e^{iz\xi}d\xi.$$

By virtue of Paley–Wiener theorem and (3.25), $\varphi(x,k)$ has the following representation

$$\varphi(x,k) = \frac{\sin kx}{k} + \int_0^x K(x,y)\frac{\sin ky}{k}dy. \tag{3.26}$$

We insert this expression to the equation (3.17). Then K is shown to satisfy the equation

$$(\partial_y^2 - \partial_x^2 + V(x))K(x,y) = 0. \tag{3.27}$$

The crucial fact is

$$2\frac{d}{dx}K(x,x) = V(x). \tag{3.28}$$

One can further derive the following equation

$$K(x,y) + \Omega(x,y) + \int_0^x K(x,t)\Omega(t,y)dt = 0, \quad x > y, \tag{3.29}$$

where $\Omega(x,y)$ is a function constructed from the S-matrix and informations of bound states. This is called the *Gel'fand–Levitan equation.*

Thus the scenario of the reconstruction of $V(x)$ is as follows. From the scattering matrix and the bound states, one constructs $\Omega(x,y)$. Solving (3.29) one gets $K(x,y)$. The potential $V(x)$ is obtained by (3.28).

3.6. *What is the hidden mechanism?*

This is truly an ingenious trick and it is not easy to find the key fact behind their theory. It is Kay and Moses[43] who studied an algebraic aspect of the Gel'fand–Levitan method.

Let H_0 and H be two self-adjoint operators. A unitary operator U from $\mathcal{H}_{ac}(H_0)$ to $\mathcal{H}_{ac}(H)$ is said to *intertwine* H_0 and H if it satisfies

$$HU = UH_0. \tag{3.30}$$

An important example of intertwining operator for $H_0 = -\Delta$ and $H = -\Delta + V(x)$ is the wave operator

$$W_{\pm} = \operatorname*{s-lim}_{t\to\pm\infty} e^{itH}e^{-itH_0}. \tag{3.31}$$

Another important example of intertwining operator is the so called *spectral representation.* Let A be self-adjoint and I be its absolutely continuous spectrum, $I = \sigma(A|_{\mathcal{H}_{ac}(A)})$. A unitary operator T from $\mathcal{H}_{ac}(A)$ to $L^2(I; \mathbf{h})$, $\mathbf{h}$ being an auxiliaray Hilbert space, is called a spectral representation of A if

$$(TAu)(\lambda) = \lambda(Tu)(\lambda), \quad \lambda \in I, \quad u \in \mathcal{H}_{ac}(A).$$

For $H_0 = -d^2/dx^2$ and $H = -d^2/dx^2 + V(x)$ on $(0,\infty)$, the above (generalized) sine transformations $\mathcal{F}_0$ and $\mathcal{F}$ are spectral representations. It is obvious that $U = \mathcal{F}^*\mathcal{F}_0$ intertwines H_0 and H.

It is also evident that if U intertwines $H_0 = -d^2/dx^2$ and $H = -d^2/dx^2 + V(x)$, then its kernel $U(x,y)$ satisfies

$$(-\partial_x^2 + V(x))U(x,y) = -\partial_y^2 U(x,y).$$

The spectral representation for the one-dimensional Schrödinger operator H has the following distinguished property. Let $\psi^{(+)}(x,k)$ be as in (3.21). Let

$$U(x,y) = \int_0^\infty \psi^{(+)}(x,k)\frac{\sin ky}{k}\rho(k)dk, \quad \rho(k) = \frac{2k^2}{\pi} \tag{3.32}$$

be the integral kernel of the intertwining operator $\mathcal{F}^*\mathcal{F}_0$. In view of the formula (3.26), we have

$$U(x,y) = \delta(x-y) + \eta(x-y)K(x,y),$$

where $\eta(t)$ is the Heaviside function. Namely, U is an integral operator of *Volterra* type. The key fact discovered by Kay–Moses is the following theorem.

Theorem 3.3: Let $H_0 = -d^2/dx^2$ be defined on $(0, \infty)$ with Dirichlet boundary condition, and let $H = H_0 + Q$ be a self-adjoint perturbation of H_0. Suppose $U = I + K$ intertwines H_0 and H and that U is Volterra, i.e. $K(x, y) = 0$ if $x < y$. Then Q is an operator of multiplication by

$$q(x) = 2\frac{d}{dx}K(x, x). \tag{3.33}$$

For $x > y$, the following equation holds

$$(\partial_x^2 - \partial_y^2)K(x, y) = q(x)K(x, y). \tag{3.34}$$

Proof. Let us content ourselves by the formal proof, since we are mainly interested in the algebraic or operator theoretical aspect of the Gel'fand–Levitan theory.

Since $U(x, y) = \delta(x - y) + \eta(x - y)K(x, y)$ satisfies the wave equation

$$(\partial_x^2 - \partial_y^2)U(x, y) = \int_0^\infty Q(x, z)U(z, y)dz,$$

we have

$$\int_0^\infty Q(x, z)U(z, y)dz = 2\delta(x - y)\frac{d}{dx}K(x, x) + \eta(x - y)(\partial_x^2 - \partial_y^2)K(x, y).$$

Since U is Volterra, so is U^{-1}. Hence by multiplying U^{-1} to the above equation we have

$$Q(x, y) = 2\delta(x - y)\frac{d}{dx}K(x, x) + \eta(x - y)C(x, y).$$

However, since Q is self-adjoint, $C(x, y) = 0$, which proves

$$Q(x, y) = 2\delta(x - y)\frac{d}{dx}K(x, x).$$

This means that Q is the operator of multiplication by $2\frac{d}{dx}K(x, x)$. □

Let us summarize the above arguments. In the one-dimensional case, the generalized eigenfunction $\varphi(x, k)$ of the Schrödinger operator $H = -d^2/dx^2 + V(x)$ has the triangular expression (3.26). This makes the intertwining operator $\mathcal{F}^*\mathcal{F}_0$ into Volterra type. The potential $V(x)$ is reconstructed from the kernel of this Volterra operator.

An excellent exposition of the one-dimensional inverse scattering is the one given by Faddeev.[20] A historical survey by Newton in the foreword of the monograph of Chadan–Sabatier[8] contains an extensive literature up to 1977.

Part II Multi-Dimensional Problem

4. n-Dimensional Borg–Levinson Theorem

The simplest multi-dimensional spectral problem is the Dirichlet problem. Let Ω be a bounded domain in $\mathbf{R}^n, n \geq 2$, with smooth boundary S. Consider the boundary value problem

$$(-\Delta + V(x))u = \lambda u \quad \text{in} \quad \Omega, \tag{4.1}$$

$$u|_S = 0, \tag{4.2}$$

where $V(x)$ is a real-function. Then there exists an infinite number of eigenvalues

$$\lambda_1(V) < \lambda_2(V) \leq \cdots .$$

The smallest eigenvalue is always simple.

As can be proved easily, even if $\lambda_i(V_1) = \lambda_i(V_2)$ for all $i \geq 1$, V_1 is not necessary equal to V_2. The one-dimensional Borg–Levinson theorem suggests that the set of normal derivatives of eigenfunctions are a candidate of the auxiliary condition to guarantee the uniqueness of the potential. Let us mention parenthetically that Gel'fand raised this problem in 1954.[25] If one considers the Neumann problem, the auxiliary condition will be the values of eigenfunctions at the boundary. However we have to be careful in choosing the eigenfunctions, since the eigenvalues are not simple in general.

Let m be the multiplicity of $\lambda_i(V)$ and $u_1, \ldots, u_m$ be a system real-valued orthonormal eigenfunctions associated with $\lambda_i(V)$. We set

$$E_i(V) = \left\{ \left(\frac{\partial u_1}{\partial \nu}, \cdots, \frac{\partial u_m}{\partial \nu} \right) \Big|_S \right\}, \tag{4.3}$$

ν being the outer unit normal to S. It is easy to see that for two such system of eigenfunctions $\{u_1, \ldots, u_m\}, \{v_1, \ldots, v_m\}$, there exists an orthogonal matrix T such that

$$\left(\frac{\partial u_1}{\partial \nu}, \cdots, \frac{\partial u_m}{\partial \nu} \right) \Big|_S = \left(\frac{\partial v_1}{\partial \nu}, \cdots, \frac{\partial v_m}{\partial \nu} \right) \Big|_S T.$$

This defines an equivalent relation $\sim$ in the space of functions on the boundary S, and for the set $E_i(V)$, there corresponds only one equivalence class, which we denote by $W_i(V)$:

$$W_i(V) = E_i(V)/\sim . \tag{4.4}$$

The following theorem is due to Nachman–Sylvester–Uhlman.[58]

Theorem 4.1: Suppose $V_1, V_2 \in C^\infty(\overline{\Omega})$ and

$$\lambda_i(V_1) = \lambda_i(V_2), \quad W_i(V_1) = W_i(V_2), \quad \forall i \geq 1.$$

Then $V_1 = V_2$.

Apparently, this theorem is a first step of generalization of the one-dimensional inverse spectral theory to the multi-dimensional case. However, unlike the one-dimensional problem, neither the map : $V \to \{\lambda_i(V)\} \times \{W_i(V)\}$ is an isomorphism, nor $\{W_i(V)\}$ is a system of coordinates of the manifold of isospectral potentials. This can be seen most easily, perhaps, in the following theorem.[32]

Theorem 4.2: Let $V_1, V_2 \in C^\infty(\overline{\Omega})$. Suppose there exists $N > 0$ such that

$$\lambda_i(V_1) = \lambda_i(V_2), \quad W_i(V_1) = W_i(V_2), \quad \forall i \geq N.$$

Then $V_1 = V_2$.

In the one-dimensional case, it is possible to deform continuously the potential $V(x)$ by varying $W_1(V)$ keeping all $\lambda_i(V)$ fixed. This is not true in the multi-dimensional case.

Let us briefly explain the proof of Theorem 4.2. We use the Dirichlet–Neumann map $N(\lambda, V)$:

$$N(\lambda, V)f = \frac{\partial u}{\partial \nu}\Big|_S, \tag{4.5}$$

where u is a solution to the Dirichlet problem

$$(-\Delta + V - \lambda)u = 0 \quad \text{in } \Omega, \tag{4.6}$$

$$u|_S = f. \tag{4.7}$$

We are assuming that $\lambda \notin \sigma_p(-\Delta_D + V)$. $N(\lambda, V)$ has, formally, the integral kernel

$$N(\lambda, V; x, y) = \sum_{i=1}^{\infty} \frac{1}{\lambda_i - \lambda} \frac{\partial u_i}{\partial \nu}(x) \frac{\partial u_i}{\partial \nu}(y), \tag{4.8}$$

u_i being the eigenfunctions associated with λ_i.

Let (,) and $\langle \, , \, \rangle$ be the inner product of $L^2(\Omega)$ and $L^2(S)$, respectively. Letting $\varphi_{\lambda,\omega}(x) = e^{i\sqrt{\lambda}\omega\cdot x}, \omega \in S^{n-1}$, we put

$$S(\lambda, \theta, \omega; V) = \langle N(\lambda, V)\varphi_{\lambda,\omega}, \overline{\varphi_{\lambda,-\theta}} \rangle. \tag{4.9}$$

Then, by integration by parts one can show

$$S(\lambda,\theta,\omega;V) = -\frac{\lambda}{2}(\theta-\omega)^2\int_\Omega e^{-i\sqrt{\lambda}(\theta-\omega)\cdot x}dx + \int_\Omega e^{-i\sqrt{\lambda}(\theta-\omega)\cdot x}V(x)dx - (R(\lambda)V\varphi_{\lambda,\omega}, V\overline{\varphi_{\lambda,-\theta}}), \quad (4.10)$$

where $R(\lambda) = (-\Delta_D + V - \lambda)^{-1}$. From $S(\lambda,\theta,\omega;V)$, one can reconstruct $V(x)$. In fact for $0 \neq \xi \in \mathbf{R}^n$, choose $\eta \in S^{n-1}$ such that $\eta \perp \xi$. For a large parameter N, we put

$$\begin{aligned} \theta_N &= C_N\eta + \xi/(2N), \quad C_N = (1-|\xi|^2/(4N^2))^{1/2}, \\ \omega_N &= C_N\eta - \xi/(2N), \\ \sqrt{t_N} &= N + i. \end{aligned}$$

Then we have

$$\lim_{N\to\infty} S(t_N,\theta_N,\omega_N;V) = -\frac{|\xi|^2}{2}\int_\Omega e^{-ix\cdot\xi}dx + \int_\Omega e^{-ix\cdot\xi}V(x)dx. \quad (4.11)$$

This proves Theorem 4.1, since $S(\lambda,\theta,\omega;V)$ is determined by $\{\lambda_i(V)\}$ and $\{W_i(V)\}$. Moreover from the formal formula (4.8), one can show that if V_1 and V_2 satisfy the assumption of Theorem 4.2,

$$\|N(\lambda,V_1) - N(\lambda,V_2)\|_{\mathbf{B}(L^2(\partial\Omega))} \leq C/|\lambda|,$$

for large $|\lambda|$. This and (4.11) imply $\hat{V}_1(\xi) = \hat{V}_2(\xi)$, which proves Theorem 4.2.

The expression (4.10) is very close to the S-matrix in potential scattering, and the above proof is inspired by the Born approximation in scattering theory, which will be discussed later.

5. Generalized Gel'fand Problem

Let us recall the problem raised by Gel'fand.

Gel'fand problem: *Do the eigenvalues and the eigenfunctions at the boundary determine the potential part of the following Neumann problem uniquely ?*

$$\begin{cases} (-\Delta + V)u = \lambda u & \text{in } \Omega, \\ \dfrac{\partial u}{\partial \nu} = 0 & \text{on } \partial\Omega. \end{cases}$$

As has been seen in the previous section, the answer is affirmative. Belishev and Kurylev observed this problem from a more general view point.

Let M be an m-dimensional Riemannian manifold with boundary $S = \partial M$, equipped with Riemannian metric $g_{jk}dx^j dx^k$. Let $g = \det(g_{jk})$. In $L^2(M, \sqrt{g}dx)$, consider the following differential operator

$$A = -\frac{1}{\sqrt{g}}(\partial_j + ib_j)\sqrt{g}g^{jk}\mu(\partial_k + ib_k) + q \tag{5.1}$$

with boundary condition

$$(\partial_\nu + ib_\nu + \sigma)u|_S = 0, \tag{5.2}$$

where $b = (b_1, \ldots, b_m), \mu, q, \sigma$ are real-valued and $\mu \geq \mu_0$ for a constant $\mu_0 > 0$. The ∂_ν is the normal derivative and b_ν is the normal component of b at S. In the application to electromagnetism, μ is a conductivity, σ is a boundary impedance. In quantum mechanics, b and q are magnetic and electric potentials.

The operator A has the discrete spectrum

$$\lambda_1 \leq \lambda_2 \leq \cdots \to \infty.$$

Let $\varphi_k(x)$ be the associated orthonormal eigenfunctions. The boundary spectral data (BSD) of A is the triple $(S, \{\lambda_k\}, \{\varphi_k|_S\})$. Belishev–Kurylev posed the following question.

Generalized Gel'fand problem: Does BSD $(S, \{\lambda_k\}, \{\varphi_k|_S\})$ determine the manifold M and the operator A?

Let us call the triple $(M, (g_{jk}), A)$ the operator system. Belishev and Kurylev gave an affirmative answer to this question by a constructive procedure, which they call BC (boundary control) method.[45,2] Their starting point is to allow the equivalence by conformal diffeomorphism.

Definition 5.1: Two BSD $(S, \{\lambda_k\}, \{\varphi_k|_S\})$ and $(R, \{\mu_k\}, \{\psi_k|_R\})$ of the operator systems (M, g, A) and (N, h, B) are said to be equivalent if

(1) $\lambda_k = \mu_k, \quad \forall k \geq 1$,
(2) there exists a conformal diffeomorphism $\phi : S \to R$ and a real-valued function $\kappa_0 > 0$ on S such that

$$\phi^*(\psi_k|_R) = \kappa_0 \varphi_k|_S, \tag{5.3}$$

$$\phi^*(h|_R) = \kappa_0^{4/m} g|_S. \tag{5.4}$$

The conditions (5.3), (5.4) are assumed to allow the generalized gauge transformation $u \to \alpha e^{i\beta} u$.

Theorem 5.2: The BSD's $(S, \{\lambda_k\}, \{\varphi_k|_S\})$ and $(R, \{\mu_k\}, \{\psi_k|_R\})$ of the operator systems (M, g, A) and (N, h, B) are equivalent if and only if there exists a conformal diffeomorphism $\Phi : M \to N$ and a complex-valued function $\kappa \neq 0$ such that

$$B = \Phi \circ (\kappa A \kappa^{-1}) \circ \Phi^{-1},$$

$$\Phi|_S = \phi, \quad \kappa|_S = \kappa_0.$$

If $\mu = 1$ in (5.1), it is natural to call A a Schrödinger operator on M. The Schrödinger operator A admits a gauge transformation, i.e. $e^{ic} A e^{-ic}$ is again a Schrödinger operator with magnetic field $b_k + \partial_k c$.

Theorem 5.3: For a BSD $(S, \{\lambda_k\}, \{\varphi_k|_S\})$, there exists a unique operator system (M, g, A), where A is a Schrödinger operator on M, up to gauge transformation.

The original interest of Gel'fand was the determination of potential term from the Dirichlet–Neumann map of the Schrödinger operator. Let us return to this problem in the anisotropic case.

Let Ω be a bounded domain in $\mathbf{R}^m$ with smooth boundary S and consider the following boundary value problem

$$\partial_i (a^{ij}(x) \partial_j u) = 0 \quad \text{in} \quad \Omega,$$

$$u = f \quad \text{on} \quad S,$$

where $A = (a^{ij}(x))$ is a positive definite C^∞-matrix and $n = (n_1, \ldots, n_m)$ is the outer unit normal to S. Let Λ_A be the Dirichlet-Neumann map

$$\Lambda_A f = \rho n_i a^{ij} \partial_j u|_S,$$

where ρ is a positive function determined by the metric and S. If $\Phi : \Omega \to \Omega$ is a diffeomorphism which fixes the boundary and we define

$$\Phi_* A = \frac{{}^t(D\Phi) A (D\Phi)}{\det(D\Phi)} \circ \Phi^{-1},$$

then $\Lambda_{\Phi_* A} = \Lambda_A$. The converse is also true.

Theorem 5.4: Let Ω be a bounded open set in $\mathbf{R}^2$ with C^3-boundary. Suppose two C^3-metrics A and B satisfy $\Lambda_A = \Lambda_B$. Then there exists a diffeomorphism $\Phi : \Omega \to \Omega$ such that

$$B = \Phi_* A \quad \text{and} \quad \Phi|_{\partial\Omega} = I.$$

If dim $\Omega \geq 3$, this theorem is proved under the additional assumption of analyticity.

Theorem 5.5: Let Ω be a compact, simply connected real-analytic manifold of dimension ≥ 3 with real-analytic boundary. Let A and B be real-analytic metrics on Ω such that $\Lambda_A = \Lambda_B$. Suppose one of the following condition holds:

(a) Both of the metrics have the following property. For any two points $p, q \in \Omega$, there is a unique minimal geodesic joining p and q whose interior lies in Ω.

(b) Either A or B extends to a complete real analytic metric on a non-compact real-analytic manifold withoutboundary containing Ω.

Then there exists a real-analytic diffeomorphism $\Phi : \Omega \to \Omega$ such that

$$B = \Phi_* A \quad \text{and} \quad \Phi|_{\partial\Omega} = I.$$

For the proof, see Refs. 49, 70 and 56. More recent results are seen in Refs. 48 and 47.

6. Kac Problem

In 1966 M. Kac raised the following problem with an impressive title "Can one hear the shape of a drum?" and an interesting episode of D. Hilbert and H. Weyl[42]

Problem: What we can know about the geometry of a Riemannian manifold from the knowledge of eigenvalues of its Laplace–Beltrami operator?

Suppose we are given two compact Riemannian manifolds whose all eigenvalues of Laplace–Beltrami operators coincide. Are they isometric? The answer is negative. There is a counter example of 16-dimensional tori due to Milnor. Thus the next aim is to extract geometric properties as much as possible from the knowledge of eigenvalues. Kac computed the asymptotic expansion of the trace of the heat kernel of a planar domain and showed that one can know the area of the domain, the length of the boundary and in the case of polygonal region the number of holes in the domain. Kac's paper stimulated two directions of research, isospectral manifolds and spectral invariants.

6.1. *Isospectral manifolds*

These problems attracted so many people that an extensive literature has been devoted to them. Let us cite important contributions of Japanese mathematicians. Ikeda–Yamamoto[30] proved that three-dimensional isospectral lens spaces are isometric. Ikeda[29] constructed non-isometric isospectral lens spaces of dimension ≥ 5. Urakawa[72] constructed non-congruent regions in Euclidean space with the same Dirichlet (and Neumann) eigenvalues. Sunada[69] developed a general method of constructing isospectral manifolds. An exposition of Sunada's theory is given by Bérard.[4] For more details on this subject, see a review article of Urakawa.[73]

Let us cite an example of two non-congruent polygonal regions having same Dirichlet and Neumann eigenvalues due to Chapman.[9]

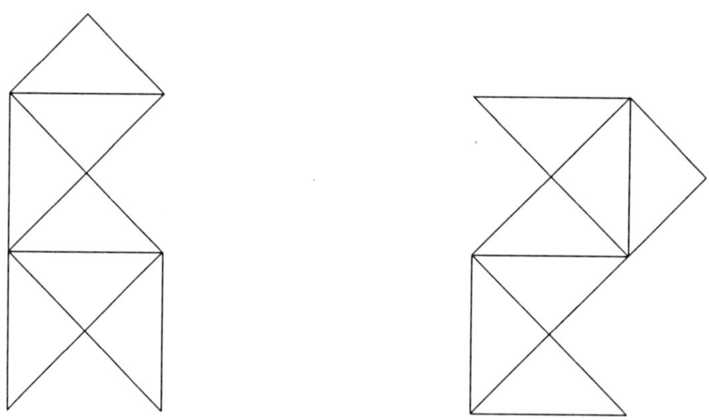

Ω_1 Ω_2

6.2. *Spectral invariants*

Spectral invariants are the quantities determined only by the spectrum. The trace of the heat kernel is often employed to compute them. Let M be a compact n-dimensional Riemannian manifold without boundary. Let

$$Z(t) = \mathrm{tr}\, e^{t\Delta} = \sum_{j=1}^{\infty} e^{-t\lambda_j},$$

where $\lambda_1 \le \lambda_2 \le \cdots$ are eigenvalues of the Laplace-Beltrami operator $-\Delta$ on M. Mckean–Singer [50] proved that as $t \to 0$,

$$(4\pi t)^{n/2} Z(t) = \mathrm{Vol}(M) + \frac{t}{3}\int_M K + \frac{t^2}{180}\int_M (10A - B + 2C) + \cdots,$$

where K is the scalar curvature and A, B, C are polynomials of the curvature tensor. If $n = 2$, this formula reads

$$Z(t) = \frac{|M|}{4\pi t} + \frac{E}{t} + \frac{\pi t}{60}\int_M K^2 + \cdots,$$

where $|M|$ = the area of M, $E = (2\pi)^{-1}\int_M K$ = the Euler characteristics of M.

If M has a boundary S, the above formula is modified as follows. Let Δ_D and Δ_N be the Dirichlet and Neumann Laplacians on M. Let

$$Z_+(t) = \mathrm{tr}\; e^{t\Delta_N}, \quad Z_-(t) = \mathrm{tr}\; e^{t\Delta_D}.$$

Then as $t \to 0$,

$$(4\pi t)^{n/2} Z_\pm(t) = \mathrm{Vol}(M) \pm \frac{\sqrt{\pi t}}{2}|S| + \frac{t}{3}\int_M K - \frac{t}{6}\int_S J + \cdots,$$

where $|S|$ = the surface area of S and J = the mean curvature.

We also have the complete asymptotic expansion. Let M be a compact manifold and let $H = -\Delta + V$. Then

$$(4\pi t)^{n/2}\mathrm{tr}\; e^{-tH} \simeq \sum_{j=0}^{\infty} t^j a_j(V),$$

where $a_j(V)$'s are integrals of certain functions over M which are universal polynomials in the covariant derivatives of V and of the curvature tensor of M.

By this asymptotic expansion one can prove (see e.g. Brüning,[6] Brooks–Perry–Petersen[5]) that the set of Riemaniann metrics on M isospectral to (M, g) is compact with respect to C^∞-topology on Riemaniann metrics. See also a survey of Perry.[64]

7. Overdeterminacy

7.1. *High energy Born approximation*

Let us return to the inverse scattering problem in $\mathbf{R}^n, n \ge 2$, for the Schrödinger operator $H = -\Delta + V(x)$. If the potential $V(x)$ satisfies

$$|V(x)| \le C(1 + |x|)^{-1-\epsilon}, \quad \epsilon > 0, \tag{7.1}$$

the scattering operator S in (3.3) is well-defined and unitary on $L^2(\mathbf{R}^n)$. One can also allow certain local singularities for V, which is omitted for the sake of simplicity. To write down the scattering amplitude, we need the limit of $R(z) = (H - z)^{-1}$ when z approaches to $\lambda \in \sigma_{cont}(H) = [0, \infty)$. For $s \in \mathbf{R}$, let $L^{2,s}$ be the function space defined by

$$u \in L^{2,s} \Longleftrightarrow \|u\|_s^2 = \int_{\mathbf{R}^n} (1 + |x|)^{2s} |u(x)|^2 dx < \infty. \tag{7.2}$$

Then if $s > 1/2$ and $\lambda > 0$, there exists a limit

$$R(\lambda \pm i0) = \lim_{\epsilon \downarrow 0} R(\lambda \pm i\epsilon) \in \mathbf{B}(L^{2,s}; L^{2,-s}). \tag{7.3}$$

Here for Banach spaces X and Y, $\mathbf{B}(X;Y)$ denotes the set of all bounded operators from X to Y.

We basically assume that

$$|V(x)| \leq C(1 + |x|)^{-n-\epsilon}, \quad \epsilon > 0. \tag{7.4}$$

Then the scattering amplitude is a continuous function of $\lambda > 0, \theta, \theta' \in S^{n-1}$ and is written as, up to a constant depending only on $\lambda > 0$

$$\begin{aligned} A(\lambda; \theta, \theta') = &\int_{\mathbf{R}^n} e^{-i\sqrt{\lambda}(\theta - \theta')\cdot x} V(x) dx \\ &- \int_{\mathbf{R}^n} e^{-i\sqrt{\lambda}\theta\cdot x} V(x) R(\lambda + i0)(V(\cdot) e^{i\sqrt{\lambda}\theta'\cdot}) dx. \end{aligned} \tag{7.5}$$

The first rigorous mathematical result for the multi-dimensional inverse scattering is due to Faddeev.[19] For $0 \neq \xi \in \mathbf{R}^n$, choose $\eta \in S^{n-1}$ such that $\eta \perp \xi$ and put

$$\theta = \left(1 - \frac{|\xi|^2}{4\lambda}\right)^{1/2} \eta + \frac{\xi}{2\sqrt{\lambda}}, \tag{7.6}$$

$$\theta' = \left(1 - \frac{|\xi|^2}{4\lambda}\right)^{1/2} \eta - \frac{\xi}{2\sqrt{\lambda}}. \tag{7.7}$$

Theorem 7.1: Let $n \geq 2$ and θ, θ' be as above. Then we have

$$\lim_{\lambda \to \infty} A(\lambda; \theta, \theta') = \int_{\mathbf{R}^n} e^{-ix\cdot\xi} V(x) dx.$$

In fact, this theorem follows from (7.5) and the high-energy resolvent estimate

$$\|R(\lambda \pm i0)\|_{\mathbf{B}(L^{2,s}; L^{2,-s})} \leq C/\sqrt{\lambda}, \tag{7.8}$$

for $s > 1/2$ and $\lambda > \lambda_0 > 0$.

This is usually called the Born approximation, since it neglects the second-term of the righthand side of (7.5). This method is a powerful tool to derive the uniqueness of the perturbation term with given scattering amplitude for large $\lambda > 0$. It is extended to general short-range potentials ($V(x) = O(|x|^{-1-\epsilon})$) by Saito,[68] long-range potentials by Isozaki–Kitada[38] ($V(x) = O(|x|^{-1/2-\epsilon})$), magnetic Schrödinger operators by Nicoleau,[62] Dirac operators by Ito[39] and even for N-body Schrödinger operators by Wang.[74]

7.2. *Time-dependent inverse scattering*

Enss[14] invented a beautiful method for proving the asymptotic completeness of wave operators by localizing solutions of Schrödinger equations along the orbit of scattering particles in classical mechanics. He also used this idea to prove the uniqueness of the potential with a given scattering operator S. This method is simple, has a wide range of applicability and can be extended easily to N-body problems as well as time-dependent potentials. See Refs. 15, 77 and 40.

7.3. *Overdeterminacy*

The above Theorem 7.1 already reveals the characteristic feature of the multi-dimensional inverse scattering problem. Unlike the one-dimensional case, the potential is determined by the scattering matrix, moreover only its high-energy part is necessary to reconstruct $V(x)$. This apparently convenient fact is at the same time a cause of difficulties in multi-dimensional inverse problem.

In the one-dimensional case, both of the potential and the S-matrix are functions of one variable. The Gel'fand–Levitan–Marchenko theory gives a necessary and sufficient condition for a function $S(k)$ to be the scattering matrix of a Schrödinger operator $-d^2/dx^2 + V(x)$.

In the n-dimensional case, the scattering amplitude $A(\lambda; \theta, \theta')$ is a function of $(2n-1)$-parameters, while the potential $V(x)$ is a function of n-variables. This overdeterminacy requires a sort of compatibility condition for a function $A(\lambda; \theta, \theta')$ of $(2n-1)$-parameters to be the scattering amplitude of associated with a Schrödinger operator $-\Delta + V(x)$. To find this compatibility condition or at least to seek a necessary condition which weakens this overdeterminacy is the very question of multi-dimensional inverse problems.

7.4. *Inverse back scattering*

The back scattering amplitude $A(\lambda;\theta,-\theta)$ is a function of n-parameters. Therefore to reconstruct the potential $V(x)$ from $A(\lambda;\theta,-\theta)$ seems to be a natural attempt. This was studied by Moses[54] and Prosser[65] in a rather formal manner. Eskin and Ralston[17] continued this direction and studied the bijectivity of the map

$$V(x) \to A(|\xi|/2;\hat{\xi},-\hat{\xi}), \quad \hat{\xi} = \xi/|\xi|.$$

They show that this map is a local diffeomorphism on a certain open set of the Banach space of real potentials.

Part III Inverse Scattering in n-Dimensions

8. Key Idea of Faddeev

8.1. *Spectral representation*

The remaining part of this paper is devoted to discussing the Faddeev theory of inverse scattering for multi-dimensional Schrödinger operators and its developments. We basically assume (7.4).

Consider the equation

$$(-\Delta + V - \lambda)u = f \in L^{2,s}, \tag{8.1}$$

for $s > 1/2, \lambda > 0$. A solution u of (8.1) is said to satisfy the outgoing radiation condition if for some $0 < \alpha < 1/2 < s$,

$$u \in L^{2,-s}, \quad (\frac{\partial}{\partial r} - i\sqrt{\lambda})u \in L^{2,-\alpha}. \tag{8.2}$$

An outgoing spherical wave $r^{-(n-1)/2}e^{i\sqrt{\lambda}r}$ satisfies (8.2). It is known that the solution of (8.1) satisfying (8.2) is unique and is given by $u = R(\lambda+i0)f$. Moreover this u admits an asymptotic expansion

$$u \simeq r^{-(n-1)/2}e^{i\sqrt{\lambda}r}\psi_0(\hat{x}), \quad \hat{x} = x/r \tag{8.3}$$

as $r \to \infty$. Therefore the solution $\varphi(x,E,\omega)$ of (3.1) having the asymptotic expansion (3.2) is given by

$$\varphi(x,E,\omega) = e^{i\sqrt{E}\omega\cdot x} - R(E+i0)(V(\cdot)e^{i\sqrt{E}\omega\cdot}). \tag{8.4}$$

Let us call this $\varphi(x,E,\omega)$ the physical eigenfunction of H. We also put

$$\Phi(x,\xi) = \varphi(x,|\xi|^2,\xi/|\xi|). \tag{8.5}$$

Using this Φ, we define a spectral representation for H by

$$(\mathcal{F}u)(\xi) = (2\pi)^{-n/2} \int_{\mathbf{R}^n} \overline{\Phi(x,\xi)} u(x) dx. \tag{8.6}$$

When $V = 0$, this reduces to the usual Fourier transformation

$$(\mathcal{F}_0 u)(\xi) = (2\pi)^{-n/2} \int_{\mathbf{R}^n} e^{-ix\cdot\xi} u(x) dx. \tag{8.7}$$

The operator $\mathcal{F}$ defined for $u \in C_0^\infty(\mathbf{R}^n)$ is uniquely extended to a bounded operator on $L^2(\mathbf{R}^n)$. Let $\mathcal{H}_{ac}(H)$ be the absolutely continuous subspace for H, and let $\mathcal{H}_{pp}(H)$ the closure of the linear hull of eigenvectors of H. We then have the orthogonal decomposition

$$L^2(\mathbf{R}^n) = \mathcal{H}_{ac}(H) \oplus \mathcal{H}_{pp}(H). \tag{8.8}$$

For $u \in \mathcal{H}_{pp}(H)$, $\mathcal{F}u = 0$. $\mathcal{F}$ is unitary from $\mathcal{H}_{ac}(H)$ onto $L^2(\mathbf{R}^n)$. For $u \in \mathcal{H}_{ac}(H) = E((0,\infty))L^2(\mathbf{R}^n)$, where $E(\lambda)$ denotes the spectral measure for H, the following inversion formula holds

$$u = \operatorname*{s-lim}_{N\to\infty} (2\pi)^{-n/2} \int_{\frac{1}{N}<|\xi|<N} \Phi(x,\xi) \mathcal{F}u(\xi) d\xi. \tag{8.9}$$

8.2. *Higher dimensional Volterra operator*

The Volterra integral operator played an important role in the 1-dimensional problem. Therefore it would be useful to generalize the notion of Volterra operator to higher dimension. Let $\gamma \in S^{n-1}$ be arbitrarily fixed. An integral operator K is said to be triangular with respect to γ if its kernel $K(x,y)$ satisfies

$$K(x,y) = 0 \quad \text{if} \quad x\cdot\gamma < y\cdot\gamma. \tag{8.10}$$

$T = I + K$ is called the Volterra type with respect to γ if K is triangular with respect to γ.

The following theorem is proved in the same way as in Theorem 3.3.

Theorem 8.1: Let $L_0 = -\Delta$ on $\mathbf{R}^n$, and let $L = L_0 + Q$ be a self-adjoint perturbation of L_0. Suppose $U = I + K$ intertwines L_0 and L. If K is triangular with respect to $(1, 0, \ldots, 0)$, Q is the multiplication operator with respect to x_1, and its kernel is give by

$$Q(x_1, x', y') = 2\frac{d}{dx_1} K(x_1, x', x_1, y').$$

Moreover for $x_1 > y_1$

$$(\Delta_x - \Delta_y)K(x,y) = \int_{\mathbf{R}^{n-1}} Q(x_1,x',z')K(x_1,z',y_1,y')dz'$$

holds.

We have now arrived at a crucial point of Gel'fand–Levitan theory. In the one-dimensional case, we constructed the generalized sine transformation $\mathcal{F}$ by using the solution of (3.17) and (3.18). Then the intertwining operator $\mathcal{F}^*\mathcal{F}_0$ becomes the Volterra type, and the potential $V(x)$ is reconstructed from its kernel.

This is not the case for the multi-dimensional problem. If one constructs the generalized Fourier transformation by using the physical eigenfunction, the intertwining operator $\mathcal{F}^*\mathcal{F}_0$ does not have a direction with respect to which it becomes Volterra. One can easily convince oneself by thinking of the fact that in the one-dimension one is on the real line, the straight line, which has clearly a distinguished direction, while in the three-dimensions there is no special direction.

What shall we do? The idea is to abandon the physical eigenfunction and to look for non physical eigenfunction by which the intertwining operator becomes Volterra. Kay and Moses tried to find a good non-physical eigenfunction. But they could not succeed in getting it. It is Faddeev who found an essential idea.[21,22,23]

8.3. *Faddeev's Green operator*

For a solution φ of $(-\Delta + V - E)\varphi = 0$, take $\zeta = (\zeta_1, \ldots, \zeta_n) \in \mathbf{C}^n$ such that $\zeta^2 = \sum_{j=1}^n \zeta_j^2 = E$ and let $\varphi = e^{ix\cdot\zeta}(1+v)$. Then v satisfies

$$(-\Delta - 2i\zeta\cdot\nabla + V)v = -V(x).$$

With this in mind let us start with the equation

$$(-\Delta - 2i\zeta\cdot\nabla + V)u = f. \tag{8.11}$$

We first consider the case $V = 0$. Among infinite number of solutions of (8.11), we select the one written by the Fourier transformation

$$u(x) = (2\pi)^{-n/2}\int_{\mathbf{R}^n} \frac{e^{ix\cdot\xi}}{\xi^2 + 2\zeta\cdot\xi}\hat{f}(\xi)d\xi =: \widetilde{G}(\zeta)f. \tag{8.12}$$

This is the Green operator introduced by Faddeev. Here we note that the integrand of (8.12) is absolutely convergent if $f \in \mathcal{S}$ = the Schwartz space of rapidly decreasing functions, and Im $\zeta \neq 0$. In fact the zeros of the denominator form a sphere of codimension 2 and hence the denominator can be rewritten near zeros in the form $\xi_1 + i\xi_2$ if we choose suitable coordinates. This yields the absolute convergence.

For the moment, let us proceed formally. Let $u = (1 + \widetilde{G}(\zeta)V)^{-1}\widetilde{G}(\zeta)f$, which is a solution to (8.11). Any $\zeta \in \mathbf{C}^n$ with Im $\zeta \neq 0$ can be uniquely written as $\zeta = \eta + z\gamma$, where $\eta \in \mathbf{R}^n, \gamma \in S^{n-1}, \eta \cdot \gamma = 0$ and $z \in \mathbf{C}_+ = \{\text{Im } z > 0\}$. This z will play an important role. We let z tend to $t \in \mathbf{R}$. Then $\widetilde{G}(\zeta)$ converges to $\widetilde{G}(k)$, where $k = \eta + t\gamma \in \mathbf{R}^n$ with $k^2 = E$. We have thus obtained the following solution

$$\psi = e^{ik\cdot x} - e^{ik\cdot x}(1 + \widetilde{G}(k)V)^{-1}\widetilde{G}(k)V \tag{8.13}$$

of $(-\Delta + V - E)\psi = 0$. This is a non-physical eigenfunction. This is different from φ which behaves like (3.2) and was utilized to define the S-matrix. An important feature of Faddeev's Green operator is that $\widetilde{G}(\eta + z\gamma)$ is analytic in $z \in \mathbf{C}_+$. Therefore the second term of the right hand side of (8.13) is the boundary value of a function analytic in the upper-half plane. Therefore by the Paley–Wiener theorem, it has the following expression

$$\psi = e^{ik\cdot x} - \int_{\gamma\cdot x}^{\infty} A_\gamma(x, \eta, s)e^{ist}ds. \tag{8.14}$$

Theorem 8.2: *(Paley–Wiener)* A function $f(x) \in L^2(\mathbf{R})$ is a boundary value of an analytic function $f(x+iy)$ in the upper-half plane such that

$$\sup_{y>0} \int_{-\infty}^{\infty} |f(x+iy)|^2 dy < \infty,$$

if and only if

$$f(x) = \frac{1}{\sqrt{2\pi}} \int_0^\infty \hat{f}(\lambda)e^{i\lambda x}d\lambda.$$

The formula (8.14) is a multi-dimensional counter part of (3.26). With this eigenfunction ψ, one can construct a spectral representation of H and the associated intertwining operator for H_0 and H. Since ψ has the triangular expression, this intertwining operator will be Volterra. One can thus expect an analogy to the one-dimensional Gel'fand–Levitan theory.

Faddeev proceeds as follows. The basic strategy is to replace the usual Green operator by the direction dependent Green operator as above. Let

$R_\gamma(E,t)$ be the direction dependent Green operator for $H = -\Delta + V$. (We shall explain it precisely later.) Using the non-physical eigenfunction

$$\Psi_\gamma(x,E,\theta) = e^{i\sqrt{E}\theta\cdot x} - R_\gamma(E,\sqrt{E}\theta\cdot\gamma)(V(\cdot)e^{i\sqrt{E}\theta\cdot}), \tag{8.15}$$

we define a new scattering amplitude

$$\widetilde{A}_\gamma(E,\theta,\theta') = (2\pi)^{-n}2^{-1}E^{(n-2)/2}\int_{\mathbf{R}^n} e^{-i\sqrt{E}\theta\cdot x}V(x)\Psi_\gamma(x,E,\theta')dx. \tag{8.16}$$

We put

$$Q_\gamma^{(\pm)}(E,\theta,\theta') = 2\pi i F(\pm\gamma\cdot(\theta-\theta')\geq 0)\widetilde{A}_\gamma(E,\theta,\theta'), \tag{8.17}$$

where $F(\cdots)$ denotes the characteristic function of the set $\{\cdots\}$. Let $Q_\gamma^{(\pm)}(E)$ be the integral operator with kernel $Q_\gamma(E,\theta,\theta')$. Then the main results are as follows:

(I) *Factorization of the S-matrix.*

$$\hat{S}(E) = (1 - Q_\gamma^{(-)}(E))(1 + Q_\gamma^{(+)}(E))^{-1}. \tag{8.18}$$

(II) *Volterra operator.* Let $\Phi_\gamma(x,\xi) = \Psi_\gamma(x,|\xi|^2,\xi/|\xi|)$ and put

$$U_\gamma(x,y) = (2\pi)^{-n}\int_{\mathbf{R}^n}\Phi_\gamma(x,\xi)e^{-iy\cdot\xi}d\xi. \tag{8.19}$$

Then we have

$$U_\gamma(x,y) = \delta(x-y) - K_\gamma(x,y), \tag{8.20}$$

$$K_\gamma(x,y) = 0 \quad \text{if} \quad x\cdot\gamma > y\cdot\gamma. \tag{8.21}$$

(III) *Gel'fand–Levitan equation.* For $x\cdot\gamma < y\cdot\gamma$, we have

$$K_\gamma(x,y) + \Omega_\gamma(x,y) + \int_{(x-y)\cdot\gamma<0} K_\gamma(x,z)\Omega_\gamma(z,y)dz = 0, \tag{8.22}$$

where $\Omega_\gamma(x,y)$ is a function constructed from the scattering data.

The scenario of the reconstruction of $V(x)$ will be as follows. From the scattering matrix $\hat{S}(E)$, construct $Q_\gamma^{(\pm)}(E)$, and then $\Omega_\gamma(x,y)$. Solve the Gel'fand–Levitan equation to get $K_\gamma(x,y)$. Since $1 - K_\gamma$ is a Volterra operator intertwining H_0 and H, the potential $V(x)$ will be obtained from $K_\gamma(x,y)$.

Faddeev goes further. He observes that the analyticity of $\widetilde{G}(\eta + z\gamma)$ plays a key role to guarantee that $V(x)$, reconstructed from $\widetilde{A}_\gamma$, is independent of the artificially introduced direction γ. This will be crucial in the characterization of the physical scattering amplitude.

9. Changing Green Operators

In the following sections, we examine the Faddeev theory in detail. The first question we address is "What occurs when one changes the Green operator in scattering theory?" The stationary scattering theory is composed of the usual Green operator $(-\Delta - E - i0)^{-1}$ of the Laplacian. Since there are many Green operators, it would be worthwhile to consider the effect of changing Green operators.

We shall discuss in a general setting. Let H_0 be a self-adjoint operator in a Hilbert space $\mathcal{H}$. Let I be the spectrum of H_0. Suppose there are two Banach spaces $\mathcal{H}_\pm$ such that

$$\mathcal{H}_+ \subset \mathcal{H} \subset \mathcal{H}_-, \tag{9.1}$$

and the inclusions from $\mathcal{H}_+$ to $\mathcal{H}$, from $\mathcal{H}$ to $\mathcal{H}_-$ are dense and continuous. Moreover we assume that the inner product $(\ ,\ )$ on $\mathcal{H}$ is uniquely extended to a sesquilinear form on $\mathcal{H}_- \times \mathcal{H}_+$. We assume that for any

$$\lambda \in I_{int} = \{\text{the interior points of } I\},$$

there exists a strong limit

$$R_0(\lambda + i0) = (H_0 - \lambda \mp i0)^{-1} \in \mathbf{B}(\mathcal{H}_+; \mathcal{H}_-) \tag{9.2}$$

and $R_0(\lambda \pm i0)$ is a $\mathbf{B}(\mathcal{H}_+; \mathcal{H}_-)$-valued continuous function of $\lambda \in I_{int}$. Suppose there exists an auxiliary Hilbert space h such that for any $\lambda \in I_{int}$, there exists a bounded operator $\mathcal{F}_0(\lambda) \in \mathbf{B}(\mathcal{H}_+; h)$ satisfying

$$\mathcal{F}_0(\lambda) H_0 u = \lambda \mathcal{F}_0(\lambda) u, \quad \forall u \in D(H_0) \cap \mathcal{H}_+. \tag{9.3}$$

We finally assume that the operator

$$\mathcal{F}_0 u(\lambda) = \mathcal{F}_0(\lambda) u$$

defined on $\mathcal{H}_+$ is uniquely extended to be a unitary operator from $\mathcal{H}$ to $L^2(I, h; d\lambda)$.

Example. Let $H_0 = -\Delta$ in $\mathcal{H} = L^2(\mathbf{R}^n)$, and $\mathcal{H}_\pm = L^{2,\pm s}$ with $s > 1/2$. Let $I = [0, \infty), h = L^2(S^{n-1})$ and

$$\mathcal{F}_0(\lambda) u = (2\pi)^{-n/2} 2^{-1/2} \lambda^{(n-2)/4} \int_{\mathbf{R}^n} e^{-i\sqrt{\lambda}\omega\cdot x} u(x) dx. \tag{9.4}$$

Let V be a bounded self-adjoint operator on $\mathcal{H}$ such that $V \in \mathbf{B}(\mathcal{H}_+; \mathcal{H}_+)$, and let $H = H_0 + V$.

Definition 9.1: Let $E \in I_{int}$. An operator $G^{(0)} \in \mathbf{B}(\mathcal{H}_+;\mathcal{H}_-)$ is said to be a Green operator of $H_0 - E$ if

$$(H_0 - E)G^{(0)} = I \quad \text{on} \quad \mathcal{H}_+. \tag{9.5}$$

An operator $G \in \mathbf{B}(\mathcal{H}_+;\mathcal{H}_-)$ is called a perturbed Green operator associated with $G^{(0)}$ if it satisfies

$$G = G^{(0)} - G^{(0)}VG = G^{(0)} - GVG^{(0)}. \tag{9.6}$$

Note that $(H - E)G = I$ on $\mathcal{H}_+$ by virtue of (9.5) and (9.6).

With a perturbed Green operator G, we define the scattering amplitude associated with G by

$$A = \mathcal{F}_0(E)(V - VGV)\mathcal{F}_0(E)^*. \tag{9.7}$$

Now suppose that we are given two Green operators $G_1^{(0)}, G_2^{(0)}$ for $H_0 - E$. Let G_1, G_2 be the associated perturbed Green operators. Let A_1, A_2 be the scattering amplitudes associated with G_1, G_2. What is the relationship between A_1 and A_2?

Theorem 9.2: Suppose there exists $M \in \mathbf{B}(h;h)$ such that

$$G_2^{(0)} - G_1^{(0)} = \mathcal{F}_0(E)^* M \mathcal{F}_0(E). \tag{9.8}$$

Then we have

$$A_2 = A_1 - A_1 M A_2. \tag{9.9}$$

The above theorem follows from the equation

$$G_2 = G_1 + (1 - G_1V)T(1 - VG_2), \tag{9.10}$$

where $T = G_2^{(0)} - G_1^{(0)}$.

Let us next discuss the solvability of the equation (9.9). We need the following assumptions:

$$G_1^{(0)}V, G_2^{(0)}V : \mathcal{H}_- \to \mathcal{H}_- \text{ are compact,}$$

$$A_1 : h \to h \text{ is compact.}$$

Theorem 9.3: Under the above assumptions, the equation (9.9) is uniquely solvable with respect to A_2:

$$A_2 = (1 + A_1 M)^{-1} A_1.$$

Proof: The proof again follows from simple algebraic manipulations. Letting $\tilde{K} = (1 - G_1V)TV$, we have

$$1 + G_2^{(0)}V = (1 + G_1^{(0)}V)(1 + \tilde{K}). \tag{9.11}$$

The existence of the perturbed Green operator G_j is equivalent to $-1 \notin \sigma_p(G_j^{(0)}V)$. Here for an operator A in a Hilbert space $\mathcal{H}$, $\sigma_p(A)$ means the set of eigenvalues of A. Putting

$$S_1 = (1 - G_1V)\mathcal{F}_0(E)^*M, \quad S_2 = \mathcal{F}_0(E)V,$$

we have

$$\tilde{K} = S_1S_2, \quad A_1M = S_2S_1.$$

Since $\sigma_p(S_1S_2) \setminus \{0\} = \sigma_p(S_2S_1) \setminus \{0\}$, we have

$$-1 \notin \sigma_p(\tilde{K}) \Longleftrightarrow -1 \notin \sigma_p(A_1M). \tag{9.12}$$

Theorem 9.3 follows from (9.11) and (9.12). □

Theorem 9.3 means that for a certain pair of Green operators there is a linear equation between the corresponding scattering amplitudes, which is solvable.

Now the question is: What kind of property of Green operator is useful in inverse scattering? According to Faddeev, it is *analyticity*. We elucidate it in the next sections.

10. Direction Dependent Green Operators

We summarize various properties of Faddeev's Green operator in this section. It is a little simpler to consider the following operator

$$G_{\gamma,0}(\lambda, z)f(x) = (2\pi)^{-n/2} \int_{\mathbf{R}^n} \frac{e^{ix\cdot\xi}}{\xi^2 + 2z\gamma\cdot\xi - \lambda^2} \hat{f}(\xi)d\xi, \tag{10.1}$$

where $\gamma \in S^{n-1}, \lambda > 0, z \in \mathbf{C}_+$. If $\zeta = \eta + z\gamma, \eta \in \mathbf{R}^n, \eta \cdot \gamma = 0$, we have

$$e^{-ix\cdot\eta}G_{\gamma,0}(|\eta|^2, z)e^{ix\cdot\eta} = \widetilde{G}(\zeta). \tag{10.2}$$

Theorem 10.1: Let $s > 1/2$.

(1) As a $\mathbf{B}(L^{2,s}; L^{2,-s})$-valued function, $G_{\gamma,0}(\lambda, z)$ is continuous with respect to $\lambda \geq 0, \gamma \in S^{n-1}, z \in \overline{\mathbf{C}_+}$ except for $(\lambda, z) = (0, 0)$.
(2) $G_{\gamma,0}(\lambda, z)$ is a $\mathbf{B}(L^{2,s}; L^{2,-s})$-valued analytic function of $z \in \mathbf{C}_+$.

(3) For any $\epsilon_0 > 0$, there exists $C > 0$ such that for $\lambda + |z| \geq \epsilon_0$,

$$\|G_{\gamma,0}(\lambda, z)\|_{\mathbf{B}(L^{2,s};L^{2,-s})} \leq C(\lambda + |z|)^{-1}.$$

(4) For $t \in \mathbf{R}$, let $\widetilde{R}_{\gamma,0}(\lambda, t) = e^{it\gamma\cdot x}G_{\gamma,0}(\lambda, t)e^{-it\gamma\cdot x}$. Then

$$(-\Delta - \lambda^2 - t^2)\widetilde{R}_{\gamma,0}(\lambda, t) = 1.$$

For the proof, see Ref. 75.

The most important features of the Green operator of Faddeev are the analyticity in $z \in \mathbf{C}_+$ and the following formula

$$\widetilde{R}_{\gamma,0}(\lambda, t) = R_0(E - i0)M_\gamma^{(+)}(t) + R_0(E + i0)M_\gamma^{(-)}(t), \tag{10.3}$$

where $E = \lambda^2 + t^2$, $R_0(E \pm i0) = (-\Delta - E \mp i0)^{-1}$, and

$$M_\gamma^{(\pm)}(t) = (\mathcal{F}_{x\to\xi})^{-1}F(\pm\gamma\cdot(\xi - t\gamma) \geq 0)\mathcal{F}_{x\to\xi}. \tag{10.4}$$

The equation (10.4) is intuitively obvious, since

$$\widetilde{R}_{\gamma,0}(\lambda, t)f = (2\pi)^{-n/2}\int_{\mathbf{R}^n}\frac{e^{ix\cdot\xi}}{\xi^2 + 2i0\gamma\cdot(\xi - t\gamma) - E}\hat{f}(\xi)d\xi,$$

$$R_0(E \pm i0)f = (2\pi)^{-n/2}\int_{\mathbf{R}^n}\frac{e^{ix\cdot\xi}}{\xi^2 - (E \pm i0)}\hat{f}(\xi)d\xi,$$

Namely, $\widetilde{R}_{\gamma,0}(\lambda, t)$ is outgoing in the half-space $\gamma\cdot\xi < t$ and incoming in the half-space $\gamma\cdot\xi > t$.

The analyticity of $G_{\gamma,0}(\lambda, z)$ in $z \in \mathbf{C}_+$ follows from the following lemma (see Lemma 3.2 of Ref. 36).

Lemma 10.2: *Let D be an open set in $\mathbf{C}$. Let $p(\xi, z)$ be a $\mathbf{C}$-valued function which is smooth in $\xi \in \mathbf{R}^n$ and analytic in $z \in D$. Let $M_z = \{\xi \in \mathbf{R}^n; p(\xi, z) = 0\}$ and assume that for $z \in D$, $\nabla_\xi \mathrm{Re}\, p(\xi, z)$ and $\nabla_\xi \mathrm{Im}\, p(\xi, z)$ are linearly independent on M_z. Then the distribution $S(z)$ defined by*

$$S(z)f = \int_{\mathbf{R}^n}\frac{f(\xi)}{p(\xi, z)}d\xi, \quad f(\xi) \in C_0^\infty(\mathbf{R}^n)$$

satisfies

$$\overline{\partial_z}S(z)f = \pi\int_{\mathbf{R}^n} f(\xi)\overline{\partial_z p(\xi, z)}\delta(p(\xi, z))d\xi,$$

where

$$\int_{\mathbf{R}^n} g(\xi)\delta(p(\xi, z))d\xi = \int_{M_z} g(\xi)dM_z,$$

and dM_z is the induced measure on M_z.

Let $E = \lambda^2 + t^2$. Then the integral kernel of $G_{\gamma,0}(\lambda, t)$ is formally written as

$$(2\pi)^{-n}\int \frac{e^{i(x-y)\cdot\xi}}{\xi^2 + 2(t+i0)\gamma\cdot\xi + t^2 - E}d\xi.$$

To find an analytic continuation with respect to t of this kernel is an important problem. Apparently it should be

$$(2\pi)^{-n}\int \frac{e^{i(x-y)\cdot\xi}}{\xi^2 + 2z\gamma\cdot\xi + z^2 - E}d\xi.$$

However this is not analytic! Eskin–Ralston[18] found the analytic continuation of the above operator. They formulated it by passing to the Fourier transformation. Let us show its expression in x-space.

The Green operator of Eskin–Ralston consists of two parts:

$$U_{\gamma,0}(E,z) = V_{\gamma,0}(E,z) + W_{\gamma,0}(E,z). \tag{10.5}$$

Let us first explain $V_{\gamma,0}(E,z)$. Let

$$D_\epsilon = \{z \in \mathbf{C}_+; |\mathrm{Re} z| < \epsilon/2\}.$$

Let $\varphi_1(t) \in C^\infty(\mathbf{R})$ be such that $\varphi_1(t) = 1$ if $|t| > 2\epsilon$, $\varphi_1(t) = 0$ if $|t| < \epsilon$. For $z \in D_\epsilon$, $V_{\gamma,0}(E,z)$ is defined by

$$V_{\gamma,0}(E,z)f = (2\pi)^{-n/2}\int_{\mathbf{R}^n} \frac{e^{i(x-y)\cdot\xi}\varphi_1(\gamma\cdot\xi)}{\xi^2 + 2z\gamma\cdot\xi + z^2 - E}\hat{f}(\xi)d\xi. \tag{10.6}$$

This is $\mathbf{B}(L^2; L^2)$-valued analytic in $z \in D_\epsilon$.

To explain $W_{\gamma,0}(E,z)$, let $\gamma = (1, 0, \ldots, 0)$. We write $x \in \mathbf{R}^n$ as $x = (x_1, x')$, $x' \in \mathbf{R}^{n-1}$, and let $\Delta' = \sum_{j=2}^n (\partial/\partial x_j)^2$. For $a \in \mathbf{R}$, let

$$\mathcal{H}'_a = \left\{ f; \int_{\mathbf{R}^{n-1}} e^{2a|x'|}|f(x')|^2 dx' < \infty \right\},$$

$$\mathcal{H}_a = \left\{ f; \int_{\mathbf{R}^n} e^{2a|x|}|f(x)|^2 dx < \infty \right\}.$$

It is well-known that for any $\delta > 0$, $(-\Delta' - z)^{-1}$ defined on $\mathbf{C}_\pm$ has an analytic continuation across the positive real axis $(0, \infty)$ into the region $\{z; \pm\mathrm{Im}\sqrt{z} > -\delta\}$ as a $\mathbf{B}(\mathcal{H}'_\delta; \mathcal{H}'_{-\delta})$-valued function. Denoting this operator by $r_\pm(z)$, we define

$$\begin{aligned} W_{\gamma,0}(E,z) = (\mathcal{F}_{x_1\to\xi_1})^{-1}\{ & r_+(E - (\xi_1 + z)^2)F(\xi_1 < 0) \\ & + r_-(E - (\xi_1 + z)^2)F(\xi_1 > 0)\}\varphi_0(\xi_1)\mathcal{F}_{x_1\to\xi_1}, \end{aligned} \tag{10.7}$$

where $\varphi_0(t) = 1 - \varphi_1(t)$.

Theorem 10.3: Let $E > 0$.

(1) For any $\delta > 0$, there exists $\epsilon > 0$ such that $U_{\gamma,0}(E,z)$ is $\mathbf{B}(\mathcal{H}_\delta;\mathcal{H}_{-\delta})$-valued analytic on D_ϵ.

(2) $U_{\gamma,0}(E,z)$ has a continuous boundary value for $z \in \overline{D_\epsilon} \cap \mathbf{R}$, and for $t \in (-\epsilon/2, \epsilon/2)$,

$$U_{\gamma,0}(E,t) = G_{\gamma,0}(\sqrt{E-t^2}, t).$$

(3) For $\tau > 0$

$$U_{\gamma,0}(E,i\tau) = G_{\gamma,0}(\sqrt{E+\tau^2}, i\tau).$$

(4) For $0 < s < 1$

$$\|U_{\gamma,0}(E,i\tau)\|_{\mathbf{B}(L^{2,s};L^{2,s-1})} \leq C/\tau, \quad \tau > 1.$$

(5) Let $R_{\gamma,0}(E,t) = e^{it\gamma\cdot x}U_{\gamma,0}(E,t)e^{-it\gamma\cdot x}$. Then

$$(-\Delta - E)R_{\gamma,0}(E,t) = 1.$$

For the proof see Isozaki.[34] As is seen from (2) and (3) an important role of the Green operator of Eskin–Ralston is to connect the Green operator of Faddeev on the real axis $G_{\gamma,0}(\sqrt{E-t^2}, t)$ to that on the imaginary axis $G_{\gamma,0}(\sqrt{E+\tau^2}, i\tau)$ via the analytic continuation.

11. Inverse Scattering at a Fixed Energy

We shall discuss in this section the reconstruction of the potential $V(x)$ from the scattering amplitude at a fixed energy $E > 0$. This is a physically reasonable problem, and demonstrates the utility of direction dependent Green operators.

11.1. *Perturbed Green operators*

We first construct the direction dependent Green operator for $H = -\Delta + V$. Assume that

$$|V(x)| \leq Ce^{-\delta_0|x|}, \tag{11.1}$$

for some $\delta_0, C > 0$. Then for $\delta < \delta_0/2$, $U_{\gamma,0}(E,z)V$ is compact on $\mathcal{H}_{-\delta}$. We define the set of *exceptional points*, $\mathcal{E}_\gamma(E)$, to be the set of $z \in \overline{D_\epsilon}$ such that $-1 \in \sigma_p(U_{\gamma,0}(E,z)V)$. It is easy to see that $\mathcal{E}_\gamma(E)$ is independent of $\delta < \delta_0/2$.

Lemma 11.1: *$\mathcal{E}_\gamma(E) \cap \mathbf{C}_+$ is discrete and there exists $C > 0$ such that $\{i\tau \in \mathcal{E}_\gamma(E); \tau > C\} = \emptyset$. Moreover $\mathcal{E}_\gamma(E) \cap \mathbf{R}$ is a closed set of measure zero.*

Since $R_{\gamma,0}(E,t) = e^{it\gamma\cdot x}U_{\gamma,0}(E,t)e^{-it\gamma\cdot x}$, $t \in \mathcal{E}_\gamma(E)$ is equivalent to $-1 \in \sigma_p(R_{\gamma,0}(E,t)V)$. We define for $E > 0$ and $t \in (-\epsilon/2, \epsilon/2) \setminus \mathcal{E}_\gamma(E)$,

$$R_\gamma(E,t) = (1 + R_{\gamma,0}(E,t)V)^{-1}R_{\gamma,0}(E,t). \tag{11.2}$$

11.2. *Faddeev scattering amplitude*

The scattering matrix has the following expression

$$\hat{S}(E) = 1 - 2\pi i A(E),$$
$$A(E) = \mathcal{F}_0(E)(V - VR(E+i0)V)\mathcal{F}_0(E)^*,$$

where $\mathcal{F}_0(E)$ is defined by (9.4), and $R(E+i0) = (H-E-i0)^{-1}$. Replacing $R(E+i0)$ by $R_\gamma(E,t)$, we define

$$A_\gamma(E,t) = \mathcal{F}_0(E)(V - VR_\gamma(E,t)V)\mathcal{F}_0(E)^*, \tag{11.3}$$

It follows from (10.3) and Theorem 10.3 [Statement (2)] that $R_{\gamma,0}(E,t)$ satisfies

$$R_{\gamma,0}(E,t) = R_0(E+i0) - T_\gamma, \tag{11.4}$$

where

$$T_\gamma = 2\pi i\mathcal{F}_0(E)^*F_\gamma(t)\mathcal{F}_0(E), \quad F_\gamma(t) = F(\gamma\cdot\theta > \frac{t}{\sqrt{E}}). \tag{11.5}$$

Therefore by Theorem 9.2, we have the following equation

$$A_\gamma(E,t) = A(E) + 2\pi i A(E)F_\gamma(t)A_\gamma(E,t). \tag{11.6}$$

For $t \in (-\epsilon/2, \epsilon/2) \setminus \mathcal{E}_\gamma(E)$, the perturbed Green operator $R_\gamma(E,t)$ exists. Therefore by virtue of Theorem 9.3, equation (11.6) is solvable with respect to $A_\gamma(E,t)$:

$$A_\gamma(E,t) = (1 - 2\pi i A(E)F_\gamma(t))^{-1}A(E), \quad t \in (-\epsilon/2, \epsilon/2) \setminus \mathcal{E}_\gamma(E). \tag{11.7}$$

Thus one can construct the Faddeev scattering amplitude from the physical scattering amplitude.

11.3. *Inverse scattering at a fixed energy*

One can now reconstruct the potential from the scattering amplitude.

Theorem 11.2: Let $n \geq 3$. Suppose $V(x)$ satisfy (11.1). Then one can reconstruct $V(x)$ uniquely from the scattering matrix of an arbitrarily fixed energy $E > 0$.

Proof: From the physical scattering amplitude $A(E)$, construct the Faddeev scattering amplitude $A_\gamma(E,t)$. Up to a constant, it has the following integral kernel

$$\int e^{-i\sqrt{E}(\theta-\theta')\cdot x}V(x)dx - \int e^{-i\sqrt{E}\theta\cdot x}V(x)R_\gamma(E,t)(V(\cdot)e^{i\sqrt{E}\theta'\cdot})dx.$$

We take $\sqrt{E}\theta = \sqrt{E-t^2}\omega + t\gamma$, $\sqrt{E}\theta' = \sqrt{E-t^2}\omega' + t\gamma$, where $\omega, \omega' \in S^{n-1}$ and $\omega \cdot \gamma = \omega' \cdot \gamma = 0$. Then the above kernel is written as

$$\begin{aligned} B_\gamma(\omega,\omega',t) = & \int e^{-i\sqrt{E-t^2}(\omega-\omega')\cdot x}V(x)dx \\ & - \int e^{-i\sqrt{E-t^2}\omega\cdot x}V(x)U_\gamma(E,t)(V(\cdot)e^{i\sqrt{E-t^2}\omega'\cdot})dx, \end{aligned}$$

where

$$U_\gamma(E,t) = e^{-it\gamma\cdot x}R_\gamma(E,t)e^{it\gamma\cdot x} = (1+U_{\gamma,0}(E,t)V)^{-1}U_{\gamma,0}(E,t).$$

$U_\gamma(E,t)$ has a unique meromorphic extension to D_ϵ and for large $\tau > 0$,

$$\|U_\gamma(E,i\tau)\|_{\mathbf{B}(L^{2,s};L^{2,s-1})} \leq C/\tau, \quad 0 < s < 1.$$

Therefore as $\tau \to \infty$

$$B_\gamma(\omega,\omega',i\tau) \simeq \int e^{-i\sqrt{E+\tau^2}(\omega-\omega')\cdot x}V(x)dx.$$

We now use the assumption $n \geq 3$. Take any $\xi \in \mathbf{R}^n$. Take $\gamma, \eta \in S^{n-1}$ such that $\xi \cdot \gamma = \xi \cdot \eta = \gamma \cdot \eta = 0$. We put

$$\omega = \omega(\tau) = (1 - \frac{\xi^2}{4\tau^2})^{1/2}\eta + \frac{\xi}{2\tau},$$

$$\omega' = \omega'(\tau) = (1 - \frac{\xi^2}{4\tau^2})^{1/2}\eta - \frac{\xi}{2\tau}.$$

Then $\sqrt{E+\tau^2}(\omega-\omega') \to \xi$. Therefore,

$$B_\gamma(\omega(\tau),\omega'(\tau),i\tau) \to \hat{V}(\xi). \qquad \square$$

11.4. *Slowly decreasing potentials*

When the potential is not exponentially decreasing, it cannot be determined from the scattering amplitude of a fixed energy (see e.g. Regge Ref. 67). However one can construct $\hat{V}(\xi)$ for $|\xi| < 2\sqrt{b}$ from the scattering amplitude for energy interval (a, b). More precisely, we have the following theorem.

Theorem 11.3: Let $n \geq 2$. Suppose $V(x)$ satisfies

$$|\partial^\alpha V(x)| \leq C(1+|x|)^{-3/2-\epsilon-|\alpha|}, \quad |\alpha| \leq n-1$$

for $\epsilon > 0$. Let S be a measurable subset of $\mathbf{R}$ with positive measure such that $b = \text{ess.sup } S > 0$. Suppose we are given the scattering amplitude $A(E)$ for all $E \in S$. Then we can reconstruct $\hat{V}(\xi)$ for all $|\xi| < 2\sqrt{b}$. If S is a half-line: $S = [E_0, \infty)$, for any $\omega, \omega' \in S^{n-1}$ such that $\omega \cdot \gamma = \omega' \cdot \gamma = 0$ and $\omega \neq \omega'$, and for a sufficiently large $\lambda > 0$, one can construct a function $C_\gamma(\lambda, t, \omega, \omega')$ such that

$$\hat{V}(\sqrt{\lambda}(\omega - \omega')) = C_\gamma(\lambda, t_0; \omega, \omega') + \text{p.v.}\frac{1}{\pi}\int_{-\infty}^{\infty} \frac{C_\gamma(\lambda, t; \omega, \omega')}{t - t_0} dt$$

holds for all $t_0 \in \mathbf{R}$.

This theorem is proved by using Faddeev's Green operator instead of Eskin–Ralston's Green operator. For the proof see Ref. 44 or 33.

12. Faddeev Theory

12.1. *Exceptional points*

Let us look closer at the Faddeev theory. By Theorem 10.3 (Statement 2), $R_{\gamma,0}(E,t) = \widetilde{R}_{\gamma,0}(\lambda,t)$ if $E = \lambda^2 + t^2$ and $|t| < \epsilon/2$. So, we re-define $R_{\gamma,0}(E,t)$ for $|t| \geq \epsilon/2$ by this formula. The perturbed Green operator is defined by

$$R_\gamma(E,t) = (1 + R_{\gamma,0}(E,t)V)^{-1} R_{\gamma,0}(E,t). \tag{12.1}$$

If we assume that $|V(x)| \leq C(1+|x|)^{-1-\epsilon_0}, \epsilon_0 > 0$, then

$$R_{\gamma,0}(E,t)V = \widetilde{R}_{\gamma,0}(\lambda,t)V = e^{it\gamma\cdot x} G_{\gamma,0}(\lambda,t) V e^{-it\gamma\cdot x}$$

is compact on $L^{2,-s}$ for $1/2 < s < (1+\epsilon_0)/2$. For $\lambda \geq 0$, the set of exceptional points $\widetilde{\mathcal{E}}_\gamma(\lambda)$ is defined to be the set of $z \in \overline{\mathbf{C}}_+$ such that $-1 \in \sigma_p(G_{\gamma,0}(\lambda, z)V)$.

Lemma 12.1: *For $\lambda \geq 0$, $\widetilde{\mathcal{E}}_\gamma(\lambda) \cap \mathbf{C}_+$ is a discrete set, and $\widetilde{\mathcal{E}}_\gamma(\lambda) \cap \{|z| > C_0\} = \emptyset$ for large $C_0 > 0$. Moreover $\widetilde{\mathcal{E}}_\gamma(\lambda) \cap \mathbf{R}$ is a closed set of measure zero.*

The existence of real exceptional points is a first barrier of the Faddeev theory. For small potentials, they do not exist. However Lavine–Nachman[46] and Khenkin–Novikov[44] proved that real exceptional points do exist if $-\Delta + V$ has bound states.

Theorem 12.2: If $\sigma_p(-\Delta + V) \neq \emptyset$, for any $\gamma \in S^{n-1}$, there exist $\lambda \geq 0$ and $t \in \mathbf{R}$ such that $-1 \in \sigma_p(G_{\gamma,0}(\lambda, t)V)$.

Let us continue our arguments under the assumption that

$$-1 \notin \sigma_p(R_{\gamma,0}(E,t)V), \quad \forall E > 0, \quad -\sqrt{E} \leq \forall t \leq \sqrt{E}. \tag{12.2}$$

This assumption is satisfied when V is sufficiently small. We let

$$\Psi_\gamma(x,E,\theta) = e^{i\sqrt{E}\theta\cdot x} - R_\gamma(E, \sqrt{E}\theta\cdot\gamma)(V(\cdot)e^{i\sqrt{E}\cdot}), \tag{12.3}$$

$$\widetilde{A}_\gamma(E,\theta,\theta') = \frac{E^{(n-2)/2}}{2(2\pi)^n}\int_{\mathbf{R}^n} e^{-i\sqrt{E}\theta\cdot x} V(x)\Psi_\gamma(x,E,\theta')dx, \tag{12.4}$$

$$Q_\gamma^{(\pm)}(E,\theta,\theta') = 2\pi i F(\pm\gamma\cdot(\theta-\theta') \geq 0)\widetilde{A}_\gamma(E,\theta,\theta'). \tag{12.5}$$

Let $Q_\gamma^{(\pm)}(E)$ be the integral operator with kernel $Q_\gamma^{(\pm)}(E,\theta,\theta')$.

12.2. *Factorization of S-matrix*

We show that

$$\hat{S}(E) = (1 - Q_\gamma^{(-)}(E))(1 + Q_\gamma^{(+)}(E))^{-1}. \tag{12.6}$$

This is proven mainly by algebraic manipulations. In fact, letting

$$L_\gamma^{(\pm)}(t) = 2\pi i F\left(\pm\gamma\cdot(\theta - \frac{t\gamma}{\sqrt{E}}) \geq 0\right) A_\gamma(E,t),$$

we have

$$\hat{S}(E)(1 + L_\gamma^{(+)}(t)) = 1 - L_\gamma^{(-)}(t). \tag{12.7}$$

The integral kernel of (12.7) reads

$$\int_{S^{n-1}} \hat{S}(E,\theta,\theta')(\delta(\theta''-\theta') + L_\gamma^{(+)}(t,\theta'',\theta'))d\theta'' = \delta(\theta-\theta') - L_\gamma^{(-)}(t,\theta,\theta').$$

Letting $t = \sqrt{E}\theta' \cdot \gamma$ in the above formula, we get

$$\hat{S}(E)(1 + Q_\gamma^{(+)}(E)) = 1 - Q_\gamma^{(-)}(E). \tag{12.8}$$

On the other hand, we can show that $(Q_\gamma^{(\pm)}(E))^2 = 0$. Therefore $1+Q_\gamma^{(+)}(E)$ is invertible, which proves (12.6).

12.3. *Volterra operator*

We shall prove (8.20) and (8.21). The proof requires involved computations using resolvent estimates. As far as the author knows, the rigorous proof has not been presented yet, and we also have to omit the detailed proof here due to the lack of space.

Let

$$\Phi_\gamma(x, \xi) = \Psi_\gamma(x, |\xi|^2, \xi/|\xi|). \tag{12.9}$$

Then we have

$$\Phi_\gamma(x, \xi) = e^{ix\cdot\xi} - \widetilde{K}_\gamma(x, \xi), \tag{12.10}$$

$$\widetilde{K}_\gamma(x, \xi) = \widetilde{R}_\gamma(|\xi'|^2, \xi \cdot \gamma)(V(\cdot)e^{i\xi\cdot}), \quad \xi' = \xi - (\xi \cdot \gamma)\gamma. \tag{12.11}$$

Here $\widetilde{R}_\gamma(\lambda, t)$ is defined by

$$\widetilde{R}(\lambda, t) = (1 + \widetilde{R}_{\gamma,0}(\lambda, t)V)^{-1}\widetilde{R}_{\gamma,0}(\lambda, t).$$

Let us put $\gamma = (1, 0, \ldots, 0)$ for the sake of simplicity. We assume that $n \geq 3$ and

$$|V(x)| \leq C(1 + |x|)^{-\rho}, \quad \rho > \max\{2, (n+1)/2\}. \tag{12.12}$$

Then one can show that if $|\xi'| \geq \epsilon_0 > 0$

$$|\widetilde{K}_\gamma(x, \xi_1, \xi')| \leq C(1 + |\xi|)^{-1}, \tag{12.13}$$

$$\left|\frac{\partial}{\partial \xi_1}\widetilde{K}_\gamma(x, \xi_1, \xi')\right| \leq C|\xi'|(1 + |\xi|)^{-1}. \tag{12.14}$$

We let

$$K_\gamma(x, y_1, \xi') = (2\pi)^{-1/2}\int_{-\infty}^{\infty} e^{-iy_1\xi_1}\widetilde{K}_\gamma(x, \xi_1, \xi')d\xi_1, \tag{12.15}$$

where the integral is taken in the limit in the mean. One can then show that

$$(1 + |y_1|)K_\gamma(x, y_1, \xi') \in L^2(\mathbf{R}_{y_1}).$$

This implies that

$$K_\gamma(x, y_1, \xi') \in L^1(\mathbf{R}_{y_1}). \tag{12.16}$$

Noting that

$$K_\gamma(x, y_1, \xi') = (2\pi)^{-1/2} \int_{-\infty}^{\infty} e^{-i(y_1 - x_1)\xi_1} G_\gamma(|\xi'|^2, \xi_1)(V(x)e^{ix'\cdot\xi'})d\xi_1,$$

and that $G_\gamma(\lambda, z)$ is analytic in $z \in \mathbf{C}_+$, we have by the Paley–Wiener theorem, $K_\gamma(x, y_1, \xi') = 0$ if $y_1 - x_1 < 0$. Therefore

$$\Psi_\gamma(x, \xi) = e^{ix\cdot\xi} - (2\pi)^{-1/2} \int_{x_1}^{\infty} e^{iy_1\xi_1} K_\gamma(x, y_1, \xi')dy_1. \tag{12.17}$$

This proves (8.20).

12.4. *Gel'fand–Levitan equation*

We have no longer sufficient estimates to guarantee the validity of the remaining arguments. Therefore the following arguments are formal, although very interesting.

Let us introduce three types of integral operators:

$$T_0 f(\xi) = (2\pi)^{-n/2} \int_{\mathbf{R}^n} e^{-ix\cdot\xi} f(x)dx, \tag{12.18}$$

$$T f(\xi) = (2\pi)^{-n/2} \int_{\mathbf{R}^n} \overline{\Phi(x, \xi)} f(x)dx, \tag{12.19}$$

$$T_\gamma f(\xi) = (2\pi)^{-n/2} \int_{\mathbf{R}^n} \overline{\Phi_\gamma(x, \xi)} f(x)dx, \tag{12.20}$$

where $\Phi(x, \xi) = e^{ix\cdot\xi} - R(|\xi|^2 + i0)(V(\cdot)e^{i\cdot\xi})$. We define

$$Q_\gamma^{(+)} f(\xi) = \left(Q_\gamma^{(+)}(|\xi|^2) f(|\xi|\cdot)\right)(\xi/|\xi|). \tag{12.21}$$

Then we can show

$$T^* = T_\gamma^*(1 - Q_\gamma^{(+)}). \tag{12.22}$$

Now let us derive the Gel'fand–Levitan equation (8.22). For the sake of simplicity we assume that $\sigma_p(-\Delta + V) = \emptyset$. In view of (8.20), we let

$$U_\gamma = T_\gamma^* T_0 = 1 + K_\gamma.$$

From $T^*T = 1$ and (12.22), we have

$$T_\gamma^*(1 + \widetilde{Q}_\gamma)T_\gamma = 1, \tag{12.23}$$

where

$$\widetilde{Q}_\gamma = Q_\gamma^{(+)}Q_\gamma^{(+)^*} - Q_\gamma^{(+)} - Q_\gamma^{(+)^*}.$$

Replacing T_γ in (12.23) by $T_0 U_\gamma^* = T_0(1 + K_\gamma^*)$, we have

$$1 + K_\gamma + \Omega_\gamma + K_\gamma \Omega_\gamma = (1 + K_\gamma^*)^{-1},$$

where

$$\Omega_\gamma = T_0^* \widetilde{Q}_\gamma T_0.$$

Now $K_\gamma^*(x,y)$ is supported in $\{(x-y)\cdot\gamma \geq 0\}$. Therefore, letting $(1 + K_\gamma^*)^{-1} = 1 + C_\gamma$, we see that $C_\gamma(x,y)$ is also supported in $\{(x-y)\cdot\gamma \geq 0\}$. Then for $(x-y)\cdot\gamma < 0$ we have

$$K_\gamma(x,y) + \Omega_\gamma(x,y) + \int_{(x-y)\cdot\gamma<0} K_\gamma(x,z)\Omega_\gamma(z,y)dz = 0.$$

This is the Gel'fand–Levitan equation.

Little is known about the solvability of this equation. When the scattering amplitude is small, $\Omega_\gamma(x,y)$ is also small. Therefore the above equation might be solved with respect to K_γ. By differentiating $K_\gamma(x,y)$, one might get $V(x)$. However, the estimates of this $V(x)$ (e.g. the spatial decay) would be hard to prove.

Putting these difficulties aside, Faddeev further stepped into the characterization problem of the scattering amplitude. According the above procedure, starting from the scattering amplitude, one can get $K_\gamma(x,y)$. By differentiating this $K_\gamma(x,y)$, one can get an integral kernel depending on γ containing the δ-function (see Theorem 8.1). What guarantees that it is the operator of multiplication independent of γ? Faddeev observed that it is the analyticity of $B_\gamma(\omega,\omega',t)$, which appeared in the proof of Theorem 11.2, with respect to t.

Newton[61] analyzed Faddeev's arguments and encountered the above difficulty. He tried to find another route to obtain $V(x)$ based on the dispersion relation, and pointed out that in this case a sort of miraculous condition must be imposed (Ref. 61, p.150).

13. $\overline{\partial}$-Approach

In 1980's $\overline{\partial}$-approach was introduced as a new view point of the inverse scattering by Beals–Coifman[1] and Nachman–Ablowitz.[57] In this approach Faddeev's method is rewritten as follows. Let $\widetilde{G}(\zeta)$ be as in (8.12).

Lemma 13.1: *Let $\overline{\partial_j} = \partial/\partial\overline{\zeta_j}$. Then for Im $\zeta_j \neq 0$*

$$\overline{\partial_j}\widetilde{G}(\zeta)f = (2\pi)^{1-n/2}\int_{\mathbf{R}^n} e^{ix\cdot\xi}\hat{f}(\xi)\xi_j\delta(\xi^2 + 2\zeta\cdot\xi)d\xi. \tag{13.1}$$

and $\overline{\partial_j}\widetilde{G}(\zeta) \in \mathbf{B}(L^{2,s}; L^{2,-s}), s > 1$.

Let us rewrite the Faddeev scattering amplitude. We define for $\zeta = \eta + i\tau\gamma$,

$$\widetilde{G}_V(\zeta) = (1 + \widetilde{G}(\zeta)V)^{-1}\widetilde{G}(\zeta). \tag{13.2}$$

If $i\tau \notin \mathcal{E}_\gamma(E)$ it satisfies

$$\widetilde{G}_V(\eta + i\tau\gamma) = e^{-ix\cdot\eta}U_\gamma(E, i\tau)e^{ix\cdot\eta}. \tag{13.3}$$

For $\omega, \omega' \in S^{n-1}$ satisfying $\omega\cdot\gamma = \omega'\cdot\gamma = 0$, we let $\eta = \sqrt{E+\tau^2}\omega, \eta' = \sqrt{E+\tau^2}\omega'$. Then $B_\gamma(\omega, \omega', i\tau)$ has the following expression

$$B_\gamma(\omega, \omega'; i\tau) = \int e^{-i(\eta-\eta')\cdot x}V(x)dx - \int e^{-i(\eta-\eta')\cdot x}V(x)\widetilde{G}_V(\zeta)V dx,$$

where $\zeta = \eta' + i\tau\gamma$. Let us note that $\xi = \eta - \eta'$ satisfies $\xi^2 + 2\zeta\cdot\xi = 0$. Starting from the scattering amplitude $A(E)$, we have thus constructed

$$T(\xi, \zeta) = \int e^{-ix\cdot\xi}V(x)dx - \int e^{-ix\cdot\xi}V(x)\widetilde{G}_V(\zeta)V dx, \tag{13.4}$$

on the set $\{(\xi, \zeta)\}$ where $\xi \in \mathbf{R}^n$ and $\zeta \in \mathbf{C}^n$ satisfy $\zeta^2 = E, |\zeta| > C, \text{Im}\,\zeta \neq 0$ and $\xi^2 + 2\zeta\cdot\xi = 0$, C being a large constant. This set has a structure of fibred space and each fibre

$$\mathcal{V}_\xi = \{\zeta \in \mathbf{C}^n;\ \zeta^2 = E,\ |\zeta| > C,\ \text{Im}\,\zeta \neq 0,\ \xi^2 + 2\zeta\cdot\xi = 0\}. \tag{13.5}$$

is a complex manifold of dimension $2n - 4$. On this complex manifold, the function $T(\xi, \zeta)$ satisfies a $\overline{\partial}$-equation.

Theorem 13.2: As a one-form on $\mathcal{V}_\xi$, we have

$$\overline{\partial}T(\xi, \zeta) = \sum_{j=1}^{n} A_j(\xi, \zeta)d\overline{\zeta_j},$$

$$A_j(\xi, \zeta) = -(2\pi)^{1-n/2}\int_{\mathbf{R}^n} T(\xi - \eta, \zeta + \eta)T(\eta, \zeta)\eta_j\delta(\eta^2 + 2\zeta\cdot\eta)d\eta.$$

There are two important applications of Theorem 13.2. Recall the generalized Cauchy formula

$$f(z) = \frac{1}{2\pi i}\int_{\partial D}\frac{f(\zeta)}{\zeta - z}d\zeta - \frac{1}{2\pi i}\int_{D}\frac{\overline{\partial} f(\zeta)}{\zeta - z}d\overline{\zeta}\wedge d\zeta,$$

where D is a domain in $\mathbf{C}$. Using a similar formula on the complex manifold $\mathcal{V}_\xi$, Nachman[55] derived the following representation formula of the potential by means of $T(\xi,\zeta)$:

$$\hat{V}(\xi) = T(\xi,\zeta_0) + \int_{\partial\mathcal{V}_\xi} T(\xi,\zeta)K(\zeta,\zeta_0) + \int_{\mathcal{V}_\xi}\sum_{j=1}^{n} A_j(\xi,\zeta)d\overline{\zeta_j}\wedge K(\zeta,\zeta_0), \tag{13.6}$$

where $\zeta_0 \in \mathcal{V}_\xi$ and $K(\zeta,\zeta_0)$ is a suitable $(2n-5)$-form on $\mathcal{V}_\xi$.

Another important application is the characterization of the Faddeev scattering amplitude. Namely, the $\overline{\partial}$-equation in Theorem 13.2 gives a necessary and sufficient condition for a function $T(\xi,\zeta)$ defined on the fibred space $\{(\xi,\zeta);\, \zeta^2 = E,\, \xi^2+2\zeta\cdot\xi = 0\}$ to be the Faddeev scattering amplitude associated with some Schrödinger operator $-\Delta+V(x)$. This interesting fact was fully discussed by Beals–Coifman[1] and Khekin–Novikov.[44] The advantage compared to Faddeev's characterization is that in the $\overline{\partial}$-approach, one can get the precise estimates of $T(\xi,\zeta)$ and also specify the associated class of potentials. See also Weder.[76] However, the characterization of the physical scattering amplitude itself is still open, though it is linked with the Faddeev scattering amplitude through the equation (11.6).

14. Inverse Conductivity Problem

It is an interesting fact that inverse eigenvalue problems are reduced to inverse scattering problems. Let Ω be a bounded domain in $\mathbf{R}^n$ and consider the following boundary value problem

$$\nabla\cdot(c(x)\nabla u) = 0 \quad \text{in} \quad \Omega, \tag{14.1}$$

$$u = f \quad \text{on} \quad \partial\Omega,$$

where $c(x) \geq c_0 > 0$. In electromagnetism, $c(x)$ corresponds to the conductivity. Let Λ_c be the Dirichlet–Neumann map

$$\Lambda_c f = c(x)\frac{\partial u}{\partial \nu},$$

ν being the outer unit normal to the boundary. A. P. Calderón posed the following problem:

Question: *Does Λ_c determine c?*

By the substitution $v = c^{1/2}u$, the above problem is transferred to

$$-\Delta v + qv = 0 \quad \text{in} \quad \Omega, \tag{14.2}$$

$$v = c^{1/2}f \quad \text{on} \quad \partial\Omega,$$

where $q = c^{-1/2}\Delta c^{1/2}$. Let Λ be the D-N map

$$\Lambda g = \frac{\partial v}{\partial \nu},$$

where v is a solution to (14.2) satisfying $v = g$ on $\partial\Omega$.

Theorem 14.1: Let $n \geq 2$. Suppose $q \in L^\infty(\Omega)$ and $0 \notin \sigma_p(-\Delta_D + q)$. Then q is uniquely determined form the knowledge of the D–N map.

The above theorem is proved by Sylvester–Uhlmann for $n \geq 3$ (Ref. 71) in the category of pure boundary value problem by using the method of complex geometrical optics. Here we discuss its relation to the scattering problem.

The following observation is useful to see how the interior boundary value problem and the scattering problem are related.

Take a constant $E > 0$ and let

$$V(x) = \begin{cases} q(x) + E & x \in \Omega \\ 0 & x \notin \Omega. \end{cases}$$

Let ψ^+ be a solution to the Schrödinger equation

$$(-\Delta + V(x) - E)\psi^+ = 0 \quad \text{in} \quad \mathbf{R}^n$$

satisfying

$$\psi^+ \simeq e^{i\sqrt{E}\omega\cdot x} + \frac{e^{i\sqrt{E}r}}{r^{(n-1)/2}} f(E;\theta,\omega) \quad r = |x| \to \infty, \quad \theta = x/r.$$

Then by Green's formula we have

$$\begin{aligned}\int_{\partial\Omega} e^{-i\sqrt{E}\theta\cdot x}\left(\frac{\partial}{\partial\nu} + i\sqrt{E}\theta\cdot\nu\right)\psi^+ dS &= \int_\Omega e^{-i\sqrt{E}\theta\cdot x}(\Delta + E)\psi^+ dx \\ &= A(E;\theta,\omega),\end{aligned}$$

where $A(E;\theta,\omega)$ is the scattering amplitude associated with $-\Delta + V$. Since ψ^+ satisfies (14.2) we have

$$\frac{\partial}{\partial\nu}\psi^+ = \Lambda\psi^+. \tag{14.3}$$

We follow Isakov–Nachman.[31] Let $\Omega_e = \mathbf{R}^n \setminus \Omega$. Then the D–N map Λ_e for the exterior Dirichlet problem is defined in the same way as in the interior problem:

$$\Lambda_e f = \frac{\partial u_e}{\partial \nu},$$

where u_e is a unique solution to

$$(-\Delta - E)u_e = 0 \quad \text{in} \quad \Omega_e,$$

$$u_e = f \in H^{3/2}(\partial\Omega) \quad \text{on} \quad \partial\Omega$$

satisfying the outgoing radiation condition

$$(\frac{\partial}{\partial r} - i\sqrt{E})u_e \in L^{2,-\alpha}(\Omega_e)$$

for some $0 < \alpha < 1/2$.

Another tool is the simple layer potential associated with $-\Delta + V$. Let $R^{(+)}(E,x,y)$ be the resolvent kernel of $(-\Delta + V - E - i0)^{-1}$, and let

$$K(E)f(x) = \int_{\partial\Omega} R^{(+)}(E,x,y)f(y)dS_y. \tag{14.4}$$

We define $B(E) \in \mathbf{B}(H^{1/2}(\partial\Omega); H^{3/2}(\partial\Omega))$ by

$$B(E)f = K(E)f|_{\partial\Omega}. \tag{14.5}$$

The relations between these operators are summarized in the following lemma.

Lemma 14.2: *Suppose E is not an eigenvalue of the interior problem for $-\Delta + V$. Let Λ_{V-E} and Λ_e be the D-N maps of $-\Delta + V - E$ in Ω and $-\Delta - E$ in Ω_e. Then*
(1) $\Lambda_{V-E} - \Lambda_e$ is an isomorphism from $H^{3/2}(\partial\Omega)$ to $H^{1/2}(\partial\Omega)$.
(2) $(\Lambda_{V-E} - \Lambda_e)^{-1} = B(E)$.

Let $\varphi^e_\pm(x,E,\omega)$ be the generalized eigenfunctions for the exterior Dirichlet problem. Namely,

$$(-\Delta - E)\varphi^e_\pm = 0 \quad \text{in} \quad \Omega_e,$$

$$\varphi^e_\pm = e^{i\sqrt{E}\omega\cdot x} \quad \text{on} \quad \partial\Omega,$$

$$\varphi^e_\pm \simeq e^{i\sqrt{E}\omega\cdot x} + \frac{e^{\pm i\sqrt{E}r}}{r^{(n-1)/2}} f_\pm(E,\hat{x},\omega) \quad r \to \infty.$$

One can then show:

Theorem 14.3:

$$\int_{\partial\Omega} \overline{\frac{\partial}{\partial \nu}\varphi^e_-(x,E,\theta)}B(E)\varphi^e_+(\cdot,E,\omega)dS_x = C(E)(A(E,\theta,\omega) - A_{ext}(E,\theta,\omega)),$$

where $A(E,\theta,\omega)$ and $A_{ext}(E,\theta,\omega)$ are the scattering amplitudes of $-\Delta+V$ in $\mathbf{R}^n$ and $-\Delta$ in Ω_e, respectively.

By the above two lemmas, one can construct the scattering amplitude $A(E,\theta,\omega)$ from the knowledge of the D–N map of the interior boundary value problem. By Theorem 11.2, one can reconstruct $V(x)$ if $n \geq 3$.

Conversely, the inverse scattering problem for the compactly supported potential can be reduced to the inverse boundary value problem. In fact, Nachman[55] directly constructed a counterpart of Faddeev scattering amplitude from the knowledge of the D–N map. The potential $V(x)$ is then reconstructed by the formula (13.6).

The two-dimensional problem is much harder, because in this case the problem is no longer overdetermined. Nachman[56] proved Theorem 14.1 when q comes from the conductivity problem, i.e. $q = c^{-1/2}\Delta c^{1/2}$. He used the $\overline{\partial}$-operator method and the theory of quasi-analytic functions.

15. Other Applications

Direction dependent Green operators are now used in various problems other than Schrödinger operators, for example Dirac equations by Isozaki[34] and Goto,[27] Maxwell equations by Ola–Päivärinta–Sommersalo[63] and also elastic equations by Nakamura–Uhlmann.[59]

Let us also mention the application to wave equation in layered media

$$\partial_t^2 u = c(x,y)^2\Delta_{x,y}u, \quad x \in \mathbf{R}^n, \quad y \in \mathbf{R},$$

where $\Delta_{x,y} = \sum_{i=1}^n (\partial/\partial x_i)^2 + (\partial/\partial y)^2$. In this case the unperturbed equation is $c_0(y)^2\Delta_{x,y}$ and the coefficient $c_0(y)$ is piecewise constant with discontinuities at the interfaces. Even in this case one can accommodate Faddeev's approach. See Isozaki,[35] Weder,[78] Guillot–Ralston[28] and Beltita.[3]

In Ref. 3, the coefficients are assumed to satisfy

$$|c_0(y) - c_\pm| \leq Ce^{-\alpha|y|}, \quad \text{for} \quad \pm y > 0,$$

$$|c(x,y) - c_0(y)| \leq Ce^{-\alpha(|x|+|y|)},$$

for a constant $\alpha > 0$. This allows any number of interface for the background medium. One can then reconstruct the sound speed $c(x, y)$ from the knowledge of the scattering matrix at a fixed energy. In the works of Refs. 35 and 3, to construct the direction dependent Green operator the commutator calculus is used, which was developed in the study of N-body Schrödinger operators and has now become a basis of spectral and scattering theory.

Let us finally mention that Melin[53] is preparing a multi-dimensional inverse scattering theory based on the ultra-hyperbolic equation

$$(\Delta_x - \Delta_y)U = V(x)U.$$

His starting point is a fundamental solution to the free ultra-hyperbolic equation $(\Delta_x - \Delta_y)U = 0$, and has an advantage that it directly constructs the intertwining operator with triangular kernel. We also remark that Ref. 37 proposes a new approach to the inverse boundary value problem in Euclidean space by imbedding the problem to hyperbolic manifolds.

Although there remain difficulties in the characterization of scattering amplitudes, Faddeev's method gives a deep insight to the multi-dimensional inverse scattering problem. This theory seems to have much room to accept highly developed tools of partial differential equations and spectral theory for further developments.

Bibliography

1. R. Beals and R. R. Coifman, Mutlidimensional inverse scatterings and non-linear differential equations, *Proc. Symp. Pure Math.* **43** (1985), 45–70.
2. M. I. Belishev and Y. V. Kurylev, To the reconstruction of a Riemannian manifold via its boundary spectral data (BC-method), *Commun. in P.D.E.* **17** (1992), 767–804.
3. I. Beltita, Inverse scattering in a layered medium, *Commun. in P.D.E.* **26** (2001), 1739–1786.
4. P. H. Bérard, Variétés riemaniennes isospectrales non isométriques, *Séminaire Bourbaki* **705** (1988/89).
5. R. Brooks, P. Perry and P. Petersen, Compactness and finiteness theorems for isospectral manifolds, *J. Reine Angew. Math.* **426** (1992), 67–89.
6. J. Brüning, On the compactness of isospectral potentials, *Commun. in P.D.E.* **9** (1984), 687–698.
7. A. P. Calderón, On an inverse boundary value problem, *Seminar on Numerical Analysis and its Applications to Continuum Physics*, Soc. Brasileria de Matematica, Rio de Janeiro, (1980), 65–73.
8. K. Chadan and P. C. Sabatier, *Inverse Problems in Quantum Scattering Theory*, 2nd ed. (Springer, New York, 1989).

9. S. J. Chapman, Drums that sound the same, *Amer. Math. Monthly*, **102** (1995), 124–138.
10. T. Christiansen, Scattering theory for perturbed stratified media, *J. Anal. Math.* **76** (1998), 1–43.
11. M. M. Crum, Associated Sturm–Liouville operators, *Quart. J. Math. Oxford* (2), **6** (1955), 121–127.
12. G. Darboux, Sur une proposition relative aux équations linéaires, *C. R. Acad. Sci. Paris* **94** (1882), 1456–1459.
13. P. Deift, Application of a commutation formula, *Duke Math. J.* **45** (1978), 267–310.
14. V. Enss, Asymptotic completeness for quantum mechanical potential scattering I. Short-range potentials, *Commun. Math. Phys.* **61** (1978), 285–291.
15. V. Enss and R. Weder, The geometrical approach to multidimensional inverse scattering, *J. Math. Phys.* **36** (1995), 3902–3921.
16. G. Eskin, The inverse scattering problem in two dimensions at fixed energy, *Commun. in P. D. E.* **26** (2001), 1055–1090.
17. G. Eskin and J. Ralston, The inverse backscattering problem in three dimensions, *Commun. Math. Phys.* **124** (1989), 169–215.
18. G. Eskin and J. Ralston, Inverse scattering problem for the Schrödinger equation with magnetic potential at a fixed energy, *Commun. Math. Phys.* **173** (1995), 199–224.
19. L. D. Faddeev, Uniqueness of the inverse scattering problem, *Vestnik Leningrad Univ.* **11** (1956), 126–130.
20. L. D. Faddeev. The inverse problem in the quantum theory of scattering, *J. Math. Phys.* **4** (1963), 72–104.
21. L. D. Faddeev, Increasing solutions of the Schrödinger equations, *Sov. Phys. Dokl.* **10** (1966), 1033–1035.
22. L. D. Faddeev, Factorization of the S-matrix for the multidimensional Schrödinger operator, *Sov. Phys. Dokl.* **11** (1966), 209–211.
23. L. D. Faddeev, Inverse problem of quantum scattering theory, *J. Sov. Math.* **5** (1976), 334–396.
24. L. D. Faddeev, *40 Years in Mathematical Physics*, World Scientific Series in 20th Century Mathematics (World Scientific, 1995).
25. I. M. Gel'fand, Some aspects of functional analysis and algebra, *Proc. Inter. Congr. Math.* **1** (1954), 253–277.
26. I. M. Gel'fand and B. M. Levitan, On the determination of a differential equation from its spectral function, *Izv. Akad. Nauk USSR, Ser. Mat.* **15**, (1951), 309–360.
27. M. Goto, Inverse scattering for Dirac operators with electro-magnetic potentials at a fixed energy, preprint (1999).
28. J. C. Guillot and J. Ralston, Inverse scattering at fixed energy for layered media, *J. Math. Pures et Appl.* **78** (1999), 27–48.
29. A. Ikeda, On lens spaces which are isospectral but not isometric, *Ann. Scient. Éc. Norm. Sup.* **13** (1980), 303–315.
30. A. Ikeda and Y. Yamamoto, On the spectra of 3-dimensional lens spaces, *Osaka J. Math.* **16** (1979), 447–469.

31. V. Isakov and A. Nachman, Global uniqueness for a two-dimensional semi-linear elliptic inverse problem, *Trans. A.M.S.* **347** (1995), 3375–3390.
32. H. Isozaki, Some remarks on the multi-dimensional Borg–Levinson theorem, *J. Math. Kyoto Univ.* **31** (1991), 743–753.
33. H. Isozaki, Multi-dimensional inverse scattering theory for Schrödinger operators, *Reviews in Math. Phys.* **8** (1996), 591–622.
34. H. Isozaki, Inverse scattering theory for Dirac operators, *Ann. l'I. H. P. Physique Théorique* **66** (1997), 237–270.
35. H. Isozaki, Inverse scattering theory for wave equations in stratified media, *J. Diff. Eq.* **138** (1997), 19–54.
36. H. Isozaki, Multi-dimensional inverse scattering theory, *Analytic Extension Formulas and their Applications*, eds. Saitoh et al, (Fukuoka, 1999/Kyoto, 2000) (Kluwer Academic Publishers, 2001).
37. H. Isozaki, Inverse scattering on hyperbolic manifolds and its applications to inverse boundary value problems in Euclidean space, *preprint* (2002).
38. H. Isozaki and H. Kitada, Scattering matrices for two-body Schrödinger operators, *Scientific Papers of the College of Arts and Sciences, The University of Tokyo*, **35** (1985), 81–107.
39. H. T. Ito, High-energy behavior of the scattering amplitude for a Dirac operator, *Publ. RIMS. Kyoto Univ.* **31** (1995), 1107–1133.
40. H. T. Ito, An inverse scattering problem for Dirac equations with time-dependent electromagnetic potentials, *Publ. RIMS. Kyoto Univ.* **34** (1998), 355–381.
41. C. G. J. Jacobi, Zur Theorier der Variationsrechnung und der Differentialgleichungen, *J. Reine Angew. Math.* **17** (1837), 68–82.
42. M. Kac, Can one hear the shape of a drum? *Amer. Math. Monthly*, **73** (1966), 1–23.
43. I. Kay and H. E. Moses, The determination of the scattering potential from the spectral measure function, I. II. III, *Nuovo Cimento* **2** (1955), 917–961.
44. G. M. Khenkin and R. G. Novikov, The $\overline{\partial}$-equation in the multi-dimensional inverse scattering problem, *Russian Math. Surveys* **42** (1987), 109–180. (1956), 66–84, 276–304.
45. Y. V. Kurylev, A multi-dimensional Gel'fand–Levitan inverse boundary problem, *Differential Equations and Mathematical Physics, Proc. Inter. Conf. Univ. of Alabama at Birmingham, 1994*, ed. I.Knowles, 117–131. (Int. Press, 1995).
46. R. Lavine and A. Nachman, On the inverse scattering transform for the n-dimensional Schrödinger operator, in *Topics in Soliton Theory and Exactly Solvable Nonlinear Equations*, edited by M.Ablowitz, B.Fuchssteiner and M.Kruskal, Oberwolfach (1986), (World Scientific, 1987).
47. M. Lassas, M. Taylor and G. Uhlmann, The Dirichlet-to-Neumann map for complete Riemannian manifolds with boundary, *preprint* (2002).
48. M. Lassas and G. Uhlmann, On determining a Riemannian manifold from the Dirichlet-to-Neumann map, *Ann. Scient. Éc. Norm. Sup.* **34** (2001), 771–787.
49. J. Lee and G. Uhlmann, Determining anisotropic real-analytic conductivity by boundary measurements, *Comm. Pure Appl. Math.* **42** (1989), 1097–1112.

50. H. P. Mckean and I. M. Singer, Curvature and the eigenvalues of the Laplacian, *J. Diff. Geom.* **1** (1967), 43–69.
51. V.A.Marchenko, *Strum–Liouville Operators and Applications*, Operator Theory : Advances and Applications Vol 22, (Birkhäuser, 1986).
52. A. Martin, Construction of the scattering amplitude from differential cross section, *Nuovo Cimento* **59** A (1969), 131–151.
53. A. Melin, The Faddeev approach to inverse scattering, in *Partial Differential Equations and Mathematical Physics*, eds. L. Hörmander and A. Melin, Progress in non-linear differential equations and their applications, pp 226–245, (Birkhäuser, 1996).
54. H. E. Moses, Calculation of the scattering potential from reflection coefficients, *Phys. Rev.* (2) **102** (1956), 559–567.
55. A. Nachman, Reconstruction from boundary measurements, *Ann. Math.* **128** (1988), 531–576.
56. A. Nachman, Global uniqueness for a two-dimensional inverse boundary value problem, *Ann. Math.* **143** (1996), 71–96.
57. A. Nachman and M. Ablowitz, A multidimensional inverse scattering method, *Studie in Appl. Math.* **71** (1984), 243–250.
58. A. Nachman, J. Sylvester and G. Uhlman, An n-dimensional Borg–Levinson theorem, *Commun. Math. Phys.* **115** (1988), 595–605.
59. G. Nakamura and G. Uhlmann, Global uniqueness for an inverse boundary problem arising in elasticity, *Invent. Math.* **118** (1994), 457–474.
60. R. G. Newton, *Scattering Theory of Waves and Particles*, 2nd edition, (Springer, 1982).
61. R. G. Newton, *Inverse Schrödinger Scattering in Three Dimensions*, (Springer-Verlag, 1989).
62. F. Nicoleau, A stationary approach to inverse scattering for Schrödinger operators with first order perturbations, *Commun. in P. D. E.* **22** (1997), 527–553.
63. P. Ola, L. Päivärinta and E. Sommersalo, An inverse boundary value problem in electrodynamics, *Duke Math. J.* **70** (1993), 617–653.
64. P. A. Perry, Inverse spectral problems on compact Riemannian manifolds, in *Schödinger Operators*, Eds. H. Holden, A. Jensen, *Lecture Notes in Physics* 345, (Springer-Verlag, 1989).
65. R. T. Prosser, Formal solutions of inverse scattering problems III, *J. Math. Phys.* **21** (1980), 2648–2653.
66. J. Pöschel and E. Trubowitz, *Inverse Spectral Theory*, (Academic Press, 1987).
67. T. Regge, Introduction to complex orbital moments, *Nuovo Cimento* **14** (1959), 951–976.
68. Y. Saito, An asymptotic behavior of the S-matrix and the inverse scattering problem, *J. Math. Phys.* **25** (1984), 3105–3111.
69. T. Sunada, Riemannian coverings and isospectral manifolds, *Ann. Math.* **121** (1985), 169–186.
70. J. Sylvester, An anisotropic inverse boundary value problem, *Comm. Pure Appl. Math.* **43** (1990), 201–233.

71. J. Sylvester and G. Uhlmann, A global uniqueness theorem for an inverse boundary value problem, *Ann. of Math.* **125** (1987), 153–169.
72. H. Urakawa, Bounded domains which are isospectral but not congruent, *Ann. scient. Éc. Norm. Sup.* **15** (1982), 441–456.
73. H. Urakawa, Spectra of the discrete and continuous Laplacians on graphs and Riemannian manifolds, *Interdiscip. Information Sci.* **3** (1997), 95–105.
74. X. P. Wang, On the uniqueness of inverse scattering for N-body systems, *Inverse Problems* **10** (1994), 765–784.
75. R. Weder, Generalized limiting absorption method and multidimensional inverse scattering theory, *Math. Meth. in Appl. Sci.* **14** (1991), 509–524.
76. R. Weder, Characterization of the scattering data in multi-dimensional inverse scattering theory, *Inverse Problem* **7** (1991), 461–489.
77. R. Weder, Multidimensional inverse scattering in an electric field, *J. Func. Anal.* **139** (1996), 441–465.
78. R. Weder, Multidimensional inverse problems in perturbed stratified media, *J. Diff. Eq.* **52** (1999), 191–239.

CHAPTER 5

ANALYSIS OF GROUND STATES OF ATOMS INTERACTING WITH A QUANTIZED RADIATION FIELD

Fumio Hiroshima

Department of Mathematics and Physics, Setsunan University, Osaka, 572-8508, Japan

hiroshima@mpg.setsunan.ac.jp

The spectrum of nonrelativistic matters interacting with a quantized radiation field is studied. In particular ground states of the Pauli–Fierz Hamiltonian in nonrelativistic QED are investigated.

1. Preliminaries

The purpose of this paper is to review recent advances in analysis of ground states of a model in nonrelativistic quantum electrodynamics.

It may be usually the fate of a physical or mathematical theory that, after its initial success, grave difficulties become apparent. Quantum electrodynamics, however, is remarkable in showing an opposite trend. Actually profound difficulties appeared almost immediately when quantum electrodynamics was established. One of the difficulties concerns an emission of photons of very low frequencies, which yields divergences of an electromagnetic mass, a scattering cross section, etc. We will call it in this paper "infrared singularity" or "infrared divergence". In 1932, E. Fermi[50] and G. Breit[36] first introduced a nonrelativistic model coupled with a quantized radiation field. In 1938, according to a certain model describing an interaction between an electron and a quantized radiation field, W. Pauli and M. Fierz[124] recognized that the quantized radiation field reacts back on the electron to produce an electromagnetic mass. This model is today the so-called "Pauli–Fierz model" which is the main object in this paper. Actually W. Pauli and M. Fierz took the dipole approximation. See Refs. 118 and 145 for many applications of the Pauli–Fierz model to quantum optics and atomic physics.

Here we look at a typical example of the successes of the Pauli–Fierz model. Let α_μ, $\mu = 1, 2, 3$, and β are 4×4 Hermitian matrices obeying the anticommutation relations

$$\{\alpha_\mu, \alpha_\nu\} = 2\delta_{\mu\nu}1, \quad \{\alpha_\mu, \beta\} = 0, \quad \beta^2 = 1.$$

The Dirac Hamiltonian of a hydrogen-like atom is given by

$$D_{\rm hyd} := -i \sum_{\mu=1}^{3} \alpha_\mu \frac{\partial}{\partial x_\mu} + \beta M - \frac{Ze^2}{|x|},$$

where Z is an atomic number, e the charge of an electron, and M the mass of an electron. Then $D_{\rm hyd}$ has eigenvalues

$$E_{nj} = \frac{M}{\sqrt{1 + Z^2 e^4 \Big(n - (j+1/2) + \sqrt{(j+1/2)^2 - Z^2 e^4}\Big)^{-2}}},$$

where $n = 1, 2, ...,$ denotes the principal quantum number and $j = l \pm 1/2$ the total angular momentum with the angular momentum $l = 0, ..., n-1$. Eigenvalue $2S_{1/2}$ corresponds to $n = 2$, $j = 1/2$, $l = 0$, and $2P_{1/2}$ to $n = 2$, $j = 1/2$, $l = 1$. Then the Dirac theory concludes that two levels $2S_{1/2}$ and $2P_{1/2}$ in a hydrogen-like atom sit at the same energy level, i.e.

$$2S_{1/2} = 2P_{1/2}.$$

In 1947, by W. Lamb and R. Retherford,[111] it was experimentally observed, however, that

$$2S_{1/2} > 2P_{1/2}.$$

This discrepancy is called the Lamb shift. H. A. Bethe[32] regarded the Lamb shift as a manifestion of a radiation reaction, and tentatively made a non-relativistic calculation of the difference of the two levels. The resulting value was in remarkable agreement with the observation. Relativistic calculations have been made by many authors. See Ref. 131 for details. In 1948, using the Pauli–Fierz model, T. Welton[146] gave an intuitive derivation to the Lamb shift. He argued that a position-fluctuation of an electron through the radiation field will effectively modify the external potential V. The fluctuation was thought of as a Gaussian random variable Δx, then an effective potential is formally given by a mean value of $V(x + \Delta x)$;

$$\begin{aligned} V_{\rm eff}(x) : &= \langle V(x + \Delta x)\rangle_{\rm AVE} = (2\pi)^{-3/2} \int_{\mathbb{R}^3} \hat{V}(k) e^{ikx} \langle e^{ik\Delta x}\rangle_{\rm AVE} dk \\ &= (2\pi C)^{-3/2} \int_{\mathbb{R}^3} V(y) e^{-|x-y|^2/(2C)} dy, \end{aligned}$$

with a certain positive constant C. Then an electron Hamiltonian effectively turns out to be governed by a Hamiltonian with the external potential $V_{\rm eff}$ instead of V. He gave an interpretation of the Lamb shift as the difference between the spectrum of the original Hamiltonian and an effective one.

Although the Pauli–Fierz model is a nonrelativistic model, as we have illustrated above, we may say that in the realm of bound electrons the Pauli–Fierz model correctly describes the interaction between electrons and the quantized radiation field.

Mathematically the Pauli–Fierz Hamiltonian is defined as a self-adjoint operator acting in a suitable Hilbert space. Recently from a mathematical physics point of view, the spectral analysis of the Pauli–Fierz (type) Hamiltonian was considerably advanced by many authors. In particular, ground states, resonances and scattering theory have been studied according to functional analysis, stochastic analysis and the theory of operator algebras. These problems, as we will see below, can be regarded as "perturbation problems of embedded eigenvalues".

1.1. *Perturbations of embedded eigenvalues*

For the time being, we assume that an electron has no spin for the sake of brevity. The spinless Pauli–Fierz Hamiltonian $H_{\rm PF}$ is defined as a self-adjoint operator acting in the Hilbert space:

$$\mathcal{H}_{\rm rad} := L^2_{\rm as}(\mathbb{R}^{3N}) \otimes \mathcal{F}_{\rm rad}.$$

Here $L^2_{\rm as}(\mathbb{R}^{3N})$ denotes the subset of antisymmetric functions of $L^2(\mathbb{R}^{3N})$ and describes the Hilbert space of state vectors for N electrons, and $\mathcal{F}_{\rm rad}$ a Boson Fock space of photons. The Pauli–Fierz Hamiltonian is decomposed as

$$H_{\rm PF} = H_{\rm dec} + H_{\rm I}. \tag{1.1}$$

Here the first term of (1.1) is a decoupled Hamiltonian:

$$H_{\rm dec} := H_{\rm p} \otimes 1 + 1 \otimes H_{\rm rad},$$

where $H_{\rm rad}$ is the free Hamiltonian in $\mathcal{F}_{\rm rad}$ and $H_{\rm p}$ a Schrödinger Hamiltonian of the form:

$$H_{\rm p} := -\frac{1}{2M}\Delta + V$$

with external potential V and electron mass M. We suppose that $H_{\rm p}$ has point spectrum $E_0 < E_1 < E_2 < \cdots < \Sigma < \infty$ with continuum edge Σ (Fig. 1).

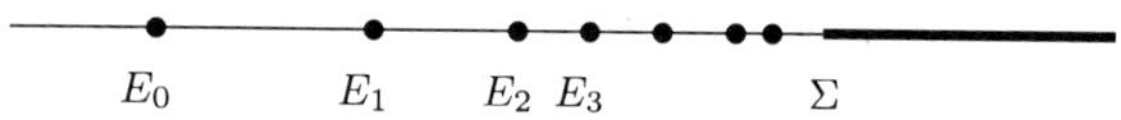

Fig. 1. The spectrum of $H_{\rm p}$

Fig. 2. The spectrum of $H_{\rm rad}$

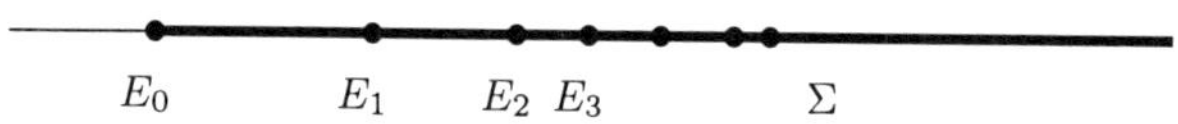

Fig. 3. The spectrum of $H_{\rm dec} = H_{\rm p} + H_{\rm rad}$

Since the spectrum of $H_{\rm rad}$ is $[0, \infty)$ and purely absolutely continuous except for zero (Fig. 2), all the point spectrum of $H_{\rm dec}$, E_j, $j = 0, 1, 2, \ldots$, are embedded in the continuum (Fig. 3).

Fock vacuum Ω is an eigenvector of $H_{\rm rad}$ associated with eigenvalue zero, i.e. $H_{\rm rad}\Omega = 0$. Let ϕ_j be an excited state of $H_{\rm p}$ with eigenvalue E_j. Then $\phi_j \otimes \Omega$ is an eigenvector of $H_{\rm dec}$ with eigenvalue E_j, i.e.

$$H_{\rm dec}(\phi_j \otimes \Omega) = E_j(\phi_j \otimes \Omega), \quad j = 0, 1, 2, \ldots$$

The perturbation of E_0 is quite different from that of E_j, $j \geq 1$. It is expected that E_j, $j \geq 1$, disappears after adding a perturbation and turns out to be a resonance pole in the lower half complex plane. Such resonances have been studied in e.g. Refs. 24–27, 29, 97–99 and 122. A coupling through a radiation field, however, enhances binding, by which E_0 may survive on the real line after adding a perturbation (see Fig. 4). This has been studied for Pauli–Fierz type models in e.g. Refs. 16–20, 24–27, 52, 53, 57, 61, 63, 72, 80, 84, 90, 140 and 142. Under some conditions, however, the absence of ground states has been also considered in e.g. Refs. 21, 39, 71 and 116.

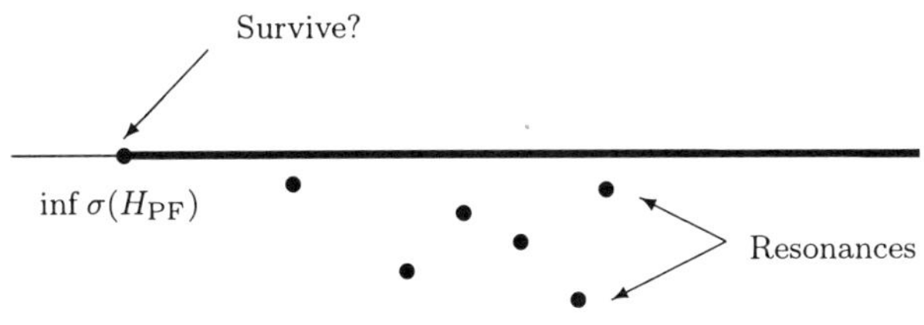

Fig. 4. The resonance poles of H_{PF}

As we have seen above, H_{PF} splits up into the sum of H_{dec} and H_{I}, and all the point spectrum of H_{dec} are embedded in the continuum. So analysis of H_{PF} may be reduced to perturbation problems of embedded eigenvalues. The problem under consideration is subdivided into two cases as in Fig. 5. In particular the main purpose of this paper is to review recent results of Problem I.

	Embedded eigenvalues	+ Perturbation
I	E_0	Survive ?
II	$E_j, \quad j \geq 1$	Resonances ?

Fig. 5. Problems I and II

1.2. *Positive temperatures*

Suppose now that the temperature is positive. We briefly sketch objects encountered in a positive temperature system. At positive temperatures, instead of H acting in $\mathcal{H}_{\mathrm{rad}}$, a self-adjoint operator acting in $\mathcal{H}_{\mathrm{rad}} \otimes \mathcal{H}_{\mathrm{rad}}$ is considered, which is called Liouvillean.

Let ρ_0 describe a normal state on the C^*-algebra of bounded operators on $L^2(\mathbb{R}^{3N})$, and ω_{KMS} a KMS state on the $*$-algebra generated by the annihilation and creation operators in $\mathfrak{F}_{\mathrm{rad}}$. The time-evolution α_t is defined by

$$\alpha_t(B) := e^{itH_{\mathrm{PF}}} B e^{-itH_{\mathrm{PF}}},$$

and the Liouvillean L_{PF} by

$$L_{\mathrm{PF}} B := i[H_{\mathrm{PF}}, B].$$

Then $\alpha_t(B)$ is expressed as

$$\alpha_t(B) = e^{itL_{\mathrm{PF}}} B.$$

By taking the Araki–Woods representation,[23] the Liouvillean can be realized as a self-adjoint operator acting in the Hilbert space

$$\mathcal{L} := \mathcal{H}_{\rm rad} \otimes \mathcal{H}_{\rm rad},$$

and can be decomposed as

$$L_{\rm PF} = L_{\rm dec} + L_{\rm I},$$

where the decoupled Liouvillean $L_{\rm dec}$ is given by

$$L_{\rm dec} := H_{\rm dec} \otimes 1 - 1 \otimes H_{\rm dec}.$$

All the point spectrum of $L_{\rm dec}$ are also embedded in the continuum (Fig. 6).

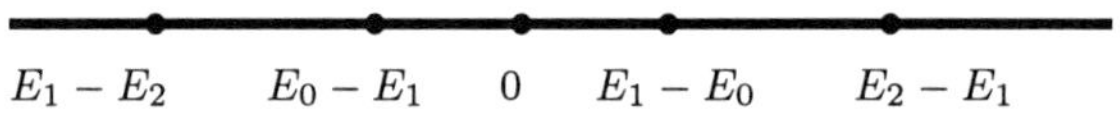

Fig. 6. The spectrum of $L_{\rm dec}$

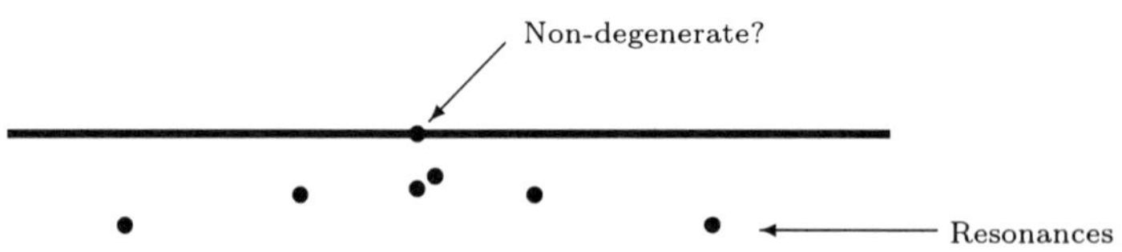

Fig. 7. The resonance poles of $L_{\rm PF}$

Then analysis of $L_{\rm PF}$ may be recognized as perturbations of embedded eigenvalues such as $H_{\rm PF}$. The perturbations of zero is quite different from those of $E_i - E_j$ $(i \neq j)$. That is to say, we expect that adding perturbation $L_{\rm I}$, embedded eigenvalues $E_i - E_j$ $(i \neq j)$, turns out to be resonance poles in the lower half complex plane, but non-degenerate eigenvalue zero survives on the real line. (See Figs. 7 and 8.) Positive temperature systems are studied, e.g., in Refs. 28, 100, 101 and 102.

	Embedded eigenvalues	+ Perturbation
III	0	Non-degenerate ?
IV	$E_i - E_j, \quad i \neq j$	Resonances ?

Fig. 8. Problems III and IV

1.3. *Organization*

This paper is organized as follows:

In Section 2 Fock space $\mathcal{F}_{\rm rad}$, creation operator $a^*(f)$, annihilation operator $a(f)$, and second quantization Γ are introduced. The radiation field quantized in the Coulomb gauge is defined, on which an ultraviolet cutoff is imposed. With these preparations we define two Hamiltonians, the Pauli–Fierz Hamiltonian including spin $H_{\rm PF}^\sigma$ and the spinless Pauli–Fierz Hamiltonian $H_{\rm PF}$. Moreover an infrared singularity caused by very low energy photons is explained.

In Section 3 we list the problems discussed in this paper.

In Section 4 we investigate the existence of a ground state. We assume that electron Hamiltonian $H_{\rm p}$ has a positive spectral gap, i.e. the distance between the bottom of the spectrum and the edge of the continuous spectrum is positive. Then we can prove the existence of a ground state of $H_{\rm PF}$ and $H_{\rm PF}^\sigma$ under no infrared cutoff assumptions in the weak coupling region. Next we study the case of a zero spectral gap but in the dipole approximation. In this case we prove that for sufficiently *large* $|e|$, $H_{\rm PF}$ also has a ground state. As a result of establishing the existence of a ground state, we may consider asymptotic fields. According to the asymptotic fields the location of the absolutely continuous spectrum of $H_{\rm PF}$ and $H_{\rm PF}^\sigma$ is specified. Furthermore we investigate a scaling limit of $H_{\rm PF}$ and derive effective Hamiltonians taking into account of radiative effects.

In Section 5 the Schrödinger representation is introduced, in which instead of $\mathcal{H}_{\rm rad}$ we work in an L^2-space with a Gaussian measure. Then the radiation field turns out to be a multiplication operator. Using this Schrödinger representation we construct a functional integral representation of

$$(F, e^{-t_1 H_{\rm PF}} T_1 e^{-t_2 H_{\rm PF}} T_2 e^{-t_3 H_{\rm PF}} \cdots T_{n-1} e^{-t_n H_{\rm PF}} G)_{\mathcal{H}_{\rm rad}}, \qquad (1.2)$$

where T_j, $j = 1, \ldots, n-1$, are suitable operators. Since the Pauli–Fierz Hamiltonian is defined through minimal coupling, we have to deal with Hilbert-space-valued stochastic integrals in the functional integral representation of (1.2).

Section 6 is devoted to prove self-adjointness of the Pauli–Fierz Hamiltonian for *arbitrary* values of the coupling constant. Using a functional integral representation of $(F, e^{-tH_{\mathrm{PF}}}G)_{\mathcal{H}_{\mathrm{rad}}}$ we construct a self-adjoint extension $\widehat{H}$ of H_{PF} for arbitrary values of the coupling constant. Finding an invariant domain of $e^{-t\widehat{H}}$ and establishing some relative bounds, we prove self-adjointness of H_{PF} and H^{σ}_{PF}.

Section 7 deals with the multiplicity of ground states. We show that H_{PF} has a unique ground state and H^{σ}_{PF} a twofold degenerate ground state. Moreover we discuss H_{PF} with a singular external potential, which gives a simple example of stable embedded eigenvalues.

In Section 8 we express the ground state energy with the help of a functional integral, and show that the charge density in the ground state of H_{PF} decays exponentially. Let Ψ_{g} be an arbitrary ground state of H_{PF}. Then we show that $\|\Psi_{\mathrm{g}}(x)\|_{\mathcal{F}_{\mathrm{rad}}}$ decays exponentially with respect to x. Moreover the effective mass of H_{PF} is considered.

In Section 9 an infinite volume Gibbs measure is introduced. We consider expectation values of suitable observables with respect to the ground state by a Gibbs measure μ on a path space, i.e. for a suitable observable $\mathcal{O}$ we want to find an associated function $f_{\mathcal{O}}$ such that

$$(\Psi_{\mathrm{g}}, \mathcal{O}\Psi_{\mathrm{g}})_{\mathcal{H}_{\mathrm{rad}}} = \int_{\mathrm{path}} f_{\mathcal{O}}(q)\mu(dq). \tag{1.3}$$

We show that μ is a limit of some sequence of probability measures.

In Section 10 we introduce the Nelson model and derive a superexponential decay of the boson density in the ground state of the Nelson model by using the Gibbs measure.

Finally the general references of this review are Refs. 60, 68, 104, 105, 127–130, 135 and 136.

2. The Pauli–Fierz Model

This section is devoted to defining the Pauli–Fierz Hamiltonian. Much of the displayed material is well known. However, some readers may be familiar with classical analysis but not the quantized version, or vice versa. So, as a first step, we derive a classical Hamiltonian with a form factor from the usual Lagrangian formalism.

2.1. *Classical Hamiltonian in the Coulomb gauge*

Here we recall a Lagrangian formalism of N electrons coupled to the electromagnetic field.[a] Let $B(x,t) \in \mathbb{R}^3$ denote the magnetic field and $E(x,t) \in \mathbb{R}^3$ the electric field. We consider the Maxwell equations with a form factor λ:

$$\dot{B} = -\nabla \times E, \tag{2.1}$$

$$\dot{E} = \nabla \times B - \sum_{j=1}^{N} e_j \lambda(\cdot - q_j)\dot{q}_j, \tag{2.2}$$

$$\nabla \cdot E = \sum_{j=1}^{N} e_j \lambda(\cdot - q_j), \tag{2.3}$$

$$\nabla \cdot B = 0. \tag{2.4}$$

Here

$$q_j = q_j(t) \in \mathbb{R}^3$$

denotes the position of the j-th electron, and e_j the charge of the j-th electron. We set

$$(j, \rho) := \left(\sum_{j=1}^{N} e_j \lambda(\cdot - q_j)\dot{q}_j \, , \quad \sum_{j=1}^{N} e_j \lambda(\cdot - q_j) \right).$$

Note that

$$\dot{\rho} + \nabla \cdot j = 0.$$

The equations of motion are given by

$$m_j \ddot{q}_j = e_j (E + \dot{q}_j \times B),$$

where m_j is the mass of the j-th electron. By (2.4) we introduce a vector potential $A = A(x,t)$ such that

$$B = \nabla \times A.$$

From (2.1) it follows that

$$\nabla \times (E + \dot{A}) = 0.$$

Furthermore a scalar potential $\phi = \phi(x,t)$ is introduced as

$$E = -\dot{A} - \nabla\phi.$$

[a]This is taken from Spohn.[143]

Then the Lagrangian is given by

$$L := \frac{1}{2}\sum_{j=1}^{N} m_j \dot{q}_j^2 + \int \mathfrak{L}_0(x)d^3x,$$

where

$$\mathfrak{L}_0 := \frac{1}{2}(E^2 - B^2) + j\cdot A - \rho\phi.$$

Take the Coulomb gauge

$$\nabla\cdot A = 0.$$

Then, Eqs. (2.1)–(2.4) turns out to be

$$\Box A = \nabla\dot{\phi} - j, \tag{2.5}$$

$$\Delta\phi = -\rho. \tag{2.6}$$

Then we see that

$$\begin{aligned}\int \frac{1}{2}(E^2 - B^2)d^3x &= \int \frac{1}{2}\left\{(\dot{A} + \nabla\phi)^2 - B^2\right\}d^3x \\ &= \int\left\{\frac{1}{2}\left(\dot{A}^2 - B^2\right) + \dot{A}\cdot\nabla\phi + \frac{1}{2}\nabla\phi\cdot\nabla\phi\right\}d^3x \\ &= \int\left\{\frac{1}{2}\left(\dot{A}^2 - B^2\right) + \frac{1}{2}\rho\phi\right\}d^3x.\end{aligned}$$

Here we used

$$\int \nabla\phi\cdot\nabla\phi d^3x = -\int \phi\Delta\phi d^3x = \int \rho\phi d^3x$$

and

$$\int \dot{A}\cdot\nabla\phi d^3x = -\int(\nabla\cdot\dot{A})\phi d^3x = 0.$$

Hence

$$\int \mathfrak{L}_0 d^3x = \int\left\{\frac{1}{2}\left(\dot{A}^2 - B^2\right) + j\cdot A - \frac{1}{2}\rho\phi\right\}d^3x.$$

By (2.6), $\frac{1}{2}\int\rho(x)\phi(x)d^3x$ is a smeared Coulomb potential,

$$\begin{aligned}\frac{1}{2}\int\rho(x)\phi(x)d^3x &= \frac{1}{2}\sum_{i,j=1}^{N}\int\frac{e_ie_j\lambda(q_i - y)\lambda(q_j - y')}{4\pi|y - y'|}dydy' \\ &=: V(q_1,\ldots,q_N).\end{aligned}$$

The canonical momenta, p_j and Π, are given by

$$p_j := \frac{\partial L}{\partial \dot{q}_j} = m_j \dot{q}_j + e_j \int A(x)\lambda(x - q_j) d^3x,$$

and

$$\Pi := \frac{\delta L}{\delta \dot{A}} = \dot{A}.$$

Hence the Hamiltonian is given by

$$\begin{aligned} H_{\text{classical}} &= \sum_{j=1}^{N} p_j \dot{q}_j + \int \Pi(x) \cdot \dot{A}(x) d^3x - L \\ &= \frac{1}{2}\sum_{j=1}^{N} m_j \dot{q}_j^2 + \frac{1}{2}\int \rho(x)\phi(x) d^3x + \frac{1}{2}\int (\dot{A}^2 + B^2) d^3x \\ &= \sum_{j=1}^{N} \frac{1}{2m_j}\left(p_j - e_j \int A(x)\lambda(x - q_j) d^3x\right)^2 + V \\ &\qquad + \frac{1}{2}\int \left\{\dot{A}^2 + (\nabla \times A)^2\right\} d^3x. \end{aligned}$$

We second-quantize $H_{\text{classical}}$ in the next section, which then defines the Pauli–Fierz Hamiltonian being the main object in this paper.

2.2. *Fock representation*

In what follows for a mathematical generalization we assume that electrons move in the d-dimensional space, and we work with the unit $\hbar = 1 = c$.

We start by introducing some basic facts of quantum field theory often used in this paper. We define the Boson Fock space over $L^2(\mathbb{R}^d)$ by

$$\mathcal{F} := \mathcal{F}(L^2(\mathbb{R}^d)) := \bigoplus_{n=0}^{\infty} [\bigotimes_s^n L^2(\mathbb{R}^d)],$$

where

$$\bigotimes_s^0 L^2(\mathbb{R}^d) := \mathbb{C},$$

and $\bigotimes_s^n L^2(\mathbb{R}^d)$ denotes the symmetric tensor product of $L^2(\mathbb{R}^d)$, i.e. $f \in \bigotimes_s^n L^2(\mathbb{R}^d)$ if and only if $f \in L^2(\mathbb{R}^{dn})$ and

$$f(k_1, \ldots, k_n) = f(k_{\sigma(1)}, \ldots, k_{\sigma(n)}),$$

for an arbitrary permutation σ of degree n. The creation operator $a^*(f)$ and the annihilation operator $a(f)$, smeared by $f \in L^2(\mathbb{R}^d)$, are defined by

$$\left(a^*(f)\Psi\right)^{(n)}(k_1,\ldots,k_n) := \frac{1}{\sqrt{n}}\sum_{j=1}^{n} f(k_j)\Psi^{(n-1)}(k_1,\ldots,\widehat{k_j},\ldots,k_n),$$

$$\left(a(f)\Psi\right)^{(n)}(k_1,\ldots,k_n) := \sqrt{n+1}\int_{\mathbb{R}^d} f(k)\Psi^{(n+1)}(k,k_1,\ldots,k_n)dk,$$

for $\Psi = \bigoplus_{n=0}^\infty \Psi^{(n)} \in \mathcal{F}$, where $\widehat{\ }$ denotes leaving out the corresponding term. $\Psi = \bigoplus_{n=0}^\infty \Psi^{(n)} \in \mathcal{F}$ is written as

$$\Psi = \{\Psi^{(0)}, \Psi^{(1)}, \ldots\}.$$

Let $a^\sharp$ stands for a^* or a. Formally we write

$$a^\sharp(f) = \int a^\sharp(k) f(k) dk.$$

Let

$$\Omega_{\rm b} := \{1,0,0,0,\ldots\} \in \mathcal{F}$$

be the Fock vacuum. It is well known that

$$\mathcal{F}_0 := \mathcal{L}\left\{a^*(f_1)\cdots a^*(f_n)\Omega, \Omega \,\middle|\, f_j \in L^2(\mathbb{R}^d), j=1,\ldots,n, n \in \mathbb{N}\right\}$$

is dense in $\mathcal{F}$, where $\mathcal{L}\{\cdots\}$ denotes the finite linear sums of vectors in $\{\cdots\}$. a^* and a leave $\mathcal{F}_0$ invariant and obey the canonical commutation relations on $\mathcal{F}_0$, i.e.

$$[a(f), a^*(g)] = (\overline{f}, g)_{L^2(\mathbb{R}^d)}, \tag{2.7}$$

$$[a^*(f), a^*(g)] = 0, \tag{2.8}$$

$$[a(f), a(g)] = 0, \tag{2.9}$$

where $(f,g)_K$ (resp. $\|f\|_K$) denotes the scalar product (resp. the norm) of Hilbert space K. We omit K in $(f,g)_K$ unless confusion may arise. Note that $(f,g)_K$ is linear in g and antilinear in f. Moreover $a^\sharp$ satisfies that

$$(a(f)\Psi, \Phi)_{\mathcal{F}} = (\Psi, a^*(\overline{f})\Phi)_{\mathcal{F}}$$

for $\Psi, \Phi \in \mathcal{F}_0$. We define

$$\mathcal{F}_{\rm rad} = \mathcal{F}_{\rm rad}(L^2(\mathbb{R}^d)) := \underbrace{\mathcal{F}\otimes\cdots\otimes\mathcal{F}}_{d-1}.$$

We identify $\mathcal{F}_{\rm rad}$ with

$$\mathcal{F}_{\rm rad} \cong \bigoplus_{n=0}^\infty \left[\bigotimes_s^n \left(\oplus^{d-1} L^2(\mathbb{R}^d)\right)\right] := \bigoplus_{n=0}^\infty \mathcal{F}_{\rm rad}^{(n)}.$$

The vacuum of $\mathcal{F}_{\rm rad}$ is defined by

$$\Omega := \{1, 0, 0, 0, \ldots\}.$$

Set

$$\mathcal{F}^0_{\rm rad} := \underbrace{\mathcal{F}_0 \widehat{\otimes} \cdots \widehat{\otimes} \mathcal{F}_0}_{d-1},$$

where $\widehat{\otimes}$ denotes an algebraic tensor product.

$$a^{\sharp r}(f) : \mathcal{F}_{\rm rad} \to \mathcal{F}_{\rm rad}$$

is defined by

$$a^{\sharp r}(f) := \underbrace{1 \otimes \cdots \otimes \overset{\text{the } r_{\rm th}}{a^{\sharp}(f)} \otimes \cdots \otimes 1}_{d-1}.$$

Formally we write

$$a^{\sharp r}(f) = \int a^{\sharp r}(k) f(k) dk.$$

From the canonical commutation relations (2.7)–(2.9) it follows that

$$\begin{aligned} [a^r(f), a^{*s}(g)] &= (\overline{f}, g)_{L^2(\mathbb{R}^d)} \delta_{rs}, \\ [a^{*r}(f), a^{*s}(g)] &= 0, \\ [a^r(f), a^s(g)] &= 0, \end{aligned}$$

on $\mathcal{F}^0_{\rm rad}$. Since $a^{\sharp r}$ is closable, we denote its closed extension by the same symbol $a^{\sharp r}$. Vectors

$$e^r(k) := (e^r_1(k), \ldots, e^r_d(k)), \quad r = 1, \ldots, d-1,$$

denote $d-1$ possible orthonormal polarization vectors perpendicular to k, i.e.

$$e^r(k) \cdot e^s(k) = \delta_{rs}, \quad e^r(k) \cdot k = 0, \quad \text{a.e. } k \in \mathbb{R}^d.$$

Note that[b]

$$d_{\mu\nu}(k) := e^r_\mu(k) e^r_\nu(k) = \delta_{\mu\nu} - \frac{k_\mu k_\nu}{|k|^2} = \left(1 - \frac{k}{|k|} \otimes \frac{k}{|k|}\right)_{\mu\nu}.$$

We define the quantized radiation field, $A_\mu(\hat{\lambda})$, by

$$A_\mu(\hat{\lambda}) := A_\mu(\hat{\lambda}, x) := \frac{1}{\sqrt{2}} \int e^r_\mu(k) \left\{ a^{*r}(k) e^{-ikx} \hat{\lambda}(-k) + a^r(k) e^{ikx} \hat{\lambda}(k) \right\} dk,$$

[b]The summation over repeated indices is automatically understood (the Einstein rule).

and $\Pi_\mu(\hat{\lambda})$ by

$$\Pi_\mu(\hat{\lambda}) := \Pi(\hat{\lambda}, x) := \frac{i}{\sqrt{2}} \int e^r_\mu(k) \left\{ a^{*r}(k) e^{-ikx} \hat{\lambda}(-k) - a^r(k) e^{ikx} \hat{\lambda}(k) \right\} dk,$$

where $\widehat{g}$ denotes the Fourier transform of g. Note that

$$\mathrm{div} A(\hat{\lambda}) = [p_\mu, A_\mu(\hat{\lambda})] = 0$$

on some dense domain, since $e^r(k) \cdot k = 0$. It is easily checked that

$$[A_\mu(\hat{\lambda}), \Pi_\nu(\hat{\rho})] = i \int_{\mathbb{R}^d} d_{\mu\nu}(k) \hat{\lambda}(k) \hat{\rho}(k) dk,$$
$$[A_\mu(\hat{\lambda}), A_\nu(\hat{\rho})] = [\Pi_\mu(\hat{\lambda}), \Pi_\nu(\hat{\rho})] = 0,$$

on $\mathfrak{F}^0_{\mathrm{rad}}$ and

$$(A_\mu(\hat{\lambda})\Omega, A_\nu(\hat{\rho})\Omega) = \frac{1}{2}(d_{\mu\nu}\hat{\lambda}, \hat{\rho}) = (\Pi_\mu(\hat{\lambda})\Omega, \Pi_\nu(\hat{\rho})\Omega).$$

Throughout this paper we assume that

$$\hat{\lambda}(-k) = \overline{\hat{\lambda}(k)}, \tag{2.10}$$

i.e. λ is a real-valued function. This assumption ensures that both $A_\mu(\hat{\lambda})$ and $\Pi_\nu(\hat{\lambda})$ are symmetric operators in $\mathfrak{F}_{\mathrm{rad}}$.

2.3. *Second quantization*

Let h be a self-adjoint operator acting in $L^2(\mathbb{R}^d)$. Define

$$\Gamma_{\mathrm{rad}}(e^{ith}) : \mathfrak{F}_{\mathrm{rad}} \to \mathfrak{F}_{\mathrm{rad}}, \quad t \in \mathbb{R},$$

by

$$\Gamma_{\mathrm{rad}}(e^{ith})\Omega := \Omega,$$
$$\Gamma_{\mathrm{rad}}(e^{ith}) a^{*r_1}(f_1) \cdots a^{*r_n}(f_n)\Omega := a^{*r_1}(e^{ith} f_1) \cdots a^{*r_n}(e^{ith} f_n)\Omega.$$

It is seen that $\Gamma_{\mathrm{rad}}(e^{ith})$ is a strongly continuous one-parameter unitary group on $\mathfrak{F}_{\mathrm{rad}}$. Thus there exists a self-adjoint operator $d\Gamma_{\mathrm{rad}}(h)$ in $\mathfrak{F}_{\mathrm{rad}}$ such that

$$\Gamma_{\mathrm{rad}}(e^{ith}) = e^{itd\Gamma_{\mathrm{rad}}(h)}, \quad t \in \mathbb{R}.$$

We call $d\Gamma_{\mathrm{rad}}(h)$ the second quantization of h. Actually $d\Gamma_{\mathrm{rad}}(h)$ acts as

$$d\Gamma_{\mathrm{rad}}(h)\Omega = 0,$$
$$d\Gamma_{\mathrm{rad}}(h) a^{*r_1}(f_1) \cdots a^{*r_n}(f_n)\Omega = \sum_{j=1}^n a^{*r_1}(f_1) \cdots a^{*r_j}(hf_j) \cdots a^{*r_n}(f_n)\Omega,$$

i.e.

$$d\Gamma_{\rm rad}(h) = \bigoplus_{n=1}^{\infty}\sum_{j=1}^{n} \underbrace{1 \otimes \cdots \otimes \overset{\text{the } j\text{th}}{h} \otimes \cdots \otimes 1}_{n}.$$

The dispersion relation is defined by

$$\omega(k) := |k|.$$

Define the free Hamiltonian in $\mathcal{F}_{\rm rad}$ by

$$H_{\rm rad} := d\Gamma_{\rm rad}(\omega).$$

Formally $H_{\rm rad}$ is written as

$$H_{\rm rad} = \int \omega(k) a^{*r}(k) a^r(k) dk.$$

Let $\sigma(T)$ denote the spectrum of T, and $\sigma_{\#}(T)$ ($\# =$ ess, disc, p, ac) the essential, discrete, point, absolutely continuous spectrum of T, respectively. It is known that

$$\begin{aligned}\sigma(H_{\rm rad}) &= [0,\infty),\\ \sigma_{\rm p}(H_{\rm rad}) &= \{0\},\\ \sigma_{\rm ess}(H_{\rm rad}) &= [0,\infty),\end{aligned}$$

and zero is a non-degenerate eigenvalue with

$$H_{\rm rad}\Omega = 0.$$

It is a direct calculation to show that

$$\begin{aligned} e^{itd\Gamma_{\rm rad}(h)} a^{*r}(f) e^{-itd\Gamma_{\rm rad}(h)} &= a^{*r}(e^{ith} f),\\ e^{itd\Gamma_{\rm rad}(h)} a^{r}(f) e^{-itd\Gamma_{\rm rad}(h)} &= a^{r}(e^{-ith} f).\end{aligned}$$

Define the number operator by

$$N_{\rm rad} := d\Gamma_{\rm rad}(1),$$

which satisfies that

$$e^{i(\pi/2)N_{\rm rad}} a^{*r}(f) e^{-i(\pi/2)N_{\rm rad}} = i a^{*r}(f), \tag{2.11}$$

$$e^{i(\pi/2)N_{\rm rad}} a^{r}(f) e^{-i(\pi/2)N_{\rm rad}} = -i a^{r}(f). \tag{2.12}$$

From (2.11) and (2.12) it follows that

$$e^{i(\pi/2)N_{\rm rad}} A_\mu(\hat{\lambda}) e^{-i(\pi/2)N_{\rm rad}} = \Pi_\mu(\hat{\lambda}). \tag{2.13}$$

2.4. *The Pauli–Fierz Hamiltonian*

Let

$$\mathcal{H}_{\rm rad} := L^2_{\rm as}(\mathbb{R}^{dN}) \otimes \mathcal{F}_{\rm rad}.$$

We identify $\mathcal{H}_{\rm rad}$ with the set of $\mathcal{F}_{\rm rad}$-valued $L^2_{\rm as}$-functions on $\mathbb{R}^{dN}$, i.e.

$$\mathcal{H}_{\rm rad} \cong \int_{\mathbb{R}^{dN}}^{\oplus} \mathcal{F}_{\rm rad} dx.$$

Define the quantized radiation field by the constant fiber direct integral[130] of $A_\mu(\hat{\lambda}, x^j)$, i.e.

$$A_{j,\mu} := \int_{\mathbb{R}^{dN}}^{\oplus} A_\mu(\hat{\lambda}, x^j) dx.$$

Formally for $\Psi \in \mathcal{H}_{\rm rad}$,

$$(A_{j,\mu}\Psi)(x^1, \ldots, x^N) = A_\mu(\hat{\lambda}, x^j)\Psi(x^1, \ldots, x^N).$$

$\hat{\lambda}$ serves as an ultraviolet cutoff function. Let $\sigma_1, \ldots, \sigma_d$ denote $2^{[d/2]} \times 2^{[d/2]}$ Hermitian matrices satisfying the anti-commutation relations:

$$\{\sigma_\mu, \sigma_\nu\} := \sigma_\mu\sigma_\nu + \sigma_\nu\sigma_\mu = 2\delta_{\mu\nu}1,$$

where $[z]$ denotes the integer part of z. The state space of N electrons interacting with a quantized radiation field is given by

$$\mathcal{H}_\sigma := \left(\otimes^N \mathbb{C}^{2^{[d/2]}}\right) \otimes \mathcal{H}_{\rm rad} \cong L^2_{\rm as}(\mathbb{R}^{dN}; \mathbb{C}^{2^{[d/2]}}) \otimes \mathcal{F}_{\rm rad}.$$

Then the Pauli–Fierz Hamiltonian is defined as a symmetric operator acting in $\mathcal{H}_\sigma$ by

$$\begin{aligned} H^\sigma_{\rm PF} &:= \sum_{j=1}^N \frac{1}{2m_j} \left\{\sigma^j \otimes (p_j \otimes 1 - e_j A_j)\right\}^2 + 1 \otimes V \otimes 1 + 1 \otimes 1 \otimes H_{\rm rad} \\ &= 1 \otimes H_{\rm PF} + H_{\rm spin}, \end{aligned}$$

where

$$\begin{aligned} p_j &:= (-i\frac{\partial}{\partial x^j_1}, \ldots, -i\frac{\partial}{\partial x^j_d}), \quad j = 1, \ldots, N, \\ \sigma^j &:= (\sigma^j_1, \ldots, \sigma^j_d), \quad j = 1, \ldots, N, \end{aligned}$$

with

$$\sigma^j_\mu := \underbrace{1 \otimes \cdots \otimes \overset{\text{the } j\text{th}}{\sigma_\mu} \otimes \cdots \otimes 1}_{N}.$$

The spin interaction is given by

$$H_{\mathrm{spin}} := -\sum_{j=1}^{N} \frac{e_j}{2m_j} \sum_{1\le\mu<\nu\le d} (-i)\sigma_\mu^j \sigma_\nu^j \otimes B_{\mu\nu}^j(\hat{\lambda}),$$

with

$$B_{\mu\nu}^j(\hat{\lambda}) := \int_{\mathbb{R}^{dN}}^{\oplus} B_{\mu\nu}(\hat{\lambda}, x^j)dx^j,$$
$$B_{\mu\nu}(\hat{\lambda}, x) := \partial_\mu A_\nu(\hat{\lambda}, x) - \partial_\nu A_\mu(\hat{\lambda}, x),$$

and the spinless Pauli–Fierz Hamiltonian H_{PF} is defined by

$$H_{\mathrm{PF}} := \sum_{j=1}^{N} \frac{1}{2m_j}(p_j \otimes 1 - e_j A_j)^2 + V \otimes 1 + 1 \otimes H_{\mathrm{rad}}.$$

The decoupled Hamiltonian H_{dec} is defined by H_{PF}^{σ} with $e = 0$, i.e.

$$H_{\mathrm{dec}} := H_{\mathrm{p}} \otimes 1 + 1 \otimes H_{\mathrm{rad}},$$

where

$$H_{\mathrm{p}} := \sum_{j=1}^{N} \frac{1}{2m_j} p_j^2 + V.$$

2.5. *Examples and infrared singularities*

We give a physically reasonable example of the Pauli–Fierz Hamiltonian. A typical example is $d = 3$, and

$$\sigma := (\sigma_1, \sigma_2, \sigma_3)$$

with the 2×2 Pauli matrices defined by

$$\sigma_1 := \begin{pmatrix} 0 & 1 \\ 1 & 0 \end{pmatrix}, \quad \sigma_2 := \begin{pmatrix} 0 & -i \\ i & 0 \end{pmatrix}, \quad \sigma_3 := \begin{pmatrix} 1 & 0 \\ 0 & -1 \end{pmatrix}.$$

Then it follows that

$$\sigma_1\sigma_2 = i\sigma_3, \quad \sigma_2\sigma_3 = i\sigma_1, \quad \sigma_3\sigma_1 = i\sigma_2,$$

and

$$H_{\mathrm{spin}} = -\sum_{j=1}^{N} \frac{e_j}{2m_j} \sigma_\mu^j \otimes B_\mu^j(\hat{\lambda}),$$

where

$$B_\mu^j(\hat{\lambda}) := \int_{\mathbb{R}^{3N}}^{\oplus} B_\mu(\hat{\lambda}, x^j) dx,$$
$$B_\mu(\hat{\lambda}, x) := (\nabla_x \times A(\hat{\lambda}, x))_\mu$$
$$= \frac{-i}{\sqrt{2}} \int (k \times e^r(k))_\mu \left\{ \hat{\lambda}(-k) e^{-ikx} a^{*r}(k) - \hat{\lambda}(k) e^{ikx} a^r(k) \right\} dk.$$

$\hat{\lambda}$ is given by

$$\hat{\lambda}(k) = \frac{\hat{\rho}(k)}{\sqrt{(2\pi)^3 \omega(k)}},$$

where ρ describes the charge density, i.e.

$$\int_{\mathbb{R}^3} \rho(x) dx = 1, \quad \rho(x) \geq 0. \tag{2.14}$$

A typical example of an external potential is

$$V_\rho(x^1, \ldots, x^N) = \sum_{1 \leq i < j \leq N} \int_{\mathbb{R}^3 \times \mathbb{R}^3} \frac{e_i e_j \rho(x^i - y) \rho(x^j - y')}{4\pi |y - y'|} dy dy'$$
$$- \sum_{j=1}^{N} \sum_{i=1}^{Z} \int_{\mathbb{R}^3} \frac{Z_i e_j \rho(x^j - y)}{4\pi |y - y^i|} dy,$$

or

$$V(x^1, \ldots, x^N) = \sum_{i \neq j}^{N} \frac{e_i e_j}{4\pi |x^i - x^j|} - \sum_{j=1}^{N} \sum_{i=1}^{Z} \frac{Z_i e_j}{4\pi |x^j - y^i|},$$

where $y^i \in \mathbb{R}^3$, $i = 1, \ldots, Z$, denotes the position of the ith static nucleus and Z_i their charge. The former potential V_ρ is a Coulomb potential smeared by charge density ρ. From (2.14) it follows that

$$\hat{\rho}(0) = \frac{1}{\sqrt{(2\pi)^3}} \neq 0.$$

Then $|\hat{\rho}(k)|^2/|k|^3$ has a singularity at $k = 0$ and

$$\int_{\mathbb{R}^3} \frac{\hat{\lambda}(k)^2}{\omega(k)^2} dk = \frac{1}{(2\pi)^3} \int_{\mathbb{R}^3} \frac{|\hat{\rho}(k)|^2}{|k|^3} dk = \infty. \tag{2.15}$$

This is called *the infrared divergence* or *the infrared singularity* . In contrast to (2.15) the following condition

$$\int_{\mathbb{R}^3} \frac{\hat{\lambda}(k)^2}{\omega(k)^2} dk < \infty \tag{2.16}$$

$\mathcal{H}_\sigma$	$L^2(\mathbb{R}^3;\mathbb{C}^2)\otimes\mathcal{F}_{\rm rad}$
$H_{\rm p}$	$-\frac{1}{2M}\Delta+V$
$H_{\rm rad}$	$\int \omega(k)a^{*r}(k)a^r(k)dk$
$H_{\rm dec}$	$H_{\rm p}+H_{\rm rad}$
CCR	$[a^r(k),a^{*s}(k')]=\delta(k-k')\delta_{rs}$
$A_\mu(\hat\lambda)$	$\frac{1}{\sqrt2}\int e^r_\mu(k)\left\{a^{*r}(k)\hat\lambda(-k)e^{-ikx}+a^r(k)\hat\lambda(k)e^{ikx}\right\}dk$
$\Pi_\mu(\hat\lambda)$	$\frac{i}{\sqrt2}\int e^r_\mu(k)\left\{a^{*r}(k)\hat\lambda(-k)e^{-ikx}-a^r(k)\hat\lambda(k)e^{ikx}\right\}dk$
CCR	$[A_\mu(\hat\lambda),\Pi_\nu(\hat\rho)]=i\int_{\mathbb{R}^3}d_{\mu\nu}k\hat\lambda(-k)\hat\rho(k)dk$
$e^r(k)$	$e^r(k)\cdot k=0,\quad e^r(k)\cdot e^s(k)=\delta_{rs}$
$B_\mu(\hat\lambda)$	$\frac{-i}{\sqrt2}\int (k\times e^r(k))_\mu\left\{a^{*r}(k)\hat\lambda(-k)e^{-ikx}-a^r(k)\hat\lambda(k)e^{ikx}\right\}dk$
$H^\sigma_{\rm PF}$	$\frac{1}{2M}\left(p-eA(\hat\lambda)\right)^2+V+H_{\rm rad}-\frac{e}{2M}\sigma B(\hat\lambda)$

Fig. 9. The Pauli–Fierz model for $N=1$ and $d=3$

is called *the infrared cutoff condition* or *the infrared regularity condition.* We have two typical choices satisfying (2.16).

(I) We replace $\omega(k)$ with $\omega_{\rm ph}(k)=\sqrt{|k|^2+m_{\rm ph}^2}$ with some $m_{\rm ph}>0$.
(II) We set $\hat\rho(0)=0$.

Note that assumption **(II)** actually breaks (2.14).

3. The List of Problems

In what follows we assume that $N = 1$ for simplicity. Set

$$m_1 = M, \quad e_1 = e, \quad p_1 = p,$$
$$A_1 = A = A(\hat{\lambda}), \quad B^1_{\mu\nu} = B_{\mu\nu} = B_{\mu\nu}(\hat{\lambda}), \quad B^1_\mu = B_\mu = B_\mu(\hat{\lambda}).$$

We set $p^2 := -\Delta$. We abbreviate $1 \otimes X$ and $X \otimes 1$ by simply X unless confusion may arise. Then

$$\mathcal{H}_\sigma = L^2(\mathbb{R}^d; \mathbb{C}^{2^{[d/2]}}) \otimes \mathcal{F}_{\mathrm{rad}},$$
$$\mathcal{H}_{\mathrm{rad}} = L^2(\mathbb{R}^d) \otimes \mathcal{F}_{\mathrm{rad}},$$

and Hamiltonians under consideration turn out to be

$$H^\sigma_{\mathrm{PF}} = \frac{1}{2M}\{\sigma(p - eA)\}^2 + V + H_{\mathrm{rad}},$$
$$= H_{\mathrm{PF}} - \frac{e}{2M}\sum_{\mu<\nu}(-i)\sigma_\mu\sigma_\nu B_{\mu\nu},$$
$$H_{\mathrm{PF}} = \frac{1}{2M}(p - eA)^2 + V + H_{\mathrm{rad}}.$$

In particular in the case of $d = 3$

$$H^\sigma_{\mathrm{PF}} = H_{\mathrm{PF}} - \frac{e}{2M}\sigma_\mu B_\mu.$$

Here we list the problems discussed in this paper.

I. Self-adjointness

A dynamics of a quantum system is defined by the unitary time-evolution generated by the Pauli–Fierz Hamiltonian. So, our first natural task is to establish the self-adjointness of the Pauli–Fierz Hamiltonian. The decoupled Hamiltonian is written as

$$H_{\mathrm{dec}} = H_{\mathrm{p}} + H_{\mathrm{rad}}.$$

We put

$$H^\sigma_{\mathrm{PF}} = H_{\mathrm{dec}} + eH_{\mathrm{I}}.$$

As we see in Section 4.2, H_{I} is relatively bounded with respect to H_{dec}. Thus, for $|e| \ll 1$, H^σ_{PF} is self-adjoint on $D(H_{\mathrm{dec}}) = D(H_{\mathrm{p}}) \cap D(H_{\mathrm{rad}})$ by the Kato-Rellich theorem. The problem is more subtle for large $|e|$. To our best knowledge a general way to define a self-adjoint extension of H^σ_{PF} for all values of coupling constants is to take the Friedrichs extension since H^σ_{PF} consists of a quadratic part and a potential V. It must be noted that no extra conditions on $\hat{\lambda}$ is needed to define the Friedrichs extension. This extension, however, does not give information of the domain, and there still remains the possibility that several distinct self-adjoint extensions exist. To show self-adjointness for arbitrary e we need a completely different technique from weak coupling cases. By applying a functional integral representation of matrix elements of $e^{-tH_{\rm PF}}$, we shall obtain self-adjointness of $H_{\rm PF}$ for arbitrary $e \in \mathbb{R}$.

II. The existence of a ground state

Let us define

$$E := \inf \sigma(H_{\rm PF}^{\sigma}).$$

Then ground states of $H_{\rm PF}^{\sigma}$ are defined by vectors, $\Psi_{\rm g}$, such as

$$H_{\rm PF}^{\sigma}\Psi_{\rm g} = E\Psi_{\rm g}.$$

Since the bottom of the spectrum of $H_{\rm dec}$ is embedded in the continuous spectrum, it is a priori not clear whether a ground state still exists after adding the perturbation $eH_{\rm I}$. We let

$$\sigma(H_{\rm p}) = \{E_j\}_{j=0}^{\infty} \cup [\Sigma, \infty), \quad E_0 \le E_1 \le \cdots < \Sigma \le \infty. \tag{3.1}$$

Then

$$\sigma(H_{\rm dec}) = [E_0, \infty), \quad \sigma_{\rm p}(H_{\rm dec}) = \{E_j\}_{j=0}^{\infty}.$$

Thus all the point spectrum of $H_{\rm dec}$ are embedded in the continuum. We will see that, under a certain condition, the ground state survives after adding $eH_{\rm I}$ (see Section 4.3). It is worthwhile to point out that no infrared cutoff condition are needed in the proof.

A more difficult, but physically reasonable, question is about enhanced binding. One may expect that a coupling with a radiation field enhances binding. We assume instead of (3.1) that

$$\sigma(H_{\rm p}) = [0, \infty), \quad \sigma_{\rm p}(H_{\rm p}) = \emptyset.$$

Then

$$\sigma(H_{\rm dec}) = [0, \infty), \quad \sigma_{\rm p}(H_{\rm dec}) = \emptyset.$$

In particular $H_{\rm dec}$ has no ground states. Our question is as follows. "Does there exist a ground state of $H_{\rm dec} + eH_{\rm I}$ for some $|e| > 0$?" We answer this question in the affirmative for a suitable external potential V. On a heuristic level one argues that a coupling with a radiation field amounts to renormalizing bare mass M to an effective mass

$$M_{\text{eff}}(e^2) = M + \delta M(e^2)$$

with $\delta M(e^2)$ increasing in e^2. Thus effectively instead of $H_{\text{p}} = -(1/2M)\Delta + V$ we should consider

$$-\frac{1}{2M_{\text{eff}}(e^2)}\Delta + V. \tag{3.2}$$

Hence a bound state could be produced for a sufficiently large $|e|$. See Fig. 10. Most likely (3.2) has no sharp mathematical meaning. However we will see an associated phenomenon of the Pauli–Fierz Hamiltonian in Section 4.4.

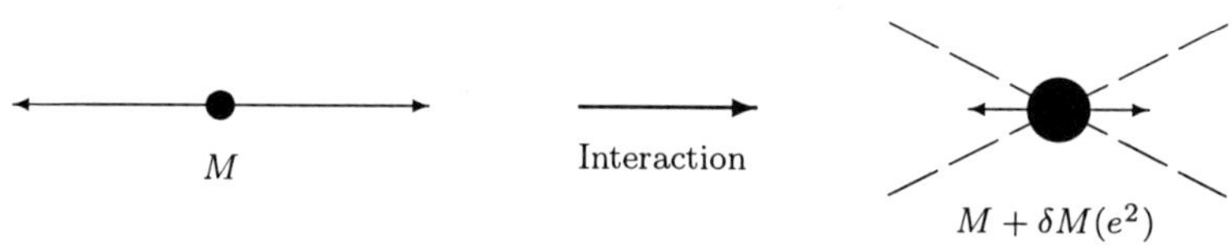

Fig. 10. Effective mass

III. Scaling limits

Formally it may be expected that the radiation field modifies the external potential V, or the mass M of an electron. Then effectively instead of $H_{\text{p}} = -(1/2M)\Delta + V$ we should consider

$$-\frac{1}{2M}\Delta + V_{\text{eff}}, \tag{3.3}$$

or

$$-\frac{1}{2M_{\text{eff}}}\Delta + V. \tag{3.4}$$

We want to derive (3.3) and (3.4) from scaling limits of H_{PF}, and to see explicit forms of V_{eff} and M_{eff}. The scaling under consideration is as follows

$$H_{\text{PF}}(\kappa) := H_{\text{p}} - \kappa^l \left(\frac{e}{M} p \cdot A\right) + \kappa^2 \left(\frac{e^2}{2M} A^2 + H_{\text{rad}}\right),$$

where κ is a scaling parameter going to infinity, and $l \le 2$. We shall prove that in the dipole approximation (3.3) can be derived from a scaling limit, $\kappa \to \infty$, of $H_{\text{PF}}(\kappa)$ subtracting a renormalization term for $l = 2$, and (3.4) for $l = 1$. In the case of $l = 1$ this scaling limit is a formal analogy of the so-called "weak coupling limit".[41]

IV. The multiplicity of ground states

Let K be a self-adjoint operator acting in a Hilbert space $\mathcal{K}$. Let P be the projection onto the subspace spanned by ground states of K. The multiplicity of ground states is identical with $\mathrm{Tr}P$. Since one method to prove the existence of a ground state of H_{PF} and H_{PF}^{σ} is *not* constructive, it does not enable us to estimate the multiplicity of ground states. Assume that the electron is spinless. To estimate the multiplicity of ground states of H_{PF} we extend the Perron–Frobenius argument.[62] The method is as follows: We can prove that $e^{-tH_{\mathrm{PF}}}$ is unitarily equivalent with some positivity improving semigroup, which is established by employing a functional integral representation of $(F, e^{-tH_{\mathrm{PF}}}G)$. One important corollary is that the ground state of H_{PF} is unique. Our proof does not depend on the existence of ground states of H_{p}. Therefore we can infer that the ground state appearing in enhanced binding through a coupling with a radiation field is also unique. Next we assume that an electron has spin and dimension $d = 3$. Suppose that H_{p} has a twofold degenerate ground state spanned by

$$\begin{pmatrix} f \\ 0 \end{pmatrix} \quad \text{and} \quad \begin{pmatrix} 0 \\ f \end{pmatrix}.$$

Then for $e = 0$, H_{PF}^{σ} also has a twofold degenerate ground state spanned by

$$\begin{pmatrix} f \\ 0 \end{pmatrix} \otimes \Omega \quad \text{and} \quad \begin{pmatrix} 0 \\ f \end{pmatrix} \otimes \Omega.$$

The problem is whether H_{PF}^{σ} for $e \neq 0$ still has a twofold degenerate ground state or not. In this case the functional integral method is *not* available, since we can not discuss the positivity argument of $e^{-tH_{\mathrm{PF}}^{\sigma}}$ by the spin interaction. Let P_e, $e \in \mathbb{R}$, be the projection onto the subspace spanned by ground states of H_{PF}^{σ} for coupling constant e. We prove that

$$P_0 P_e P_0 = a P_0$$

for some $a > 0$. This algebraic relation implies that $\mathrm{Tr}P_e \geq 2$. In addition to this we prove $\mathrm{Tr}P_e < 3$ for small $|e|$ by the pull-through formula. We conclude that the ground state of H_{PF}^{σ} is still twofold degenerate for sufficiently small $|e|$.

V. Ground state energies

The ground state energy

$$E = E(e^2) = \inf \sigma(H_{\rm PF})$$

is a function of e^2. By means of the following (i)–(iv):

(i) the existence of a ground state $\Psi_{\rm g}$,
(ii) the overlap, $(f \otimes \Omega, \Psi_{\rm g})_{\mathcal{H}_{\rm rad}} \neq 0$, for $f \geq 0$ but $f \neq 0$,
(iii) a functional integral representation of $(f \otimes \Omega, e^{-tH_{\rm PF}} f \otimes \Omega)_{\mathcal{H}_{\rm rad}}$,
(iv) the formula

$$E(e^2) = -\lim_{t \to \infty} \frac{1}{t} \log(f \otimes \Omega, e^{-tH_{\rm PF}} f \otimes \Omega)_{\mathcal{H}_{\rm rad}},$$

we obtain that $E(\cdot)$ is continuous, concave and monotone increasing. We point out that (ii) can be derived from the positivity improving argument mentioned in **IV**.

VI. Charge density in the ground state

As is well known in quantum mechanics, eigenvectors of Schrödinger operators exponentially decay. Our question is whether the charge density in the ground state, $\Psi_{\rm g}$, decays exponentially as well. By applying the functional integral we show that there exist positive constants D_1, D_2 and j such that

$$\|\Psi_{\rm g}(x)\|_{\mathcal{F}_{\rm rad}} \leq D_1 e^{-D_2|x|^j} \tag{3.5}$$

for almost every $x \in \mathbb{R}^d$. The exponential decay (3.5) depends on external potentials. We will specify a class of external potentials for which (3.5) holds.

VII. Gibbs measures

Let $\mathcal{O}$ be a suitable observable. In order to consider its expectation value with respect to a ground state $\Psi_{\rm g}$ of $H_{\rm PF}$, by constructing a measure μ on the continuous path space $\mathrm{C}(\mathbb{R};\mathbb{R}^d)$ we want to express it as

$$(\Psi_{\rm g}, \mathcal{O}\Psi_{\rm g})_{\mathcal{H}_{\rm rad}} = \int_{\mathrm{C}(\mathbb{R};\mathbb{R}^d)} f_{\mathcal{O}}(q) d\mu(q). \tag{3.6}$$

Here $f_{\mathcal{O}}$ is a function associated with $\mathcal{O}$. Here, μ is called an infinite volume Gibbs measure. Note

$$(\Psi_{\rm g}, \mathcal{O}\Psi_{\rm g})_{\mathcal{H}_{\rm rad}} = \lim_{t \to \infty} \frac{(e^{-tH_{\rm PF}} f \otimes \Omega, \mathcal{O} e^{-tH_{\rm PF}} f \otimes \Omega)_{\mathcal{H}_{\rm rad}}}{(e^{-tH_{\rm PF}} f \otimes \Omega, e^{-tH_{\rm PF}} f \otimes \Omega)_{\mathcal{H}_{\rm rad}}} \tag{3.7}$$

for some $f \geq 0$ but $f \not\equiv 0$. With a help of a functional integral we express the right-hand side above as

$$\lim_{t\to\infty} \int_{C(\mathbb{R};\mathbb{R}^d)} f_{\mathcal{O}}(q,t) d\mu_t(q)$$

with some measure μ_t and some function $f_{\mathcal{O}}(q,t)$. If a limiting argument is established, we can get (3.6). We want to apply (3.6) to the following problems:

(1) We want to see a relationship between an infrared singularity and the divergence of $(\Psi_{\rm g}, N_{\rm rad}\Psi_{\rm g})_{\mathcal{H}_{\rm rad}}$.

(2) We want to analyze both $f(k) := (\Psi_{\rm g}, a^{*r}(k)a^r(k)\Psi_{\rm g})_{\mathcal{H}_{\rm rad}}$ in the momentum representation and $g(x) := (\Psi_{\rm g}, a^{*r}(x)a^r(x)\Psi_{\rm g})_{\mathcal{H}_{\rm rad}}$ in the position representation. Rigorously $f(k)$ and $g(x)$ are defined as kernels of distributions on $C_0^\infty(\mathbb{R}^d)$, i.e. they satisfy that, for arbitrary $h \in C_0^\infty(\mathbb{R}^d)$,

$$(\Psi_{\rm g}, d\Gamma_{\rm rad}(h)\Psi_{\rm g}) = \int_{\mathbb{R}^3} h(k)f(k)dk,$$

$$(\Psi_{\rm g}, d\Gamma_{\rm rad}(h(-i\nabla))\Psi_{\rm g}) = \int_{\mathbb{R}^3} h(-i\nabla_x)g(x)dx.$$

According to a Gibbs measure we want to express both $(\Psi_{\rm g}, a^{*r}(k)a^r(k)\Psi_{\rm g})_{\mathcal{H}_{\rm rad}}$ and $(\Psi_{\rm g}, a^{*r}(x)a^r(x)\Psi_{\rm g})_{\mathcal{H}_{\rm rad}}$, and to study their properties.

(3) We expect, as for the charge density (3.5), that the probability of a density of bosons in $\Psi_{\rm g}$ decays exponentially as $n \to \infty$, i.e.

$$\|\Psi_{\rm g}\|^2_{L^2(\mathbb{R}^d)\otimes\mathcal{F}^{(n)}_{\rm rad}} \leq De^{-\beta n}$$

with some constants D and β. This problem is reduced to whether $(\Psi_{\rm g}, e^{+\beta N_{\rm rad}}\Psi_{\rm g})_{\mathcal{H}_{\rm rad}}$ is finite or not for some $\beta > 0$.

Unfortunately Eqs. (1)–(3) are not established for the Pauli–Fierz model, since a double stochastic integral encountered in (3.7) can not be controlled. In this paper we will prove them for the Nelson model.[121]

4. The Existence of Ground States

In what follows for simplicity we set the mass of an electron as

$$M = 1.$$

4.1. *Ultraviolet cutoffs and domains*

One of a delicate problem is to define the domain of $H_{\rm PF}^\sigma$. For weak couplings, however, we will easily prove the self-adjointness of $H_{\rm PF}^\sigma$ on $D(H_{\rm p})\cap D(H_{\rm rad})$.[c]

The following fundamental inequalities are well known:

$$\|a^{*r}(f)\Psi\| \le \|f\|\|\Psi\| + \|f/\sqrt{\omega}\|\|H_{\rm rad}^{1/2}\Psi\|, \tag{4.1}$$

$$\|a^{r}(f)\Psi\| \le \|f/\sqrt{\omega}\|\|H_{\rm rad}^{1/2}\Psi\|, \tag{4.2}$$

for $\Psi\in D(H_{\rm rad}^{1/2})$, and

$$\|a^{\sharp r}(f)a^{\sharp s}(g)\Psi\| \le \left(\|f/\sqrt{\omega}\|+\|f\|\right)\left(\|g\sqrt{\omega}\|+\|g\|+\|\sqrt{\omega}g\|+\|\omega g\|\right)\|(H_{\rm rad}+1)\Psi\|, \tag{4.3}$$

for $\Psi\in D(H_{\rm rad})$. Moreover it is well known that

$$\|a^{*r}(f)\Psi\| \le \|f\|(\|\Psi\| + \|N_{\rm rad}^{1/2}\Psi\|), \tag{4.4}$$

$$\|a^{r}(f)\Psi\| \le \|f\|\|N_{\rm rad}^{1/2}\Psi\|, \tag{4.5}$$

for $\Psi\in D(N_{\rm rad}^{1/2})$. The Pauli–Fierz Hamiltonian can be written as

$$H_{\rm PF}^\sigma = H_{\rm p} + H_{\rm rad} + eH_{\rm I},$$

where

$$H_{\rm p} := -\frac{1}{2}\Delta + V,$$

$$H_{\rm I} := -p\cdot A + \frac{e}{2}A^2 + H_{\rm spin}.$$

Suppose that $\hat\lambda/\sqrt{\omega}$, $\hat\lambda$, $\omega\hat\lambda\in L^2(\mathbb{R}^d)$ and that

$$\|\Delta f\| \le a\|H_{\rm p}f\| + b\|f\| \tag{4.6}$$

for some constants a and b. Then, using Eqs. (4.1)–(4.3), we easily have

$$\|p\cdot A\Psi\| \le C_1\|(H_{\rm p}+H_{\rm rad}+1)\Psi\|, \tag{4.7}$$

$$\|A^2\Psi\| \le C_2\|(H_{\rm rad}+1)\Psi\| \tag{4.8}$$

with some constants C_1 and C_2 for $\Psi\in D(H_{\rm p})\cap D(H_{\rm rad})$. On the other side, suppose $\hat\lambda$, $\omega\hat\lambda\in L^2(\mathbb{R}^d)$. Then

$$\|p\cdot A\Psi\| \le C_3\|(H_{\rm p}+N_{\rm rad}+1)\Psi\|, \tag{4.9}$$

$$\|A^2\Psi\| \le C_4\|(N_{\rm rad}+1)\Psi\| \tag{4.10}$$

[c] $D(T)$ denotes the domain of T.

with some constants C_3 and C_4 for $\Psi \in D(H_{\rm p}) \cap D(N_{\rm rad})$. Then we have the following proposition.

Proposition 4.1: *(1) Suppose $\hat{\lambda}/\sqrt{\omega}$, $\hat{\lambda}$, $\omega\hat{\lambda} \in L^2(\mathbb{R}^d)$. Then $H_{\rm PF}^\sigma$ is well defined on $D(H_{\rm p}) \cap D(H_{\rm f})$. (2) Suppose $\hat{\lambda}$, $\omega\hat{\lambda} \in L^2(\mathbb{R}^d)$. Then $H_{\rm PF}^\sigma$ is well defined on $D(H_{\rm p}) \cap D(N_{\rm rad}) \cap D(H_{\rm f})$.*

4.2. *Self-adjointness for weak couplings*

Proposition 4.2: *Suppose that $\hat{\lambda}/\sqrt{\omega}$, $\hat{\lambda}$, $\omega\hat{\lambda} \in L^2(\mathbb{R}^d)$ and $|e|$ is sufficiently small. Moreover, assume that (4.6) holds. Then $H_{\rm PF}^\sigma$ is self-adjoint on $D(H_{\rm p}) \cap D(H_{\rm rad})$, bounded from below, and essentially self-adjoint on any core of $H_{\rm p} + H_{\rm rad}$.*

Proof: By virtue of (4.7) and (4.8), we have

$$\|H_{\rm I}\Psi\| \leq C'\|(H_{\rm p} + H_{\rm rad})\Psi\| + C''\|\Psi\| \tag{4.11}$$

with some constants C' and C''. The proposition follows from the Kato-Rellich theorem and the fact that $D(H_{\rm p} + H_{\rm rad}) = D(H_{\rm p}) \cap D(H_{\rm rad})$. □

Remark 4.3: Suppose $\hat{\lambda}/\sqrt{\omega} \notin L^2(\mathbb{R}^d)$ or $|e| \gg 1$. Then it is crucial to prove the self-adjointness or essential self-adjointness of $H_{\rm PF}^\sigma$. Actually $\hat{\lambda}/\sqrt{\omega} \notin L^2(\mathbb{R}^d)$ breaks the relative boundedness (4.11), and the Kato–Rellich theorem is not available for $|e| \gg 1$. Moreover in the proof of Proposition 4.2, $H_{\rm rad}$ is regarded as a dominating part. So it is of mathematical interest to investigate the self-adjointness of

$$\frac{1}{2}(p - eA)^2 + V.$$

These problems are taken up in Section 6.3.

4.3. *Positive spectral gap*

In what follows we assume that $\hat{\lambda}/\sqrt{\omega}, \hat{\lambda}, \omega\hat{\lambda} \in L^2(\mathbb{R}^d)$ *unless otherwise stated.* In this section we will prove the existence of a ground state of $H_{\rm PF}$. An essential idea of a proof of the existence of a ground state of $H_{\rm PF}$ is due to Glimm and Jaffe,[59] Fröhlich,[52] and we learned it by Arai and Hirokawa.[18]

For $H_{\rm PF}^\sigma$ one can also prove the existence of a ground state in a similar manner. Let

$$E := \inf \sigma(H_{\rm PF}),$$

and

$$E_{\rm p} := \inf \sigma(H_{\rm p}).$$

Let us introduce classes of external potentials.

Definition 4.4: We say $V = Z + W \in V_{\rm exp}$ if and only if

(i) $\inf Z > -\infty$,
(ii) $W < 0$,
(iii) $Z \in L^1_{\rm loc}(\mathbb{R}^d)$,
(iv) $W \in L^p(\mathbb{R}^d)$ for some $p > \max\{1, d/2\}$.

Moreover we define $V(n)$, $n \geq 0$, as follows.

Definition 4.5: Suppose $V = Z + W \in V_{\rm exp}$.

(1) We say $V \in V(n)$, $n \geq 1$, if $Z(x) \geq \gamma |x|^{2n}$ outside a compact set with some positive constant γ.
(2) We say $V \in V(0)$ if $\liminf_{|x|\to\infty} Z(x) > E$.

Throughout this subsection we suppose Assumption V_n given by

Assumption 3: [Assumption V_n] We assume that

(1) $V \in V(n)$,
(2) $D(\Delta) \subset D(V)$ and $\|Vf\| \leq a\|\Delta f\| + b\|f\|$ for $f \in D(\Delta)$ with some positive constants $a < 1$ and b,
(3) there exists a positive spectral gap, i.e. $\Sigma_{\rm p} - E_{\rm p} > 0$, where

$$\Sigma_{\rm p} := \sigma_{\rm ess}(H_{\rm p}),$$

(4) $e^{-tH_{\rm p}} : L^2(\mathbb{R}^d) \longrightarrow L^\infty(\mathbb{R}^d)$ for $t > 0$ with

$$\|e^{-tH_{\rm p}} f\|_\infty \leq c_{\rm b} \|f\|_{L^2(\mathbb{R}^d)},$$

where $c_{\rm b}$ is a constant depending on t.

An example of (4) in Assumption V_n is as follows:

Example 4.6: We say $V \in K_d$ if and only if

[$d \geq 3$]:

$$\lim_{\epsilon \downarrow 0} \left(\sup_{x \in \mathbb{R}^d} \int_{|x-y|<\epsilon} |x-y|^{-(d-2)} |V(y)| dy \right) = 0,$$

$[d=2]$:

$$\lim_{\epsilon\downarrow 0}\left(\sup_{x\in\mathbb{R}^d}\int_{|x-y|<\epsilon}\log\left\{|x-y|^{-1}\right\}|V(y)|dy\right)=0,$$

$[d=1]$:

$$\sup_{x\in\mathbb{R}^d}\int_{|x-y|<1}|V(y)|dy<\infty.$$

We say $V\in K_d^{\rm loc}$ if and only if, for all $R>0$, $1_{|x|<R}V\in K_d$. Suppose that $V=V_+-V_-$ satisfies (i) $V_\pm\geq 0$, (ii) $V_+\in K_d^{\rm loc}$, and (iii) $V_-\in K_d$. Then $H_{\rm p}=-(1/2)\Delta+V$ satisfies (4) of Assumption V_n. It is known that the Coulomb potential with $d=3$ is contained in K_3. See Ref.137 for details.

Suppose $|e|\ll 1$. Then using (4.1)–(4.3), we have[d]

$$(1-|e|A-|e|B)\,H_{\rm dec}+|e|AE_{\rm p}-|e|C\leq H_{\rm PF}, \tag{4.12}$$

$$H_{\rm PF}\leq(1+|e|A+|e|B)\,H_{\rm dec}-|e|AE_{\rm p}+|e|C, \tag{4.13}$$

where A,B,C are some positive constants. Thus we get

$$|E-E_{\rm p}|\leq|e|D, \tag{4.14}$$

where $D:=BE_{\rm p}+C$. Let

$$\Gamma_a:=\left\{l=(l_1,\ldots,l_d)\in\mathbb{R}^d\,\middle|\,l_\mu=\frac{2\pi n_\mu}{a},n_\mu\in\mathbb{Z},|l_\mu|<L,\mu=1,\ldots,d\right\},$$

$$\Gamma(l,a):=[l_1,l_1+\frac{2\pi}{a})\times\cdots\times[l_d,l_d+\frac{2\pi}{a}).$$

By the map

$$\ell_2(\Gamma_a)\ni\{a_l\}_{l\in\Gamma_a}\longmapsto\left(\frac{a}{2\pi}\right)^{d/2}\sum_{l\in\Gamma_a}a_l1_{\Gamma(l,a)}(\cdot)\in L^2(\mathbb{R}^d),$$

we identify $\ell_2(\Gamma_a)$ with a subspace of $L^2(\mathbb{R}^d)$. Define

$$\mathfrak{F}^a_{\rm rad}:=\mathfrak{F}_{\rm rad}(\ell_2(\Gamma_a))\subset\mathfrak{F}_{\rm rad}.$$

Set

$$A^a_\mu(x):=\frac{1}{\sqrt{2}}\left\{a^{*r}\left(\left(\frac{a}{2\pi}\right)^{d/2}\sum_{l\in\Gamma_a}1_{\Gamma(l,a)}\hat{\lambda}(l)e^{-ilx}e^r_\mu(l)\right)\right.$$
$$\left.+a^r\left(\left(\frac{a}{2\pi}\right)^{d/2}\sum_{l\in\Gamma_a}1_{\Gamma(l,a)}\hat{\lambda}(l)e^{ilx}e^r_\mu(l)\right)\right\},$$

[d] Let A and B be self-adjoint operators in a Hilbert space K. We say that $A\leq B$ if $D(B)\subset D(A)$ and $(f,Af)\leq(f,Bf)$ for all $f\in D(B)$.

and

$$H_{\rm rad}^{m,a} := d\Gamma_{\rm rad}\left(m + \left(\frac{a}{2\pi}\right)^{d/2} \sum_{l\in\Gamma_a} |l| 1_{\Gamma(l,a)}\right).$$

Note that

$$\sigma(H_{\rm rad}^{m,a}\lceil_{\mathcal{F}_{\rm rad}^a}) = \sigma_{\rm p}(H_{\rm rad}^{m,a}\lceil_{\mathcal{F}_{\rm rad}^a}).$$

Thus a lattice Hamiltonian with an artificial mass m is defined by

$$H_{m,a} := \frac{1}{2}(p - eA^a)^2 + V + H_{\rm rad}^{m,a}.$$

Let

$$\mathcal{H}_a := L^2(\mathbb{R}^d) \otimes \mathcal{F}_{\rm rad}^a.$$

Lemma 4.7: *Let a be sufficiently large. Then $H_{m,a}$ is reduced by $\mathcal{H}_a$.*

Proof: Let $\mathcal{P}$ be the projection of $L^2(\mathbb{R}^d)$ onto $\ell_2(\Gamma_a)$ defined by

$$\mathcal{P}f = \left(\frac{a}{2\pi}\right)^{d/2} \sum_{l_a\in\Gamma_a} 1_{\Gamma(l,a)}(\cdot) f(l).$$

Define the projection P of $\mathcal{H}_{\rm rad}$ to $\mathcal{H}_a$ by

$$P := 1 \otimes \Gamma(\mathcal{P}).$$

We can easily see that

$$\begin{aligned} A_\mu^a P\Phi &= PA_\mu^a\Phi, \\ H_{\rm rad}^{m,a} P\Phi &= PH_{\rm rad}^{m,a}\Phi, \end{aligned}$$

for $\Phi = u \otimes a^{*r_1}(f_1)\cdots a^{*r_n}(f_n)\Omega$ with $u, f_1, \ldots, f_n \in C_0^\infty(\mathbb{R}^d)$. Hence

$$PH_{m,a} = H_{m,a}P$$

on a set $\mathcal{D}$ of finite linear sum of Φ's mentioned above. Since $\mathcal{D}$ is a core of $H_{m,a}$, the lemma follows. □

Let $E_{m,a} := \inf\sigma(H_{m,a})$.

Lemma 4.8: *Let a be sufficiently large. Then*

$$H_{m,a}\lceil_{\mathcal{H}_a^\perp} \geq m + E_{m,a}.$$

Proof: For instance we set $\ell_2 := \ell_2(\Gamma_a)$. Since $L^2(\mathbb{R}^d) = \ell_2 \oplus \ell_2^\perp$, it is seen that

$$\mathcal{F}_{\text{rad}} \cong \mathcal{F}_{\text{rad}}^a \otimes \mathcal{F}_{\text{rad}}(\ell_2^\perp). \tag{4.15}$$

Under the identification

$$\mathcal{H}_{\text{rad}} \cong \mathcal{H}_a \otimes \mathcal{F}_{\text{rad}}(\ell_2^\perp),$$

we have

$$H_{m,a} \cong H_{m,a}\lceil_{\mathcal{H}_a} \otimes 1 + 1 \otimes H_{\text{rad}}^{m,a}.$$

Let P_{V} be the vacuum projection of $\mathcal{F}_{\text{rad}}(\ell_2^\perp)$. Then the following identifications follow:

$$\begin{aligned}\mathcal{H}_a \otimes \mathcal{F}_{\text{rad}}(\ell_2^\perp) &= \left(\mathcal{H}_a \otimes P_{\text{V}}\mathcal{F}_{\text{rad}}(\ell_2^\perp)\right) \bigoplus \left(\mathcal{H}_a \otimes P_{\text{V}}^\perp \mathcal{F}_{\text{rad}}(\ell_2^\perp)\right)\\ &\cong (\mathcal{H}_a \otimes \mathbb{C}) \bigoplus \left(\mathcal{H}_a \otimes P_{\text{V}}^\perp \mathcal{F}_{\text{rad}}(\ell_2^\perp)\right)\\ &\cong \mathcal{H}_a \bigoplus \left(\mathcal{H}_a \otimes P_{\text{V}}^\perp \mathcal{F}_{\text{rad}}(\ell_2^\perp)\right).\end{aligned}$$

Under this identification we obtain that

$$\begin{aligned}H_{m,a} &\cong (H_{m,a}\lceil_{\mathcal{H}_a} \otimes P_{\text{V}}) \bigoplus (H_{m,a}\lceil_{\mathcal{H}_a} \otimes P_{\text{V}}^\perp)\\ &\qquad + (1 \otimes H_{\text{rad}}^{m,a} P_{\text{V}}) \bigoplus (1 \otimes H_{\text{rad}}^{m,a} P_{\text{V}}^\perp)\\ &\cong (H_{m,a}\lceil_{\mathcal{H}_a} \otimes P_{\text{V}}) \bigoplus (H_{m,a}\lceil_{\mathcal{H}_a} \otimes P_{\text{V}}^\perp + 1 \otimes H_{\text{rad}}^{m,a} P_{\text{V}}^\perp)\\ &\cong (H_{m,a}\lceil_{\mathcal{H}_a}) \bigoplus (H_{m,a}\lceil_{\mathcal{H}_a} \otimes 1 + 1 \otimes H_{\text{rad}}^{m,a})\, P_{\text{V}}^\perp.\end{aligned}$$

Hence

$$H_{m,a}\lceil_{\mathcal{H}_a^\perp} \cong (H_{m,a}\lceil_{\mathcal{H}_a} \otimes 1 + 1 \otimes H_{\text{rad}}^{m,a})\, P_{\text{V}}^\perp \geq E_{m,a} + m.$$

Thus we get the desired result. □

Lemma 4.9: *Suppose that e and m satisfy*

$$0 < m < (1 - |e|A - |e|B)(\Sigma_{\text{p}} - E_{\text{p}}) - 2|e|D.$$

Then, for sufficiently large a and L,

$$[E_{m,a}, E_{m,a} + m) \cap \sigma(H_{m,a}) \subset \sigma_{\text{disc}}(H_{m,a}).$$

Proof: Let

$$\begin{aligned}M &:= H_{m,a} - m - E_{m,a},\\ \mathcal{H}_- &:= E_{[-m,0]}^M \mathcal{H}_a,\\ \mathcal{H}_+ &:= E_{[0,\infty)}^M \mathcal{H}_a.\end{aligned}$$

Then

$$\mathcal{H} = \mathcal{H}_- \oplus \mathcal{H}_+ \oplus \mathcal{H}_a^\perp$$

and M is reduced by $\mathcal{H}_-$, $\mathcal{H}_+$ and $\mathcal{H}_a^\perp$. Moreover

$$M\lceil_{\mathcal{H}_+} \geq 0$$

is trivial and

$$M\lceil_{\mathcal{H}_a^\perp} \geq 0$$

follows from Lemma 4.8. Then it is enough to prove that $M\lceil_{\mathcal{H}_-}$ has a purely discrete spectrum in a neighborhood of $-m$. For sufficiently large a and L, Eqs. (4.12) and (4.13) hold with ω, e^r, $\hat{\lambda}$ replaced by

$$m + \left(\frac{a}{2\pi}\right)^{d/2} \sum_{l\in\Gamma_a} \omega(l) 1_{\Gamma(l,a)}(\cdot),$$

$$\left(\frac{a}{2\pi}\right)^{d/2} \sum_{l\in\Gamma_a} e^r(l) 1_{\Gamma(l,a)}(\cdot),$$

$$\left(\frac{a}{2\pi}\right)^{d/2} \sum_{l\in\Gamma_a} \hat{\lambda}(l) 1_{\Gamma(l,a)}(\cdot),$$

respectively. Moreover (4.14) is also valid with E replaced by $E_{m,a}$. Note that we can choose A, B, C, and D in (4.12)–(4.14) independent of m, a, and L. Suppose $E_{\rm p} < \Sigma_{\rm p}' < \Sigma_{\rm p}$. Let

$$l := 1 - |e|A - |e|B$$

and

$$\overline{H}_{\rm p} := H_{\rm p} - E_{\rm p}.$$

We have, by (4.12)

$$H_{m,a}\lceil_{\mathcal{H}_a} \geq lH_{\rm rad}^{m,a}\lceil_{\mathcal{H}_a} + l\overline{H}_{\rm p} + E_{\rm p} - |e|D.$$

Hence

$$\begin{aligned}
M\lceil_{\mathcal{H}_a} &\geq l\overline{H}_{\rm p} + \{lH_{\rm rad}^{m,a}\lceil_{\mathcal{H}_a} - (m + E_{m,a} - E_{\rm p} + |e|D)\} \\
&\geq l(\Sigma_{\rm p}' - E_{\rm p}) E^{\overline{H}_{\rm p}}_{[\Sigma_{\rm p}' - E_{\rm p}, \infty)} \otimes 1 \\
&\quad + (E^{\overline{H}_{\rm p}}_{[0,\Sigma_{\rm p}' - E_{\rm p})} + E^{\overline{H}_{\rm p}}_{[\Sigma_{\rm p}' - E_{\rm p}, \infty)}) \otimes \{lH_{\rm rad}^{m,a}\lceil_{\mathcal{H}_a} - (m + E_{m,a} - E_{\rm p} + |e|D)\} \\
&= M_1 + M_2 + M_3,
\end{aligned}$$

where

$$M_1 := E^{\overline{H}_{\rm p}}_{[0,\Sigma'_{\rm p}-E_{\rm p})} \otimes \{lH^{m,a}_{\rm rad}\lceil_{\mathcal{H}_a} - (m + E_{m,a} - E_{\rm p} + |e|D)\},$$

$$M_2 := E^{\overline{H}_{\rm p}}_{[\Sigma'_{\rm p}-E_{\rm p},\infty)} \otimes lH^{m,a}_{\rm rad}\lceil_{\mathcal{H}_a},$$

$$M_3 := \{l(\Sigma'_{\rm p} - E_{\rm p}) + E_{\rm p} - |e|D - m - E_{m,a}\}E^{\overline{H}_{\rm p}}_{[\Sigma'_{\rm p}-E_{\rm p},\infty)} \otimes 1.$$

Since, by (4.14),

$$E_{\rm p} - E_{m,a} \geq -|e|D,$$

one can obtain

$$l(\Sigma'_{\rm p} - E_{\rm p}) - m - E_{m,a} + E_{\rm p} - |e|D \geq l(\Sigma'_{\rm p} - E_{\rm p}) - m - 2|e|D > 0.$$

Then $M_3 \geq 0$. It is obvious that $M_2 \geq 0$. Hence we obtain that

$$M\lceil_{\mathcal{H}_a} \geq M_1 \geq E^{\overline{H}_{\rm p}}_{[0,\Sigma'_{\rm p}-E_{\rm p})} \otimes \{lH^{m,a}_{\rm rad}\lceil_{\mathcal{H}_a} - (m + 2|e|D)\}. \tag{4.16}$$

Set

$$K := E^{\overline{H}_{\rm p}}_{[0,\Sigma'_{\rm p}-E_{\rm p})} \otimes \{lH^{m,a}_{\rm rad}\lceil_{\mathcal{H}_a} - (m + 2|e|D)\}.$$

We denote by E^O_A the spectral projection of self-adjoint operator O to a Borel set $A \subset \mathbb{R}$, and define the positive and negative part of O by

$$O_- = E^O_{(-\infty,0)} O E^O_{(-\infty,0)}, \quad O_+ = E^O_{[0,\infty)} O E^O_{[0,\infty)}.$$

Let $\{\phi_n\}_{n=1}^N$ and $\{\psi_m\}_{m=1}^\infty$ be complete orthonormal systems of $\mathcal{H}_-$ and $\mathcal{H}_+$, respectively, where $N \leq \infty$. Then, by (4.16) we have

$$\begin{aligned} 0 \geq {\rm Tr} M\lceil_{\mathcal{H}_-} &= \sum_{n=1}^N (\phi_n, M\phi_n) \geq \sum_{n=1}^N (\phi_n, K\phi_n) \\ &= \sum_{n=1}^N (\phi_n, K_+\phi_n) + \sum_{n=1}^N (\phi_n, K_-\phi_n) \geq \sum_{n=1}^N (\phi_n, K_-\phi_n) \\ &\geq \sum_{n=1}^N (\phi_n, K_-\phi_n) + \sum_{m=1}^\infty (\psi_m, K_-\psi_m) = {\rm Tr} K_-. \end{aligned}$$

Note

$$K_- = E^{\overline{H}_{\rm p}}_{[0,\Sigma'_{\rm p}-E_{\rm p})} \otimes \{lH^{m,a}_{\rm rad}\lceil_{\mathcal{H}_a} - (m + 2|e|D)\}_-$$

and

$${\rm Tr} K_- = {\rm Tr} E^{\overline{H}_{\rm p}}_{[0,\Sigma'_{\rm p}-E_{\rm p})} \times {\rm Tr}\, \{lH^{m,a}_{\rm rad}\lceil_{\mathcal{H}_a} - (m + 2|e|D)\}_- .$$

Then

$$\left|\mathrm{Tr} M\lceil_{\mathcal{H}_-}\right| \le \mathrm{Tr} E^{\overline{H}_\mathrm{p}}_{[0,\Sigma'_\mathrm{p}-E_\mathrm{p})} \times \left|\mathrm{Tr}\left\{lH^{m,a}_{\mathrm{rad}}\lceil_{\mathcal{H}_a} - (m+2|e|D)\right\}_-\right| < \infty,$$

which implies that $M\lceil_{\mathcal{H}_-}$ has a purely discrete spectrum in $[-m,-\epsilon]$ with some $\epsilon > 0$ since 0 may not be discrete. Thus the lemma follows. □

We define H^m_{PF} by H_{PF} with ω replaced by

$$\omega_m := \omega + m$$

and

$$E_m := \inf \sigma(H^m_{\mathrm{PF}}).$$

Lemma 4.10: *Suppose that* e *and* m *are as in Lemma* 4.9. *Then* H^m_{PF} *has a ground state.*

Proof: Define

$$U_a := \exp\left(ix_\mu \otimes d\Gamma_{\mathrm{rad}}\left(\left(\frac{a}{2\pi}\right)^{d/2} \sum_{l\in\Gamma_a} l_\mu 1_{\Gamma(l,a)}\right)\right)$$

and

$$U := \exp\left(ix_\mu \otimes d\Gamma_{\mathrm{rad}}(k_\mu)\right).$$

Then we have

$$U_a H_{m,a} U_a^{-1} = \frac{1}{2}\left(p - d\Gamma_{\mathrm{rad}}\left(\left(\frac{a}{2\pi}\right)^{d/2} \sum_{l\in\Gamma_a} l 1_{\Gamma(l,a)}\right) - eA^a(0)\right)^2 + V + H^{m,a}_{\mathrm{rad}}.$$

Here $A^a_\mu(0)$ is defined by $A^a_\mu(x)$ with x replaced by zero. It is a direct calculation to show that

$$\lim_{L\to\infty}\lim_{a\to\infty} U_a(H_{m,a}-i)^{-1}U_a^{-1} = U(H^m_{\mathrm{PF}}-i)^{-1}U^{-1} \tag{4.17}$$

in the operator norm. Note that it is *not* clear that

$$\lim_{L\to\infty}\lim_{a\to\infty}(H_{m,a}-i)^{-1} \to (H^m_{\mathrm{PF}}-i)^{-1}$$

in the operator norm. By Lemma 4.9 and (4.17), we get

$$[E_m, E_m+m) \cap \sigma(H^m_{\mathrm{PF}}) \subset \sigma_{\mathrm{disc}}(H^m_{\mathrm{PF}}).$$

Thus the lemma follows. □

Let $\Psi^{(m)}_\mathrm{g}$ be a ground state of H^m_{PF}.

Lemma 4.11: *For arbitrary $k \geq 0$, we have $\Psi_{\mathrm{g}}^{(m)} \in D(|x|^k)$, and there exists a constant c'_k independent of m such that*

$$\||x|^k \Psi_{\mathrm{g}}^{(m)}\| \leq c'_k \|\Psi_{\mathrm{g}}^{(m)}\|.$$

Proof: See Remark 8.6. □

Lemma 4.12: *There exists a constant C independent of m such that*

$$\|N_{\mathrm{rad}}^{1/2} \Psi_{\mathrm{g}}^{(m)}\|^2 \leq |e| C \|(|x|^2 + 1)\Psi_{\mathrm{g}}^{(m)}\|^2. \tag{4.18}$$

Proof: This proof is due to Bach, Fröhlich and Sigal.[27] In this proof, C_j, $j = 0, 1, 2, \ldots$, denote sufficiently large constants. Since H_{PF}^m is self-adjoint on $D(\Delta) \cap D(H_{\mathrm{rad}}^m)$, we see that

$$\Psi_{\mathrm{g}}^{(m)} \in D(\Delta) \cap D(H_{\mathrm{rad}}^m).$$

Moreover, by the closed graph theorem one can obtain that

$$\|\Delta\Psi\| + \|H_{\mathrm{rad}}^m \Psi\| \leq C_0(\|H_{\mathrm{PF}}^m \Psi\| + \|\Psi\|). \tag{4.19}$$

The key inequalities used in this proof are as follows:

$$\|a^{\sharp r}(f)\Phi\| \leq C_1(\|f/\sqrt{\omega_m}\| + \|f\|)\|((H_{\mathrm{PF}}^m)^{1/2} + 1)\Phi\|, \tag{4.20}$$

$$\|p_\mu \Phi\| \leq C_2 \|((H_{\mathrm{PF}}^m)^{1/2} + 1)\Phi\|, \tag{4.21}$$

$$\begin{aligned}\|a^{\sharp r}(f) a^{\sharp s}(g)\Phi\| &\leq C_3(\|f/\sqrt{\omega_m}\| + \|f\|) \\ &\quad \times \big(\|g/\sqrt{\omega_m}\| + \|g\| + \|\sqrt{\omega_m} g\| + \|\omega_m g\|\big)\|(H_{\mathrm{PF}}^m + 1)\Phi\|. \end{aligned} \tag{4.22}$$

Then, we also have

$$\begin{aligned}&\|a^{\sharp r}(g) p_\mu \Phi\| \\ &\quad \leq C_4\big(\|g/\sqrt{\omega_m}\| + \|g\| + \|\sqrt{\omega_m} g\| + \|\omega_m g\|\big)\|(H_{\mathrm{PF}}^m + 1)\Phi\|. \end{aligned} \tag{4.23}$$

Inequalities (4.20)–(4.23) are derived from (4.19) and (4.1)–(4.3). For notational simplicity, we set

$$\delta(g) := \|g/\sqrt{\omega_m}\| + \|g\| + \|\sqrt{\omega_m} g\| + \|\omega_m g\|.$$

Then

$$\|a^{\sharp r}(f)\Phi\| \leq C_5 \delta(f) \|((H_{\mathrm{PF}}^m)^{1/2} + 1)\Phi\|, \tag{4.24}$$

$$\|(p_\mu - eA_\mu)\Phi\| \leq C_6 \|(H_{\mathrm{PF}}^m)^{1/2}\Phi\|, \tag{4.25}$$

$$\|a^{\sharp r}(f)(p_\mu - eA_\mu)\Phi\| \leq C_7 \delta(f)\delta(\hat{\lambda})\|(H_{\mathrm{PF}}^m + 1)\Phi\|. \tag{4.26}$$

Suppose that $\hat{\lambda}\sqrt{\omega^n} \in L^2(\mathbb{R}^d)$, $n = -1, 0, \ldots, 4$. Then from (4.24)-(4.26) it follows that

$$\|(H^m_{\rm rad})^2\Psi\| \le C_8\|((H^m_{\rm PF})^2 + 1)\Psi\|.$$

In particular

$$D((H^m_{\rm PF})^2) \subset D((H^m_{\rm rad})^2) \quad \text{and} \quad \Psi^m_{\rm g} \in D((H^m_{\rm rad})^2).$$

Moreover $[H^m_{\rm PF}, a^r(f)]\Psi^m_{\rm g}$ is well defined. To avoid troublesome domain arguments we truncate $\hat{\lambda}$ as

$$\hat{\lambda}_l(k) := \begin{cases} \hat{\lambda}(k) & 1/l < |k| < l, \\ 0 & \text{otherwise.} \end{cases}$$

Define $H(l)$ by $H^m_{\rm PF}$ with $\hat{\lambda}$ replaced by $\hat{\lambda}_l$. Let

$$g_\mu(k, x) := \frac{e}{\sqrt{2}} e^r_\mu(k) e^{-ikx} \hat{\lambda}_l(k).$$

Fix $f \in C_0^\infty(\mathbb{R}^d \setminus \{0\})$ and set

$$G_\mu := (\bar{f}, g_\mu).$$

We have

$$[H(l), a^r(f)] = -G_\mu(p_\mu - eA_\mu) - a^r(\omega_m f). \tag{4.27}$$

To remove the extra term $G_\mu(p_\mu - eA_\mu)$ in (4.27), we introduce

$$\theta := i x_\mu G_\mu.$$

We have

$$[H(l), \theta] = G_\mu(p_\mu - eA_\mu) + (-i)\vartheta, \tag{4.28}$$

where

$$\vartheta := (x_\nu \partial_\mu G_\nu)(\partial_\mu - ieA_\mu) + (1/2) x_\nu(\partial_\mu \partial_\mu G_\nu).$$

Note that $\partial_\mu \partial_\mu G_\nu = (\bar{f}, (-ik_\mu)^2 g_\nu)$ is well defined, since we truncate $\hat{\lambda}$.

Combining (4.28) with (4.27), we have[e]

$$[H(l), a^r(f) + \theta] = -a^r(\omega_m f) + (-i)\vartheta.$$

Simply we set $\Psi_{\rm g}^m = \Psi$, and $\inf \sigma(H(l)) = E_l$. It is derived that

$$\begin{aligned}
0 &\le ((a^r(f) + \theta)\Psi, (H(l) - E_l)(a^r(f) + \theta)\Psi) \\
&= ((a^r(f) + \theta)\Psi, [H(l), a^r(f) + \theta]\Psi) \\
&\quad + ((a^{*r}(\bar{f}) + \bar{\theta})(a^r(f) + \theta)\Psi, (H(l) - E_l)\Psi) \\
&:= \mathrm{I} + \mathrm{II} + \mathrm{III} + \mathrm{IV} - (a^{*r}(\omega_m f)a^r(f)\Psi, \Psi).
\end{aligned}$$

Here

$$\begin{aligned}
\mathrm{I} &:= (a^r(f)\Psi, (-i)\vartheta\Psi), \\
\mathrm{II} &:= (\theta\Psi, (-i)\vartheta\Psi), \\
\mathrm{III} &:= (\theta\Psi, -a^r(\omega_m f)\Psi), \\
\mathrm{IV} &:= ((a^{*r}(\bar{f}) + \bar{\theta})(a^r(f) + \theta)\Psi, (H(l) - E_l)\Psi).
\end{aligned}$$

By substituting $f_l/\sqrt{\omega_m}$ for f in I–IV with $\{f_l\}_{l=1}^\infty$ a complete orthonormal system in $L^2(\mathbb{R}^d)$ such that $f_l \in C_0^\infty(\mathbb{R}^d \setminus \{0\})$, and summing up from $l = 1$ to ∞, we see that

$$\begin{aligned}
&(a^{*r}(\omega_m f)a^r(f)\Psi, \Psi) \to (\Psi, N_{\rm b}\Psi), \\
&\mathrm{I} \to -(i/2)(a^r(k_\mu^2 \overline{g_\nu}/\omega_m)\Psi, x_\nu \Psi) \\
&\qquad - (a^r(k_\mu \overline{g_\nu}/\omega_m)\Psi, x_\nu(\partial_\mu - ieA_\mu)\Psi) := a, \\
&\mathrm{II} \to (1/2)(\Psi, x_{\mu'} x_\nu \Psi)(g_{\mu'}, -k_\mu^2 g_\nu/\omega_m) \\
&\qquad + (\Psi, x_{\mu'} x_\nu(\partial_\mu - ieA_\mu)\Psi)(g_{\mu'}, (-ik_\mu)g_\nu/\omega_m) := b, \\
&\mathrm{III} \to i(x_\mu \Psi, a^r(g_\mu)\Psi) := c, \\
&\mathrm{IV} \to (d\Gamma_{\rm rad}(1/\omega)\Psi, (H(l) - E_l)\Psi) \\
&\qquad + (-i)(x_\nu a^{*r}(\overline{g_\mu}/\omega_m)\Psi, (H(l) - E_l)\Psi) \\
&\qquad + i(x_\nu a^r(\overline{g_\mu}/\omega_m)\Psi, (H(l) - E_l)\Psi) \\
&\qquad + (g_\mu, \overline{g_\nu})(x_\mu x_\nu \Psi, (H(l) - E_l)\Psi) := d.
\end{aligned}$$

---

[e]Formally we have

$$\begin{aligned}
&H(l)(a^r(k) + ix_\mu g_\mu(k,x))\Psi_{\rm g}^m \\
&= E_l(a^r(k) + ix_\mu g_\mu(k,x))\Psi_{\rm g}^m - \omega_m(k)a^r(k)\Psi_{\rm g}^m - i\vartheta(k)\Psi_{\rm g}^m \\
&= (E_l - \omega_m(k))(a^r(k) + ix_\mu g_\mu(k,x))\Psi_{\rm g}^m + (i\omega_m(k)x_\mu g_\mu(k,x) - i\vartheta(k))\Psi_{\rm g}^m,
\end{aligned}$$

where $\vartheta(k)$ is a formal kernel of ϑ. Hence

$$(a^r(k) + ix_\mu g_\mu(k,x))\Psi_{\rm g}^m = (H(l) - E_l + \omega_m(k))^{-1}\left(i\omega_m(k)x_\mu g_\mu(k,x) - i\vartheta(k)\right)\Psi_{\rm g}^m.$$

This is called the pull-through formula.

Then

$$(\Psi, N_{\rm rad}\Psi) \le a+b+c+d.$$

Note that

$$\|g_\mu\| \le |e|\|\hat{\lambda}\|, \quad \|k_\mu g_\mu/\omega_m\| \le |e|\|\hat{\lambda}\|, \quad \|k_\mu^2 g_\mu/\omega_m\| \le |e|\|\omega\hat{\lambda}\|.$$

Using (4.24)–(4.26), we immediately see that

$$\begin{aligned}
|a| &\le |e|C_9(\delta(\hat{\lambda})+\delta(\hat{\lambda})^2)\||x|\Psi\|\|(H(l)+1)\Psi\|,\\
|b| &\le |e|C_{10}\||x|^2\Psi\|^2\|\hat{\lambda}\|^2 + C_9\delta(\hat{\lambda})\||x|^2\Psi\|\|(H(l)+1)\Psi\|,\\
|c| &\le |e|C_{11}\delta(\hat{\lambda})\||x|\Psi\|\|(H(l)+1)\Psi\|.
\end{aligned}$$

Moreover, noting that $\|d\Gamma_{\rm rad}(1/\omega_m)\Phi\| \le (1/m)\|N_{\rm rad}\Phi\|$, we obtain that

$$\begin{aligned}
|d| &\le |e|\left(\frac{1}{m}\|N_{\rm rad}\Psi\| + \frac{\|\hat{\lambda}\|}{m}\||x|(N_{\rm rad}+1)\Psi\| + \||x|^2\Psi\|\right)\\
&\quad \times\|(H(l)-E_l)\Psi\|\\
&:= |e|R_n\|(H(l)-E_l)\Psi\|.
\end{aligned}$$

Hence we get

$$\begin{aligned}
\|N_{\rm rad}^{1/2}\Psi\|^2 &\le |e|C_{12}\delta(\hat{\lambda})^2\left(\|(|x|^2+1)\Psi\|\|(H(l)+1)\Psi\| + \|(|x|^2+1)\Psi\|^2\right)\\
&\quad +|e|R_n\|(H(l)-E_l)\Psi\|. \qquad (4.29)
\end{aligned}$$

It is easily seen that as $l\to\infty$

$$H(l)\Psi \to H_{\rm PF}^m\Psi, \quad E_l \to E.$$

Taking $l\to\infty$ on the both sides of (4.29), we get (4.18). □

Lemma 4.13: *Suppose $|e| \ll 1$. Let*

$$Q := E_{[\epsilon,\infty)}^{\overline{H}_{\rm p}} \otimes E_{\{0\}}^{H_{\rm rad}}$$

with $\epsilon > 0$ such that $\epsilon + E_{\rm p} < \Sigma_{\rm p}$. Then there exists a constant c' independent of m such that

$$\|Q\Psi_{\rm g}^{(m)}\| \le |e|c'\|\Psi_{\rm g}^{(m)}\|.$$

Proof: Let $H_{\rm PF}^m = \overline{H}_{\rm p} + H_{\rm rad}^m + eH_{\rm I}^m + E_{\rm p}$. Since

$$(Q\Psi_{\rm g}^m, (H_{\rm PF}^m - E_m)\Psi_{\rm g}^m) = 0,$$

we obtain

$$(Q\Psi_{\rm g}^m, (\overline{H}_{\rm p} + H_{\rm rad}^m + E_{\rm p} - E_m)\Psi_{\rm g}^m) = -e(Q\Psi_{\rm g}^m, H_{\rm I}^m\Psi_{\rm g}^m).$$

It follows that

$$(Q\Psi_{\rm g}^m, (\overline{H}_{\rm p} + H_{\rm rad}^m + E_{\rm p} - E_m)\Psi_{\rm g}^m) \geq (\epsilon + E_{\rm p} - E_m)\|Q\Psi_{\rm g}^m\|^2$$

and

$$|(Q\Psi_{\rm g}^m, H_{\rm I}^m \Psi_{\rm g}^m)| \leq \|Q\Psi_{\rm g}^m\| \|H_{\rm I}^m \Psi_{\rm g}^m\|.$$

Note that E_m is continuous in m and e, and $\lim_{e\to 0} E_m = E_{\rm p}$ uniformly in m. Hence $\epsilon + E_{\rm p} - E_m > 0$ uniformly in m for sufficiently small $|e|$, and

$$\|Q\Psi_{\rm g}^m\| \leq \frac{\|H_{\rm I}^m \Psi_{\rm g}^m\|}{\epsilon + E_{\rm p} - E_m}$$

holds. One can easily see that

$$\|H_{\rm I}^m \Psi_{\rm g}^m\| \leq c''\|\Psi_{\rm g}^m\|$$

with some constant c'''. Thus the lemma follows. □

Theorem 4.14: There exists $e_* > 0$ such that for $|e| < e_*$ a ground state of $H_{\rm PF}$ exists.

Proof: Let $\Psi_{\rm g}^{(m)}$ be a normalized ground state of $H_{\rm PF}^m$. There exists a subsequence m' such that $\Psi_{\rm g}^{(m')}$ weakly converges to a vector Ψ as $m' \to \infty$. If $\Psi \neq 0$, Ψ is a ground state.[18] Let

$$P := E_{[0,\epsilon)}^{\overline{H}_{\rm p}} \otimes E_{\{0\}}^{H_{\rm rad}}. \tag{4.30}$$

Since

$$P + Q \geq 1 - N_{\rm rad},$$

we have

$$(\Psi_{\rm g}^{(m')}, P\Psi_{\rm g}^{(m')}) \geq \|\Psi_{\rm g}^{(m')}\|^2 - \|Q\Psi_{\rm g}^{(m')}\|^2 - \|N_{\rm rad}^{1/2}\Psi_{\rm g}^{(m')}\|^2.$$

From Lemma 4.11 it follows that

$$\||x|^2\Psi_{\rm g}^{(m')}\| \leq c_2'\|\Psi_{\rm g}^{(m')}\|,$$

where c_2' is independent of m. Thus by Lemma 4.12 there exists a constant a independent of m such that

$$\|N_{\rm rad}^{1/2}\Psi_{\rm g}^{(m')}\| \leq |e|a\|\Psi_{\rm g}^{(m')}\|.$$

Since P is a finite rank operator, taking $m' \to \infty$, we get by Lemma 4.13

$$(\Psi, P\Psi) \geq 1 - |e|a - e^2c'^2 > 0. \tag{4.31}$$

Thus the theorem follows. □

Corollary 4.15: *We assume that $H_{\rm p}$ has the unique ground state $\phi_{\rm p}$, and $H_{\rm PF}$ also has the unique ground state $\Psi_{\rm g}$. Then*

$$\text{s}-\lim_{e\to 0}\Psi_{\rm g} = \phi_{\rm p}\otimes\Omega.$$

Proof: Take sufficiently small $\epsilon > 0$ such that

$$P\Psi_{\rm g} = (\phi_{\rm p}\otimes\Omega, \Psi_{\rm g})\phi_{\rm p}\otimes\Omega,$$

where P is defined in (4.30). Then we have

$$\|\phi_{\rm p}\otimes\Omega - \Psi_{\rm g}\|^2 = 2 - 2\Re(\phi_{\rm p}\otimes\Omega, \Psi_{\rm g}) = 2 - 2\Re(\Psi_{\rm g}, P\Psi_{\rm g})^2.$$

Equation (4.31) yields $\lim_{e\to 0}(\Psi_{\rm g}, P\Psi_{\rm g}) = 1$. Then the corollary follows. □

4.4. *Enhanced binding*

In this subsection we assume that $H_{\rm PF}$ has *no* ground state for $e = 0$. In fact this will be the case for a sufficiently shallow V. We expect an interaction with a radiation field to enhance binding. This is precisely our main result in this subsection.

Let the electron mass M be revived in this subsection. The Pauli–Fierz Hamiltonian in the dipole approximation[f] is defined by $H_{\rm PF}$ with $A_\mu(\hat{\lambda}, x)$ replaced by

$$A_\mu(\hat{\lambda}, 0) := A_\mu(0) := \frac{1}{\sqrt{2}}\int e^r_\mu(k)\hat{\lambda}(k)\left\{a^*(k) + a^r(k)\right\}dk,$$

i.e.

$$H_{\rm dip} := \frac{1}{2M}\left(p - eA(0)\right)^2 + V + H_{\rm rad}.$$

It corresponds to neglecting collisions between an electron and photons. We suppose the following assumptions on $\hat{\lambda}$.

Assumption 4: We assume that

(1) $\hat{\lambda}$ is rotation invariant, i.e. there exists a function ψ on $\mathbb{R}$ such that $\hat{\lambda}(k) = \psi(|k|)$,

[f] Arai showed that $H_{\rm dip}$ can be diagonalized for a suitable V. Spectral properties of $H_{\rm dip}$ are established in Refs. 3–13.

(2) $\omega^{n/2}\hat{\lambda} \in L^2(\mathbb{R}^d)$ for $n = -4, 2$,
(3) $|\psi(\sqrt{s})|s^{d/2} \in L^\epsilon([0,\infty), ds)$, $0 < \epsilon < 1$, and Lipschitz continuous on $[0,\infty)$ of an order strictly less than one,
(4) $\|\hat{\lambda}\omega^{(d-2)/2}\|_\infty < \infty$ and $\|\hat{\lambda}\omega^{d/2}\|_\infty < \infty$,
(5) $\hat{\lambda}(k) \neq 0$ for $k \neq 0$.

Let K be a Hilbert space. We say that a pair of bounded operators $\{A, B\}$ is in symplectic group $\mathcal{S}_{\text{sp}}(K)$ if the following operator equation holds on $K \oplus K$:

$$\begin{pmatrix} A & B \\ \overline{B} & \overline{A} \end{pmatrix}^* \begin{pmatrix} 1 & 0 \\ 0 & -1 \end{pmatrix} \begin{pmatrix} A & B \\ \overline{B} & \overline{A} \end{pmatrix} = \begin{pmatrix} A & B \\ \overline{B} & \overline{A} \end{pmatrix} \begin{pmatrix} 1 & 0 \\ 0 & -1 \end{pmatrix} \begin{pmatrix} A & B \\ \overline{B} & \overline{A} \end{pmatrix}^* = \begin{pmatrix} 1 & 0 \\ 0 & -1 \end{pmatrix},$$

where $\overline{T}f := \overline{T\overline{f}}$.

Proposition 4.16: *There exist a vector $L = (L_1, \ldots, l_d) \in \oplus^{d-1}L^2(\mathbb{R}^d)$ and a pair of bounded operators $\mathcal{W}_\pm = (\mathcal{W}_{\pm rs})_{1\leq r,s\leq d-1}$ such that the following (1)–(4) hold:*

(1) $\{\mathcal{W}_+, \mathcal{W}_-\} \in \mathcal{S}_{\text{sp}}(\oplus^{d-1}L^2(\mathbb{R}^d))$,
(2) $\mathcal{W}_+$ *is a Hilbert–Schmidt operator on* $\oplus^{d-1}L^2(\mathbb{R}^d)$,
(3) *Operators $B^{*s}(f)$ and $B^s(f)$ defined by*

$$B^s(f) := a^{*r}(\mathcal{W}_{+rs}f) + a^r(\mathcal{W}_{-rs}f) - ep_\nu(L_\nu, f),$$
$$B^{*s}(f) := a^{*r}(\overline{\mathcal{W}}_{-rs}f) + a^r(\overline{\mathcal{W}}_{+rs}f) - ep_\nu(\overline{L}_\nu, f),$$

satisfy the canonical commutation relations:

$$[B^r(f), B^{*s}(g)] = (\bar{f}, g)\delta_{rs},$$
$$[B^{*r}(f), B^{*s}(g)] = 0,$$
$$[B^r(f), B^s(g)] = 0,$$

(4) *it follows that*

$$e^{itH_{\text{dip}}}B^{*s}(f)e^{-itH_{\text{dip}}} = B^{*s}(e^{it\omega}f),$$
$$e^{itH_{\text{dip}}}B^s(f)e^{-itH_{\text{dip}}} = B^{*s}(e^{-it\omega}f).$$

Proof: See Arai.[4–7] □

By this proposition and Berezin[31] we can diagonalize H_{dip}. See also Ruijsenaars[125,126] and Shale.[133]

Theorem 4.17: There exists a unitary operator T (a Bogoliubov transformation) such that

$$TH_{\text{dip}}T^{-1} = -\frac{1}{2M_{\text{eff}}}\Delta + H_{\text{rad}} + e^2G + V_{\text{eff}}, \tag{4.32}$$

where

$$M_{\text{eff}} := M + e^2\frac{(d-1)}{d}\|\hat{\lambda}/\sqrt{\omega}\|^2,$$

$$G := \frac{d-1}{2\pi}\int_{-\infty}^{\infty}\frac{t^2\left\|\sqrt{\omega}\hat{\lambda}/(t^2+\omega^2)\right\|^2}{M+e^2\frac{(d-1)}{d}\left\|\sqrt{\omega}\hat{\lambda}/\sqrt{t^2+\omega^2}\right\|^2}dt,$$

$$V_{\text{eff}}(x) := TVT^{-1} = V(x+\Delta x).$$

Here

$$\Delta x = \frac{1}{\sqrt{2}}\int\left\{a^*(k)\overline{K^r(k)} + a^r(k)K^r(k)\right\}dk$$

with some $K^r = (K_1^r, \ldots, K_d^r) \in \oplus^d L^2(\mathbb{R}^d)$, $r = 1, \ldots, d-1$.

Proof: We show an outline of a proof. See Ref. 90 for details. Suppose $V = 0$. Thus one can obtain that

$$[H_{\text{dip}}, p_\mu] = 0, \quad \mu = 1, 2, 3.$$

Then we can decompose H_{dip} as

$$\mathcal{H}_{\text{rad}} = \int_{\mathbb{R}^d}^{\oplus}\mathcal{F}_{\text{rad}}dp,$$

$$H_{\text{dip}} = \int_{\mathbb{R}^d}^{\oplus}H_{\text{dip}}(p)dp,$$

where

$$H_{\text{dip}}(p) = \frac{1}{2M}(p - eA(0))^2 + H_{\text{rad}}, \quad p \in \mathbb{R}^d.$$

By Proposition 4.16 we can explicitly construct a unitary operator

$$T(p) : \mathcal{F}_{\text{rad}} \to \mathcal{F}_{\text{rad}}$$

such that

$$T(p)H_{\text{dip}}(p)T(p)^{-1} = E(p) + H_{\text{rad}}$$

with some constant $E(p)$. Note that

$$\inf\sigma(H_{\text{dip}}(p)) = E(p), \quad p \in \mathbb{R}^d. \tag{4.33}$$

We calculate $E(p)$ as follows. Take a momentum lattice of a width $(2\pi)/a$ with the finite volume $(4\pi L)^d$, and then the number of lattice points is

$$J := (2[aL]+1)^d.$$

The lth lattice is denoted by Γ_l, $l = 1, \ldots, J$. A momentum lattice approximation of $H_{\rm dip}(p)$, say $H_{\rm dip}^{\rm lattice}(p)$, is defined by $\hat{\lambda}$ and ω replaced by

$$\left(\frac{a}{2\pi}\right)^{d/2} \sum_{l=1}^{J} \hat{\lambda}(l) 1_{\Gamma_l} \quad \text{and} \quad \left(\frac{a}{2\pi}\right)^{d/2} \sum_{l=1}^{J} \omega(l) 1_{\Gamma_l},$$

respectively, where 1_{Γ_l} denotes the characteristic function of Γ_l. Let

$$D := (d-1)J.$$

It is seen that there exists a unitary operator implementing

$$\mathcal{H}_{\rm rad} \cong L^2(\mathbb{R}^d) \otimes L^2(\mathbb{R}^D)$$

and

$$H_{\rm dip}^{\rm lattice}(p) \cong \frac{1}{2}(\vec{p}, \vec{p})_D + \frac{1}{2}(\vec{q}, A\vec{q})_D + \frac{1}{2M}p^2 - \frac{1}{2}(\vec{f}, A\vec{f})_D - \frac{1}{2}{\rm Tr}\sqrt{A_0} \quad (4.34)$$

where $(\cdot, \cdot)_D$ denotes the J-dimensional Euclidean scalar product, and $\{q_j, p_j\}_{j=1}^D$ the canonical pairs of $L^2(\mathbb{R}^D)$. A is given by

$$A := A_0 + \frac{e^2}{M} P,$$

where A_0 is the $D \times D$ diagonal matrix defined by

$$A_0 := \begin{pmatrix} \omega(l_1) 1_{d-1} & & & \\ & \omega(l_2) 1_{d-1} & & \\ & & \ddots & \\ & & & \omega(l_J) 1_{d-1} \end{pmatrix},$$

and $P := |\vec{v}_\mu\rangle\langle\vec{v}_\mu|$ the projection onto the subspace spanned by vectors

$$\vec{v}_\mu := \begin{pmatrix} v^\mu_{1l_1} \\ v^\mu_{2l_1} \\ \vdots \\ v^\mu_{d-1l_1} \\ \vdots \\ \vdots \\ v^\mu_{1l_J} \\ v^\mu_{2l_J} \\ \vdots \\ v^\mu_{d-1l_J} \end{pmatrix} \in \mathbb{R}^D, \quad \mu = 1, \ldots, d,$$

where

$$v_{rl}^{\mu} := \sqrt{\omega(l)}\hat{\lambda}(l)e_{\mu}^{r}(l).$$

$\vec{f}$ is defined by

$$\vec{f} := \frac{e}{M}A^{-1}p_{\mu}\vec{v}_{\mu}.$$

It is established that for $z \in \mathbb{C} \setminus \mathbb{R}$

$$\lim_{L\to\infty}\lim_{a\to\infty}(H_{\mathrm{dip}}^{\mathrm{lattice}}(p) - z)^{-1} = (H_{\mathrm{dip}}(p) - z)^{-1} \tag{4.35}$$

in the operator norm. Since (4.34) is a harmonic oscillator in $L^2(\mathbb{R}^D)$, we have by (4.33):

$$\begin{aligned}
E(p) &= \inf\sigma(H_{\mathrm{dip}}(p)) \\
&= \lim_{L\to\infty}\lim_{a\to 0}\inf\sigma(H_{\mathrm{dip}}^{\mathrm{lattice}}(p)) \\
&= \lim_{L\to\infty}\lim_{a\to 0}\left\{\frac{1}{2M}p^2 - \frac{1}{2}(\vec{f}, A\vec{f})_D + \frac{1}{2}\mathrm{Tr}\left(\sqrt{A} - \sqrt{A_0}\right)\right\}.
\end{aligned}$$

We calculate $(\vec{f}, A\vec{f})_D$ and $\mathrm{Tr}\left(\sqrt{A} - \sqrt{A_0}\right)$ as follows:

$$(\vec{f}, A\vec{f})_D = \frac{e^2\theta/M}{1 + (e^2\theta/M)}\frac{p^2}{M},$$

where

$$\theta := \frac{d-1}{d}\sum_{|l|\le 2\pi L}\frac{|\hat{\lambda}(l)|^2}{\omega(l)},$$

and

$$\begin{aligned}
\mathrm{Tr}\left(\sqrt{A} - \sqrt{A_0}\right) &= \frac{1}{\pi}\int_{-\infty}^{\infty}\mathrm{Tr}\left\{A(s^2 + A)^{-1} - A_0(s^2 + A_0)^{-1}\right\}ds \\
&= \frac{d-1}{\pi}\int_{-\infty}^{\infty}\frac{e^2 s^2/M}{1 + (e^2\xi/M)}\sum_{|l|\le 2\pi L}\frac{|\hat{\lambda}(l)|^2\omega(l)}{(s^2 + \omega(l)^2)^2}ds,
\end{aligned}$$

where

$$\xi := \frac{d-1}{d}\sum_{|l|\le 2\pi L}\frac{|\hat{\lambda}(l)|^2\omega(l)}{s^2 + \omega(l)^2}.$$

As $a \to \infty$ and then $L \to \infty$, we see that

$$\begin{aligned}
&\frac{1}{2M}p^2 - \frac{1}{2}(\vec{f}, A\vec{f})_D \to \frac{1}{2M_{\mathrm{eff}}}p^2, \\
&\frac{1}{2}\mathrm{Tr}\left(\sqrt{A} - \sqrt{A_0}\right) \to e^2 G.
\end{aligned}$$

Thus we obtain

$$E(p) = \frac{1}{2M_{\text{eff}}}p^2 + e^2 G.$$

Let

$$T := \int_{\mathbb{R}^d}^{\oplus} T(p)dp.$$

Then TVT^{-1} can be also calculated, since the explicit form of T is known. Thus the theorem follows. □

By (4.32) we see that e^2G is the ground state energy of H_{dip} with $V = 0$. Thus we have the following corollary:

Corollary 4.18: *Let*

$$\hat{\lambda}(k) = \begin{cases} 0 & |k| < \kappa, \\ 1/\sqrt{\omega(k)} & \kappa \le |k| \le \Lambda, \\ 0 & |k| > \Lambda. \end{cases}$$

Suppose that $V = 0$ *and* $M > 8\pi\kappa/3$. *Then*

$$\frac{8}{3}\left(\frac{3}{8\pi}\frac{1}{M}\right)^{1/2}\frac{\pi}{2} \le \lim_{\Lambda\to\infty}\frac{\inf\sigma(H_{\text{dip}})}{\Lambda^{3/2}} \le \frac{8}{3}\left(\frac{9}{8\pi}\frac{1}{M}\right)^{1/2}\frac{\pi}{2}.$$

Proof: See Hiroshima.[89] □

Let $D(\kappa)$ be the dilation unitary on $L^2(\mathbb{R}^d)$ defined by

$$D(\kappa)f(k) := \kappa^{d/2} f(k/\kappa).$$

We define a scaled Hamiltonian by

$$\begin{aligned} H_{\text{dip}}(\kappa) &:= \kappa^2 D(\kappa)^{-1}\left\{\frac{1}{2M}(p - eA(0))^2 + H_{\text{rad}} + \frac{1}{\kappa^2}V(x/\kappa)\right\}D(\kappa) \\ &= \frac{1}{2M}(p - \kappa e A(0))^2 + V + \kappa^2 H_{\text{rad}}. \end{aligned}$$

Lemma 4.19:

(1) *For arbitrary* $e \in \mathbb{R}$, $H_{\text{rad}} + (e^2/2M)A^2(0)$ *has a unique ground state, and*

$$\inf\sigma\left(H_{\text{rad}} + \frac{e^2}{2M}A^2(0)\right) = e^2 G.$$

(2) *Let* $V = 0$. *Then*

$$\inf\sigma(H_{\text{dip}}) = e^2 G.$$

(3) *Let P_{field} be the projection onto the one dimensional subspace spanned by the ground state of $H_{\text{rad}} + (e^2/2M)A^2(0)$. Then, for $z \in \mathbb{C} \setminus \mathbb{R}$,*

$$\text{s-}\lim_{\kappa\to\infty} \left(H_{\text{dip}}(\kappa) - e^2\kappa^2 G - z\right)^{-1} = (H_{\text{eff}} - z)^{-1} \otimes P_{\text{field}}.$$

Here

$$H_{\text{eff}} := -\frac{1}{2M_{\text{eff}}}\Delta + V.$$

Proof: See Hiroshima and Spohn.[90] □

This lemma suggests that, for $|e| \gg 1$ and sufficiently large κ, $H_{\text{dip}}(\kappa)$ may have a ground state, since $M_{\text{eff}} \gg 1$.

Fig. 11. The spectrum of $-(1/2M)\Delta + V_\kappa + H_{\text{rad}}$ with a negative shallow V

Fig. 12. The spectrum of H_κ with a negative shallow V for $|e| > e_c$

We introduce the following assumptions on V.

Assumption 5: We assume that

(1) $D(\Delta) \subset D(V)$ and $\|Vf\| \le a\|\Delta f\| + b\|f\|$ with some positive constants $a \ll 1$ and b (e.g. infinitesimally small),
(2) $V \in C^1(\mathbb{R}^d)$ and $\partial_\mu V \in L^\infty(\mathbb{R}^d)$ for $\mu = 1, \ldots, d$,
(3) there exist $\mu_0 \ge 1$ and $r > 0$ such that, for all $a \ge \mu_0$,

$$\inf \sigma\left(-(1/2M)\Delta + aV\right) \le -r,$$
$$\sigma_{\text{ess}}\left(-(1/2M)\Delta + aV\right) = [0, \infty).$$

Theorem 4.20: Let us suppose Assumption 5. Set

$$V_\kappa(x) = \frac{1}{\kappa^2}V(x/\kappa).$$

Suppose that κ is sufficiently large. Then there exists $e_c \geq 0$ such that

$$H_\kappa = \frac{1}{2M}(p - eA(0))^2 + V_\kappa + H_{\rm rad}$$

has a ground state and it is unique for arbitrary $|e| > e_c$. In particular, if $H_{\rm p}$ has a ground state, then H_κ has a ground state for all $e \in \mathbb{R}$ (Figs. 11 and 12).

Proof: We have

$$H_\kappa = \frac{1}{\kappa^2} D(\kappa) H_{\rm dip}(\kappa) D(\kappa)^{-1}.$$

By applying Lemma 4.19, we can prove that $H_{\rm dip}(\kappa)$ has a ground state for $|e| > e_c$ with some $e_c \geq 0$. Thus H_κ has a ground state for $|e| > e_c$. The uniqueness follows from Corollary 7.12. □

Remark 4.21: In Theorem 4.20 we do *not* assume the existence of a ground state of $H_{\rm p}$ for $e = 0$.

We give an example of V. Let $V \leq 0$. Set

$$N(V) := L_{0,d} \int_{\mathbb{R}^d} |MV(x)|^{d/2} dx,$$

where $L_{0,d}$ is a universal constant. It is known as the Lieb–Thirring identity[112] that

$$N(V) = \#\{\text{negative eigenvalues of } -(1/2M)\Delta + V\}.$$

Assume $N(V) < 1$. Since the scale invariance

$$N(V) = N(V_\kappa),$$

$H_{\rm p}$ with V_κ has no ground states for arbitrary $\kappa > 0$, and

$$\sigma(-(1/2M)\Delta + V_\kappa) = [0, \infty).$$

Suppose that κ is sufficiently large. Then, by Theorem 4.20, there exists e_c such that for $|e| > e_c$ a ground state of H_κ exists, but, *not* exist for $e = 0$.

4.5. *Asymptotic fields*

We have established the existence of a ground state, $\Psi_{\rm g}$, of $H_{\rm PF}$ in subsections 4.3 and 4.4. Then we can consider asymptotic fields. See Albeverio,[1] Høegh-Krohn,[93–95] Kato and Mugibayashi.[109] For instance in this subsection we suppose

$$V(x) := x^2, \quad |e| \ll 1.$$

Let

$$a_t^{\sharp r}(f) := e^{itH_{\mathrm{PF}}}e^{-itH_{\mathrm{dec}}}a^{\sharp r}(f)e^{itH_{\mathrm{dec}}}e^{-itH_{\mathrm{PF}}}, \quad r = 1, \ldots, d-1.$$

We want to consider the strong limit of $a_t^{\sharp r}$ as $t \to \pm\infty$. We will focus on s$-\lim_{t\to\infty} a_t^{*r}$ in what follows. The other statements are similar. A crucial point is that the radiation field is massless, i.e. $\omega(k) = |k|$. A massive case is studied in Høegh-Krohn.[93–95] From the definition of a_t^{*r} and fundamental limiting arguments, we have[g]

$$a_t^{*r}(f)\Psi = a_T^{*r}(f)\Psi - ie\int_T^t e^{isH_{\mathrm{PF}}}K_\mu^r(s,x,f)\,(p_\mu - eA_\mu)\,e^{-isH_{\mathrm{PF}}}\Psi ds, \tag{4.36}$$

where

$$K_\mu^r(s,x,f) := [A_\mu, a^{*r}(e^{-is\omega}f)] = \frac{1}{\sqrt{2}}\left(\overline{\hat{\lambda}e_\mu^r e^{-ikx}}, e^{-is\omega}f\right).$$

Definition 4.22: We say $f \in \mathcal{E}$ if, for all $h \in C_0^\infty(\mathbb{R}^d)$,

$$\lim_{t\to\infty} t^{\frac{d-1}{2}} \sup_{x\in\mathbb{R}^d}\left|\int_{\mathbb{R}^d} f(k)h(k)e^{ikx-it\omega(k)}dk\right| < \infty.$$

Lemma 4.23: *Suppose that $\hat{\lambda}, \partial_\mu\hat{\lambda} \in \mathcal{E}$ and $f \in C_0^\infty(\mathbb{R}^d)$. Then*

$$\mathrm{s}-\lim_{t\to\infty} a_t^{\sharp r}(f)\Psi$$

exists for $\Psi \in D(H_{\mathrm{PF}})$.

Proof: By the integral representation (4.36) it is enough to prove that

$$\left\|K_\mu^r(t,x,f)\,(p_\mu - eA_\mu)\,e^{-itH_{\mathrm{PF}}}\Psi\right\| \in L^1([T,\infty), dt). \tag{4.37}$$

By using

$$e^{it\omega(k)} = \frac{\omega(k)}{k_\mu}\frac{1}{it}\frac{\partial}{\partial k_\mu}e^{it\omega(k)}, \quad k \in \mathbb{R}^d \setminus \{0\},$$

and integrating by parts, we see that

$$\begin{aligned}&\left\|K_\mu^r(t,x,f)\,(p_\mu - eA_\mu)\,e^{-itH_{\mathrm{PF}}}\Psi\right\| \le C_1 t^{-(d-1)/2} \times \\ &\left(\|x_\mu p_\mu e^{-itH_{\mathrm{PF}}}\Psi\| + \|p_\mu e^{-itH_{\mathrm{PF}}}\Psi\| + \|x_\mu A_\mu e^{-itH_{\mathrm{PF}}}\Psi\| + \|A_\mu e^{-itH_{\mathrm{PF}}}\Psi\|\right)\end{aligned} \tag{4.38}$$

---

[g] Formally

$$a_t^{\sharp r}(f)\Psi = a_T^{\sharp r}(f)\Psi + i\int_T^t e^{isH_{\mathrm{PF}}}[-ep\cdot A + e^2A^2, a^{*r}(e^{-is\omega}f)]e^{-isH_{\mathrm{PF}}}\Psi ds.$$

with some constant C_1. Since $V(x) = x^2$, we have

$$\|x_\mu p_\mu e^{-itH_{\mathrm{PF}}}\Psi\| \leq C_2\left(\|H_{\mathrm{PF}}\Psi\| + \|\Psi\|\right) \tag{4.39}$$

with some constant C_2. The other terms in (4.38) are estimated in the similar manner as (4.39) and we have, with some constant C_3,

$$\left\|K_\mu^r(t,x,f)\,(p_\mu - eA_\mu)\,e^{-itH_{\mathrm{PF}}}\Psi\right\| \leq C_3 t^{-(d-1)/2}(\|H_{\mathrm{PF}}\Psi\| + \|\Psi\|).$$

Hence (4.37) follows. □

We define, for $\Psi \in D(H_{\mathrm{PF}})$,

$$a_\pm^{\sharp r}(f)\Psi := \mathrm{s-}\lim_{t\to\pm} a_t^{\sharp r}(f)\Psi.$$

It is immediately seen that

$$\|a_\pm^{\sharp r}(f)\Psi\| \leq C_4(\|f/\sqrt{\omega}\| + \|f\|)(\||H_{\mathrm{PF}}|^{1/2}\Psi\| + \|\Psi\|)$$

with some constant C_4. Hence we can extend $a_\pm^{\sharp r}(f)$ to f such that $f, f/\sqrt{\omega} \in L^2(\mathbb{R}^d)$. We denote the closure of $a_\pm^{\sharp r}(f)$ by the same symbol. Then

$$D(a_\pm^{\sharp r}(f)) \supset D(|H_{\mathrm{PF}}|^{1/2}).$$

Moreover it is obtained

$$[a_\pm^r(f), a_\pm^{*s}(g)] = (\bar{f}, g)\delta_{rs},$$
$$[a_\pm^{\sharp r}(f), a_\pm^{\sharp s}(g)] = 0,$$

and

$$e^{itH_{\mathrm{PF}}}a_\pm^{*r}(f)e^{-itH_{\mathrm{PF}}} = a_\pm^{*r}(e^{it\omega}f),$$
$$e^{itH_{\mathrm{PF}}}a_\pm^{r}(f)e^{-itH_{\mathrm{PF}}} = a_\pm^{r}(e^{-it\omega}f), \tag{4.40}$$

on $D(H_{\mathrm{PF}})$. It is also proven that for all f such that $f, f/\sqrt{\omega} \in L^2(\mathbb{R}^d)$,

$$a_\pm^r(f)\Psi_{\mathrm{g}} = 0.$$

We define the asymptotic Hilbert space $\mathcal{H}_{\pm\mathrm{asy}}$ by

$$\mathcal{H}_{\pm\mathrm{asy}} := \overline{\{a_\pm^{*r_1}(f_1)\cdots a_\pm^{*r_n}(f_n)\Psi_{\mathrm{g}}, \Psi_{\mathrm{g}} |}$$
$$\overline{f_j \in C_0^\infty(\mathbb{R}^d), r_j = 1, \ldots, d, j = 1, \ldots, n, n \in \mathbb{N}\}}.$$

where $\overline{\{\ldots\}}$ denotes the closure of $\{\ldots\}$. Let $W_\pm : \mathcal{H}_{\pm\mathrm{asy}} \to \mathfrak{F}_{\mathrm{rad}}$ be defined by

$$W_\pm\Psi_{\mathrm{g}} := \Omega,$$
$$W_\pm a_\pm^{*r_1}(f_1)\cdots a_\pm^{*r_n}(f_n)\Psi_{\mathrm{g}} := a^{*r_1}(f_1)\cdots a^{*r_n}(f_n)\Omega.$$

Thus $W_\pm$ can be uniquely extended to a unitary operator of $\mathcal{H}_{\pm\mathrm{asy}}$ to $\mathcal{F}_{\mathrm{rad}}$.

Theorem 4.24: We assume that H_{PF} has a ground state. Then

$$\sigma_{\mathrm{ac}}(H_{\mathrm{PF}}) = [E, \infty).$$

Proof: It is seen that $e^{itH_{\mathrm{PF}}}$ is reduced by $\mathcal{H}_{\pm\mathrm{asy}}$. Then

$$H_{\mathrm{PF}} = \left(H_{\mathrm{PF}}\lceil_{\mathcal{H}_{\pm\mathrm{asy}}}\right) \oplus \left(H_{\mathrm{PF}}\lceil_{\mathcal{H}_{\pm\mathrm{asy}}^{\perp}}\right)$$

under the identification

$$\mathcal{H}_{\mathrm{rad}} \cong \mathcal{H}_{\pm\mathrm{asy}} \oplus \mathcal{H}_{\pm\mathrm{asy}}^{\perp}.$$

By the definition of W and (4.40), we have

$$W_\pm \left(e^{itH_{\mathrm{PF}}\lceil_{\mathcal{H}_{\pm\mathrm{asy}}}}\right) W_\pm^{-1} = e^{it(H_{\mathrm{rad}}+E)}.$$

Hence

$$H_{\mathrm{PF}} \cong (H_{\mathrm{rad}} + E) \oplus \left(H_{\mathrm{PF}}\lceil_{\mathcal{H}_{\pm\mathrm{asy}}^{\perp}}\right).$$

Since $\sigma_{\mathrm{ac}}(H_{\mathrm{rad}} + E) = [E, \infty)$, the theorem follows.[h] □

4.6. *Scaling limits*

To interpret the Lamb shift[32] intuitively, Welton[146] formally derived an effective Schrödinger Hamiltonian taking into account of radiative effects. As we explained in Section 1, we expect that a position-fluctuation of an electron modifies external potential V. Then H_{p} effectively turns out to be an effective Hamiltonian given by

$$-\frac{1}{2M}\Delta + V_{\mathrm{eff}}. \tag{4.41}$$

On the other hand it was found that photons enhance bare mass M. Then M amounts to an observed mass

$$M_{\mathrm{eff}} := M + \delta M, \quad \delta M > 0,$$

through a coupling with a radiation field. Then H_{p} formally turns out to be

$$-\frac{1}{2M_{\mathrm{eff}}}\Delta + V. \tag{4.42}$$

[h] Arai[14] proved Theorem 4.24 without assuming the existence of ground states of H_{PF}.

It has been shown in Hiroshima[73,75,76,85] that scaling limits of $H_{\rm dip}$ unifies (4.41) and (4.42). Let us introduce a scaling Hamiltonian $H_{\rm dip}(\kappa)$ as follows:

$$H_{\rm dip}(\kappa) := H_{\rm p} + \kappa^l \frac{1}{M} p \cdot A(0) + \kappa^2 \left\{ \frac{1}{2M} A^2(0) + H_{\rm rad} \right\}, \quad l \leq 2.$$

Davies[41,42] considered a scaling limit $\kappa \to \infty$ of the case of $l = 1$ for a certain model.[i] He called this scaling limit *the weak coupling limit.* On the other hand Arai[11] considered the case of $l = 2$ for $H_{\rm dip}$ but without $A^2(0)$.

Performing a mass renormalization, subtracting some energy reset from $H_{\rm dip}(\kappa)$, and taking $\kappa \to \infty$ in the resolvent sense, we derive (4.41) for $l = 2$, and (4.42) for $l = 1$.

Let $G(\kappa)$ be defined by G in Theorem 4.17 with M replaced by M/κ^2.

Theorem 4.25: Suppose $\hat{\lambda}$ satisfies Assumption 4. Then there exist a self-adjoint operator $H_{\rm eff}$, a unitary U_l of $\mathcal{H}$, and $M(\kappa) \geq 0$ such that

$$\begin{aligned} \text{s-}\lim_{\kappa \to \infty} & \left(H_{\rm dip}(\kappa) - \frac{1}{2M(\kappa)} \Delta - G(\kappa) - z \right)^{-1} \\ & = U_l^{-1} \left\{ (H_{\rm eff} - z)^{-1} \otimes P_{\rm field} \right\} U_l. \end{aligned}$$

Proof: See Hiroshima.[85] □

An explicit form of $H_{\rm eff}$ is given in Fig. 13, where

$$\begin{aligned} \delta M &:= e^2 \frac{d-1}{d} \int_{\mathbb{R}^d} \frac{|\hat{\lambda}(k)|^2}{\omega(k)} dk, \\ V_{\rm eff}(x) &:= (2\pi C_{\hat{\lambda}})^{-d/2} \int_{\mathbb{R}^d} V(y) e^{-|x-y|^2/(2C_{\hat{\lambda}})} dy, \\ C_{\hat{\lambda}} &:= \frac{d-1}{2d} \frac{e^2}{M} \int_{\mathbb{R}^d} \frac{|\hat{\lambda}(k)|^2}{\omega(k)^2} \\ & \qquad \times \left(M - \frac{d-1}{2d} e^2 \int_{\mathbb{R}^d} \frac{\omega(k')|\hat{\lambda}(k')|^2}{\omega(k)^2 - \omega(k')^2 + i0} dk' \right)^{-1} dk. \end{aligned}$$

4.7. *Notes*

Binding:

As we have seen in Theorem 4.14, a ground state of $H_{\rm PF}^\sigma$ exists even for the case of

$$\|\hat{\lambda}/\omega\| = \infty. \tag{4.43}$$

[i] Hiroshima[78,79] studied a scaling limit of the Nelson model to derive a Schrödinger Hamiltonian with the Coulomb or Yukawa potential. In this scaling limit he simultaneously removed an ultraviolet cutoff. See also a recent work of Teufel.[144]

	$\frac{1}{2M(\kappa)}$	H_{eff}
$l=2$	$\kappa^2 \frac{1}{2(M+\delta M)} \frac{\delta M}{M}$	$-\frac{1}{2M}\Delta + V_{\text{eff}}$
$1<l<2$	$\kappa^{2l-2} \frac{1}{2(M+\delta M)} \frac{\delta M}{M}$	$-\frac{1}{2M}\Delta + V$
$l=1$	0	$-\frac{1}{2(M+\delta M)}\Delta + V$
$l<1$	0	$-\frac{1}{2M}\Delta + V$

Fig. 13. Effective Hamiltonians

This is due to Bach, Fröhlich and Sigal.[27] Several models, however, lose ground states under (4.43). See Section 10. Griesemer, Lieb and Loss[61] proved the existence of a ground state of H_{PF}^{σ} for arbitrary $e \in \mathbb{R}$ under some conditions on V. Arai and Hirokawa[19] gave a sufficient condition for the existence of a ground state of GSB models under (4.43).

Resonances:

The existence of resonances is considered by Bach, Fröhlich and Sigal[24–27] by a renormalization group, and Bach, Fröhlich, Sigal and Soffer by a positive commutator method. For positive temperatures, see Jakšić and Pillet.[100–102]

Localization of the number of photons:

In Section 4.3 we proved that

$$\Psi_{\text{g}} \in D(N_{\text{rad}}^{1/2}).$$

Actually by employing a pull-through formula it is shown in Hiroshima[88] that

$$\Psi_{\text{g}} \in \bigcap_{k=1}^{\infty} (N_{\text{rad}}^{k})$$

under no infrared cutoff condition. As we will see in Section 10.3, the number of bosons of a ground state of the Nelson Hamiltonian superexponentially decays under the infrared cutoff condition. As far as we know, however, superexponential or exponential decays, i.e.

$$\Psi_{\text{g}} \in D(e^{+\beta N_{\text{rad}}}),$$

have *not* yet been proven for H_{PF}. See Conjecture 2.

Enhanced binding:

In the zero spectral gap case, i.e.

$$E_{\mathrm{p}} - \Sigma_{\mathrm{p}} = 0,$$

it is not known the absence of ground states of H_κ for some $e \neq 0$ (Fig.14). Recently Hainzl, Vougalter and Vugalter[66] considered an enhanced binding of H_{PF} with *no* dipole approximation for small coupling constants. See also Arai and Kawano,[22] Catto and Hainzl,[38] Chen, Vougalter and Vugalter,[40] and Hiroshima.[89]

e	$e = 0$	$0 < \lvert e \rvert \leq e_{\mathrm{c}}$	$e_{\mathrm{c}} < \lvert e \rvert$
Ψ_{g}	Not exist	?	Exist

Fig. 14. Enhanced binding of H_κ with a negative shallow V

Scattering theory:

Scattering theory is to investigate the asymptotics of the time evolution $e^{-itH_{\mathrm{PF}}}$ as $t \to \pm\infty$. In the dipole approximation it is studied in e.g. Arai[6,7] and Spohn.[141] One expects that after long time $e^{-itH_{\mathrm{PF}}}\psi$ is in its bound states and several photons travel outwards. Namely for *some* $\psi \in \mathcal{H}_{\mathrm{rad}}$ there exist a comparison photon state $\phi \in \mathcal{F}_{\mathrm{rad}}$ and a bound state $\Psi_{\mathrm{b}} \in \mathcal{H}_{\mathrm{rad}}$ such that

$$e^{-itH_{\mathrm{PF}}}\psi \approx e^{-itE_{\mathrm{b}}}\Psi_{\mathrm{b}} \otimes_s e^{-itH_{\mathrm{rad}}}\phi \tag{4.44}$$

for $t \to \infty$. Here $H_{\mathrm{PF}}\Psi_{\mathrm{b}} = E_{\mathrm{b}}\Psi_{\mathrm{b}}$. Precisely it is formulated as follows. The wave operator[97]

$$\Omega_+ : \mathcal{H}_{\mathrm{rad}} \otimes \mathcal{F}_{\mathrm{rad}} \to \mathcal{H}_{\mathrm{rad}}$$

is defined by

$$\Omega_+ = \mathrm{s-}\lim_{t\to\infty} e^{itH_{\mathrm{PF}}} I e^{-it\widetilde{H}_{\mathrm{PF}}}(P_{\mathrm{pp}} \otimes 1),$$

where $\widetilde{H}_{\mathrm{PF}} = H_{\mathrm{PF}} \otimes 1 + 1 \otimes H_{\mathrm{rad}}$ acting in $\mathcal{H}_{\mathrm{rad}} \otimes \mathcal{F}_{\mathrm{rad}}$, P_{pp} is the projection to the space spanned by eigenvectors of H_{PF}, and

$$I : \mathcal{H}_{\mathrm{rad}} \otimes \mathcal{F}_{\mathrm{rad}} \to \mathcal{H}_{\mathrm{rad}}$$

is the so called scattering identification[56] defined by

$$I(\phi \otimes \Omega) := \phi,$$
$$I(\phi \otimes a^{*r_1}(f_1)\cdots a^{*r_1}(f_1)\Omega) := a^{*r_1}(f_1)\cdots a^{*r_1}(f_1)\phi.$$

Asymptotic completeness for Rayleigh scattering is the following statement.

$$\mathrm{Ran}\Omega_+ \supset E^{H_{\mathrm{PF}}}_{(-\infty,\Sigma_{\mathrm{p}})}\mathcal{H}_{\mathrm{rad}}. \tag{4.45}$$

Roughly speaking, (4.45) implies that for every

$$\psi \in E^{H_{\mathrm{PF}}}_{(-\infty,\Sigma_{\mathrm{p}})}\mathcal{H}_{\mathrm{rad}},$$

there exist a bound state $\Psi_{\mathrm{b}} \in \mathcal{H}_{\mathrm{rad}}$ and a photon state $\phi \in \mathcal{F}_{\mathrm{rad}}$ satisfying (4.44). Fröhlich, Griesemer and Schlein[55] established the existence of asymptotic fields for nonrelativistic and relativistic Pauli–Fierz Hamiltonians, and they proved asymptotic completeness for Rayleigh scattering for the Nelson Hamiltonian with infrared cutoffs in Fröhlich, Griesemer and Schlein.[56]

Let

$$\mathcal{K}_\pm := \left\{\Psi \in \mathcal{H}_{\mathrm{rad}} | a^r_\pm(f)\Psi = 0 \,\text{for all}\, f \in L^2(\mathbb{R}^d)\right\}.$$

It is known that

$$P_{\mathrm{pp}}\mathcal{H}_{\mathrm{rad}} \subset \mathcal{K}_\pm.$$

In contrast to this, *asymptotic completeness* is the following statement:

$$P_{\mathrm{pp}}\mathcal{H}_{\mathrm{rad}} = \mathcal{K}_\pm. \tag{4.46}$$

Gérard[58] studied an asymptotic completeness of the Nelson model. See also Dereziński and Gérard,[43] and Ammari.[2] As far as we know, explicit relationships between (4.45) and (4.46) are unknown.

Relativistic models:

Relativistic models are defined by the Dirac Hamiltonian minimally coupled to quantized radiation fields. It is studied in e.g. Arai,[15] Barbaroux, Dimassi and Guillot,[30] and Dimassi and Guillot.[44]

5. Functional Integral Representations

In this section we investigate the spinless Pauli–Fierz Hamiltonian H_{PF}. Functional integral representations of nonrelativistic quantum electrodynamics are discussed in e.g. Fefferman, Fröhlich and Graf,[49] Fröhlich and Park,[54] Gross[64] and Hiroshima.[77] Functional integral representations are useful for studying, e.g.

(1) the self-adjointness of $H_{\rm PF}$ for arbitrary $e \in \mathbb{R}$,
(2) the multiplicity of ground states,
(3) densities in a ground state,
(4) qualitative properties of ground state energies.

5.1. *Schrödinger representation*

In order to derive a functional integral representation of matrix elements of $e^{-tH_{\rm PF}}$, we take a Schrödinger representation. Note that

$$(A_\mu(\hat{\lambda})\Omega, A_\nu(\hat{\rho})\Omega) = \frac{1}{2}(d_{\mu\nu}\hat{\lambda}, \hat{\rho}),$$
$$[A_\mu(\hat{\lambda}), A_\nu(\hat{\rho})] = 0.$$

Define the quadratic form

$$q(\cdot,\cdot) : \left(\oplus^d L^2(\mathbb{R}^d)\right) \times \left(\oplus^d L^2(\mathbb{R}^d)\right) \longrightarrow \mathbb{C}$$

by

$$q(f,g) := \frac{1}{2}(d_{\mu\nu}\hat{f}_\mu, \hat{g}_\nu), \quad f, g \in \oplus^d L^2(\mathbb{R}^d).$$

In particular we set

$$q(f,f) := q(f).$$

Let (Q,ν) be a probability space[j] endowed with a mean zero Gaussian random process $\phi(f)$ indexed by *real* $f \in \oplus^d L^2(\mathbb{R}^d)$ with the covariance

$$\int_Q \phi(f)\phi(g)d\nu(\phi) = q(f,g).$$

Note that

$$\int_Q e^{\alpha\phi(f)}d\nu(\phi) = e^{(\alpha^2/2)q(f)}, \quad \alpha \in \mathbb{C}.$$

---

[j] Actually we can construct a Gaussian measure ν on "the Schwartz distribution space of transverse vector potentials",[49,77]

$$\mathcal{S}_{\rm T} := \left\{ \Psi \in \oplus^d \mathcal{S}'_{\rm real}(\mathbb{R}^d) \,\middle|\, {\rm div}\Psi = 0 \right\}$$

by the Minlos theorem (e.g. Ref. 68). Then ϕ is defined by the pairing of $\oplus^d\mathcal{S}'_{\rm real}(\mathbb{R}^d)$ and $\oplus^d\mathcal{S}_{\rm real}(\mathbb{R}^d)$,

$$\phi(f) := \langle \phi, f \rangle.$$

For $f \in \oplus^d L^2_{\rm real}(\mathbb{R}^d)$, we define

$$\phi(f) := \lim_{n\to\infty} \phi(f_n)$$

with $f_n \in \oplus^d\mathcal{S}_{\rm real}(\mathbb{R}^d)$ such that $f_n \to f$ as $n \to \infty$ in $\oplus^d L^2(\mathbb{R}^d)$.

We set for $f \in \oplus^d L^2(\mathbb{R}^d)$

$$\phi(f) := \phi(\Re f) + i\phi(\Im f),$$

where $\Re f := \bigoplus_{\mu=1}^d \Re f_\mu$ and $\Im f := \bigoplus_{\mu=1}^d \Im f_\mu$. Set

$$L_0^2(Q) := \mathcal{L}\{ :\phi(f_1)\cdots\phi(f_n):, 1 | f_j \in \oplus^d L^2(\mathbb{R}^d), j = 1, \ldots, n, n \in \mathbb{N}\},$$

where the Wick product $:\phi(f_1)\cdots\phi(f_n):$ is recursively defined by

$$\begin{aligned} &:\phi(f): = \phi(f), \\ &:\phi(f)\phi(f_1)\cdots\phi(f_n): = \phi(f) :\phi(f_1)\cdots\phi(f_n): \\ &\qquad\qquad -\frac{1}{2}(\overline{f}, f_j) :\phi(f_1)\cdots\widehat{\phi(f_j)}\cdots\phi(f_n): . \end{aligned}$$

It is known that $L_0^2(Q)$ is dense in $L^2(Q)$ and

$$(:\phi(f_1)\cdots\phi(f_n):, :\phi(g_1)\cdots\phi(g_m):) = \delta_{nm} \sum_{\pi \in \mathcal{P}_n} q(f_1, g_{\pi(1)})\cdots q(f_n, g_{\pi(n)}),$$

where $\mathcal{P}_n$ denotes the set of permutations of n degree. Let $T : L^2(\mathbb{R}^d) \to L^2(\mathbb{R}^d)$ be a contractive operator. We define the contractive operator

$$\Gamma(T) : L^2(Q) \to L^2(Q)$$

by

$$\begin{aligned} &\Gamma(T)1 = 1, \\ &\Gamma(T) :\phi(f_1)\cdots\phi(f_n): = :\phi([T]f_1)\cdots\phi([T]f_n):, \end{aligned}$$

where $[T] := \oplus^d T$. Let h be a self-adjoint operator of $L^2(\mathbb{R}^d)$. Then $\Gamma(e^{ith})$ is a strongly continuous one-parameter unitary group in t. Thus there exists a self-adjoint operator $d\Gamma(h)$ of $L^2(Q)$ such that

$$\Gamma(e^{ith}) = e^{itd\Gamma(h)}, \quad t \in \mathbb{R}.$$

$d\Gamma(h)$ is called the second quantization of h in the Schrödiger representation. The number operator in $L^2(Q)$ is defined by

$$N := d\Gamma(1).$$

We set

$$\pi(\lambda) := e^{i(\pi/2)N}\phi(\lambda)e^{-i(\pi/2)N}.$$

Let

$$\widehat{\omega} := \omega(-i\nabla).$$

We define the free Hamiltonian in $L^2(Q)$ by

$$H_{\rm f} := d\Gamma(\widehat{\omega}).$$

Set

$$\mathcal{A}_\mu(\lambda) := \phi(\bigoplus_{\nu=1}^{d} \delta_{\mu\nu}\lambda), \quad \mu = 1, \ldots, d.$$

The following proposition is well known.

Proposition 5.1: *There exists a unitary operator $\theta : \mathcal{F} \to L^2(Q)$ such that (1) $\theta\Omega = 1$, (2) $\theta^{-1} H_{\rm rad}\theta = H_{\rm f}$, (3) $\theta^{-1} A_\mu(\hat{\lambda}, x)\theta = \mathcal{A}_\mu(\lambda(\cdot - x))$ for each $x \in \mathbb{R}^d$.*

Proof: See Refs. 49 and 77. □

Remark 5.2: Let h be a nonnegative function. Then it follows that

$$\theta^{-1} d\Gamma_{\rm rad}(h)\theta = d\Gamma(h(-i\nabla)).$$

Define

$$\mathcal{H} := L^2(\mathbb{R}^d) \otimes L^2(Q).$$

We identify $\mathcal{H}$ with

$$\mathcal{H} \cong \int_{\mathbb{R}^d}^{\oplus} L^2(Q) dx.$$

i.e. $\mathcal{H}$ can be regarded as the set of $L^2(Q)$-valued L^2-functions on $\mathbb{R}^d$. Thus, for $F \in \mathcal{H}$, $F(x) \in L^2(Q)$ a.e. $x \in \mathbb{R}^d$ and

$$\int_{\mathbb{R}^d} \|F(x)\|^2_{L^2(Q)} dx = \|F\|^2_{\mathcal{H}}.$$

Set

$$\mathcal{A}_\mu := \int_{\mathbb{R}^d}^{\oplus} \mathcal{A}_\mu(\lambda(\cdot - x)) dx.$$

The Pauli–Fierz Hamiltonian in the Schrödinger representation is defined by

$$H := \frac{1}{2}(p - e\mathcal{A})^2 + V + H_{\rm f}.$$

Let

$$\Theta := \int_{\mathbb{R}^d}^{\oplus} \theta dx : \mathcal{H} \to \mathcal{H}_{\rm rad}.$$

From Proposition 5.1 it follows that:

(1) if $\hat{\lambda}/\sqrt{\omega}, \hat{\lambda}, \omega\hat{\lambda} \in L^2(\mathbb{R}^d)$, then on $D(\Delta) \cap D(H_{\rm rad})$

$$\Theta^{-1} H \Theta = H_{\rm PF}, \tag{5.1}$$

(2) if $\hat{\lambda}, \omega\hat{\lambda} \in L^2(\mathbb{R}^d)$, then on $D(\Delta) \cap D(H_{\rm rad}) \cap D(N_{\rm rad})$

$$\Theta^{-1} H \Theta = H_{\rm PF}. \tag{5.2}$$

Set

$$H_0 := \frac{1}{2} \left(p - e\mathcal{A}\right)^2.$$

5.2. *Decompositions of semigroups*

For a moment we assume that $|e| \ll 1$ and V is relatively bounded with respect to Δ with a sufficiently small relative bound. In particular H is self-adjoint. We want to derive a functional integral representation of

$$\left(\Phi, e^{-\beta_0 K} e^{-t_1 H} f_1 e^{-\beta_1 K} e^{-(t_2 - t_1)H} f_2 \cdots f_{m-1} e^{-\beta_{m-1} K} e^{-(t_m - t_{m-1})H} \Psi\right)_{\mathcal{H}},$$

where $f_j \in L^\infty(\mathbb{R}^d)$, $j = 1, \ldots, m-1$, K is a nonnegative self-adjoint operator. We start with preparing some notations. For $f, g \in \oplus^d L^2(\mathbb{R}^{d+1})$, we define

$$q_0(f, g) := \frac{1}{2} \int_{\mathbb{R}^{d+1}} d_{\mu\nu}(k) \overline{\hat{f}_\mu}(k, k_0) \hat{g}_\nu(k, k_0) dk dk_0.$$

Let (Q_0, ν_0) be a probability space endowed with a mean zero Gaussian random process $\phi_0(f)$ indexed by *real* $f \in \oplus^d L^2(\mathbb{R}^{d+1})$ with the covariance

$$\int_{Q_0} \phi_0(f) \phi_0(g) d\nu_0(\phi_0) = q_0(f, g).$$

For $f \in \oplus^d L^2(\mathbb{R}^{d+1})$, we define

$$\phi_0(f) = \phi_0(\Re f) + i\phi_0(\Im f).$$

Let

$$j_t : L^2(\mathbb{R}^d) \to L^2(\mathbb{R}^{d+1})$$

be defined by

$$\widehat{j_t f}(k, k_0) = \frac{e^{-itk_0}}{\sqrt{\pi}} \sqrt{\frac{\omega(k)}{\omega(k)^2 + |k_0|^2}} \hat{f}(k), \quad (k, k_0) \in \mathbb{R}^d \times \mathbb{R}, \quad t \in \mathbb{R}. \tag{5.3}$$

It is immediate to see that

$$(j_t f, j_s g)_{L^2(\mathbb{R}^{d+1})} = (\hat{f}, e^{-|t-s|\omega} \hat{g})_{L^2(\mathbb{R}^d)} = (f, e^{-|t-s|\widehat{\omega}} g)_{L^2(\mathbb{R}^d)}.$$

Hence

$$j_t^* j_s = e^{-|t-s|\widehat{\omega}}, \quad t, s \in \mathbb{R}. \tag{5.4}$$

Let

$$J_t : L^2(Q) \to L^2(Q_0)$$

be defined by

$$J_t 1 = 1,$$
$$J_t \ {:}\phi(f_1)\cdots\phi(f_n){:} = {:}\phi_0([j_t]f_1)\cdots\phi_0([j_t]f_n){:} \ .$$

It is easily seen that, by (5.4), J_t can be extended to an isometry of $L^2(Q)$ to $L^2(Q_0)$, and that

$$J_t^* J_s = e^{-|t-s|H_{\mathrm{f}}}, \quad t, s \in \mathbb{R}. \tag{5.5}$$

Additionally we need another Gaussian random process. For $f, g \in \oplus^d L^2(\mathbb{R}^{d+2})$, we define

$$q_1(f,g) := \frac{1}{2}\int_{\mathbb{R}^{d+2}} d_{\mu\nu}(k)\overline{\hat{f}_\mu}(k,k_0,k_1)\hat{g}_\nu(k,k_0,k_1)dkdk_0dk_1.$$

Let (Q_1, ν_1) be a probability space endowed with a mean zero Gaussian random process $\phi_1(f)$ indexed by real $f \in \oplus^d L^2(\mathbb{R}^{d+2})$ with the covariance

$$\int_{Q_1} \phi_1(f)\phi_1(g)\nu_1(d\phi_1) = q_1(f,g).$$

For $f \in \oplus^d L^2(\mathbb{R}^{d+2})$, we define

$$\phi_1(f) = \phi_1(\Re f) + i\phi_1(\Im f).$$

Suppose that h is a nonnegative multiplication operator of $L^2(\mathbb{R}^d)$. We define

$$\xi_t(h) = \xi_t : L^2(\mathbb{R}^{d+1}) \to L^2(\mathbb{R}^{d+2})$$

by

$$\widehat{\xi_t f}(k,k_0,k_1) := \frac{e^{-itk_1}}{\sqrt{\pi}}\sqrt{\frac{h(k)}{h(k)^2+|k_1|^2}}\hat{f}(k,k_0),$$

$$(k,k_0,k_1) \in \mathbb{R}^d \times \mathbb{R} \times \mathbb{R}, \quad t \in \mathbb{R}. \tag{5.6}$$

Similarly to (5.4) we have

$$\xi_t^* \xi_s = e^{-|t-s|(h(-i\nabla)\otimes 1)} \tag{5.7}$$

under the identification $L^2(\mathbb{R}^{d+1}) \cong L^2(\mathbb{R}^d) \otimes L^2(\mathbb{R})$. Define

$$\Xi_t : L^2(Q_0) \to L^2(Q_1)$$

by

$$\Xi_t 1 = 1,$$
$$\Xi_t :\!\phi_0(f_1) \cdots \phi_0(f_n)\!: = :\!\phi_1([\xi_t]f_1) \cdots \phi_1([\xi_t]f_n)\!: .$$

From the definitions of J_t and Ξ_s, the intertwining property

$$J_s e^{-td\Gamma(h(-i\nabla))} = e^{-td\Gamma(h(-i\nabla)\otimes 1)} J_s \tag{5.8}$$

follows (Fig. 16).

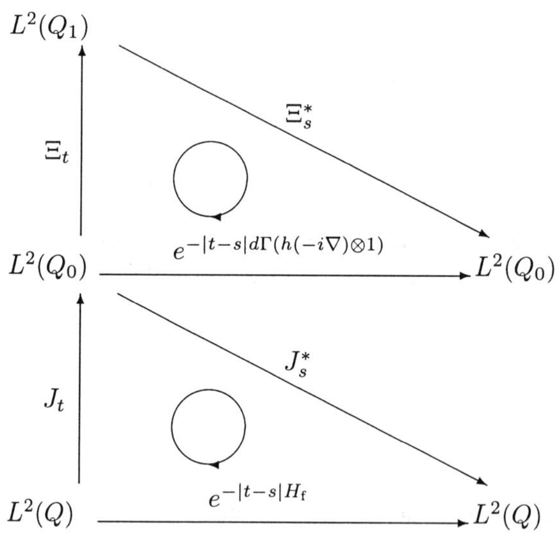

Fig. 15. Decompositions of semigroups

From (5.7) it follows that (Fig. 15)

$$\Xi_t^* \Xi_s = e^{-|t-s|d\Gamma(h(-i\nabla)\otimes 1)}. \tag{5.9}$$

We define

$$\pi_0(f) := e^{i(\pi/2)N_0} \phi_0(f) e^{-i(\pi/2)N_0},$$
$$\pi_1(g) := e^{i(\pi/2)N_1} \phi_1(g) e^{-i(\pi/2)N_1},$$

where N_0 and N_1 are the number operators in $L^2(Q_0)$ and $L^2(Q_1)$, respectively.

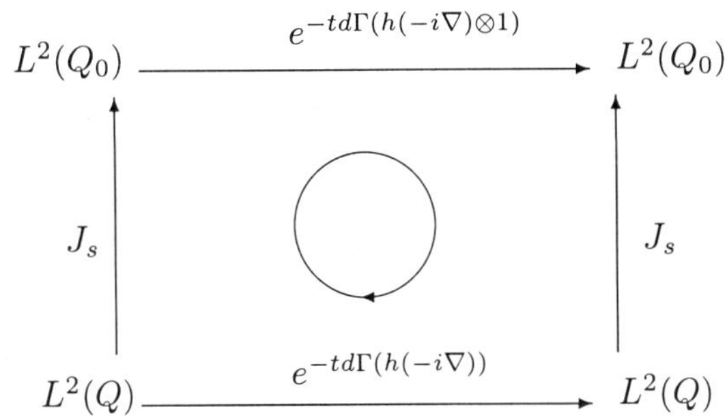

Fig. 16. Intertwining properties

5.3. *Functional integral representation I*

Let P be the Wiener measure on $\mathrm{C}([0,\infty);\mathbb{R}^d)$ with the d-dimensional Brownian motion

$$(B(t))_{t\geq 0} = (B_\mu(t))_{1\leq\mu\leq d, t\geq 0}$$

starting at 0, i.e. for $q(\cdot) = (q_1(\cdot), \ldots, q_d(\cdot)) \in \mathrm{C}([0,\infty);\mathbb{R}^d)$, it follows that

$$B_\mu(t,q) = q_\mu(t)$$

and

$$P(B(0) = 0) = 1.$$

Set

$$X_t = X_t(q,x) := B(t,q) + x, \quad t \geq 0,$$

and

$$X = X(q,x) := (X_t(q,x))_{t\geq 0}.$$

Let

$$dX := dP \otimes dx$$

be a measure on

$$\mathcal{W} := \mathrm{C}([0,\infty);\mathbb{R}^d) \times \mathbb{R}^d.$$

We fix notations. Set

$$\mathbb{E}_x(f) := \int_{\mathrm{C}([0,\infty);\mathbb{R}^d)} f(q+x)dP(q),$$

where $f \in L^1(\mathrm{C}([0,\infty);\mathbb{R}^d);dP)$, and $\mathbb{E}_x^Q$ denotes the $L^2(Q)$-valued expectation, i.e.

$$\mathbb{E}_x^Q(f_Q) := \int_{\mathrm{C}([0,\infty);\mathbb{R}^d)} f_Q(q+x)dP(q),$$

where $f_Q \in L^1(\mathrm{C}([0,\infty);\mathbb{R}^d), dP) \otimes L^2(Q)$.

Let $F(\cdot,\cdot) : \mathbb{R} \times \mathbb{R}^d \to \mathcal{K}$, where $\mathcal{K}$ is a Hilbert space. Then $\mathcal{K}$-valued stochastic integrals are defined by

$$\int_0^t F(s, B(s))dB_\mu(s)$$
$$:= \lim_{n\to\infty} \sum_{k=1}^{2^n} F\left(\frac{k-1}{2^n}t, B\left(\frac{k-1}{2^n}t\right)\right)\left\{B_\mu\left(\frac{k}{2^n}t\right) - B_\mu\left(\frac{k-1}{2^n}t\right)\right\}$$

in $L^2(\mathrm{C}([0,\infty);\mathbb{R}^d), dP) \otimes \mathcal{K}$.

Theorem 5.3: [Functional integral representation I] Let $|e|$ be sufficiently small. Suppose that $\hat{\lambda}/\sqrt{\omega}$, $\hat{\lambda}$, $\omega\hat{\lambda} \in L^2(\mathbb{R}^d)$and $\omega^2\hat{\lambda} \in L^2(\mathbb{R}^d)$. Let h be a nonnegative multiplication operator of $L^2(\mathbb{R}^d)$. Assume that $F_0, F_m \in \mathcal{H}$, and $F_1, \ldots, F_{m-1} \in L^\infty(\mathbb{R}^d) \widehat{\otimes} L_0^2(Q)$. Set

$$K := d\Gamma(h(-i\nabla))$$

and

$$\widehat{F}_j := \Xi_{\tau_j} J_{t_j} F_j.$$

Then, for $0 := t_0 \le t_1 \le t_2 \le \cdots \le t_m$ and $0 \le \tau_1 \le \tau_2 \le \cdots \le \tau_m$, we have

$$\Big(F_0, e^{-\tau_1 K} e^{-t_1 H} F_1 e^{-(\tau_2-\tau_1)K} e^{-(t_2-t_1)H} F_2 \cdots$$
$$\cdots F_{m-1} e^{-(\tau_m-\tau_{m-1})K} e^{-(t_m-t_{m-1})H} F_m\Big)$$
$$= \int_{\mathcal{W}\times Q_1} e^{-ie\phi_1(\mathcal{L}(t_0,t_1,\ldots,t_m))} \widehat{F}_0(X_0)\widehat{F}_{t_1}(X_{t_1})\cdots\widehat{F}_{t_m}(X_{t_m})dX d\nu_1(\phi_1), \tag{5.10}$$

where

$$\mathcal{L}(t_0,t_1,\ldots,t_m) := \bigoplus_{\mu=1}^d \sum_{j=1}^m \int_{t_{j-1}}^{t_j} \xi_{\tau_j} j_s \lambda(\cdot - X_s) dB_\mu(s),$$

and $\int_T^S \cdots dB_\mu(s)$ denotes an $L^2(\mathbb{R}^{d+2})$-valued stochastic integral.[k] In particular

$$(F, e^{-tH}G) = \int_{\mathcal{W}} e^{-\int_0^t V(X_s)ds} (F(X_0), \mathcal{J}_t G(X_t))_{L^2(Q)}\, dX, \tag{5.11}$$

where

$$\mathcal{J}_t := \mathcal{J}_t(X) := J_0^* e^{-ie\phi_0(\mathcal{K}(t))} J_t$$

and

$$\mathcal{K}(t) := \bigoplus_{\mu=1}^{d} \int_0^t j_s \lambda(\cdot - X_s) dB_\mu(s).$$

Proof: For instance we set $V = 0$. By the Trotter–Kato product formula[108] we have

$$e^{-tH} = \mathrm{s-}\lim_{n\to\infty} \left(e^{-(t/n)H_0} e^{-(t/n)H_\mathrm{f}} \right)^n.$$

Put $a_n := t_n - t_{n-1}$ and $b_n := \tau_n - \tau_{n-1}$. Thus

$$\begin{aligned}
&\Big(F_0, e^{-\tau_1 K} e^{-t_1 H} F_1 e^{-(\tau_2-\tau_1)K} e^{-(t_2-t_1)H} F_2 \cdots \\
&\qquad\qquad \cdots F_{m-1} e^{-(\tau_m-\tau_{m-1})K} e^{-(t_m-t_{m-1})H} F_m \Big) \\
&= \lim_{n\to\infty} \Big(F_0, e^{-b_1 K} \left(e^{-(a_1/n)H_0} e^{-(a_1/n)H_\mathrm{f}} \right)^n \\
&\quad \times F_1 e^{-b_2 K} \left(e^{-(a_2/n)H_0} e^{-(a_2/n)H_\mathrm{f}} \right)^n \cdots \\
&\quad \cdots F_{m-1} e^{-b_m K} \left(e^{-(a_m/n)H_0} e^{-(a_m/n)H_\mathrm{f}} \right)^n F_m \Big)
\end{aligned} \tag{5.12}$$

For $F(\cdot) \in \mathcal{H}$ we define

$$Q_s : \mathcal{H} \to \mathcal{H}$$

by

$$\begin{aligned}
Q_0 F(x) &:= F(x), \\
Q_s F(x) &:= \int_{\mathbb{R}^d} p_s(|x-y|) e^{-i(e/2)\phi(\lambda_{x,y})} F(y) dy, \quad s > 0,
\end{aligned}$$

[k]Note that $\lambda(\cdot - X_s) \in L^2(\mathbb{R}^d)$, $j_s\lambda(\cdot - X_s) \in L^2(\mathbb{R}^{d+1})$ and $\xi_{\tau_j} j_s \lambda(\cdot - X_s) \in L^2(\mathbb{R}^{d+2})$ for each X_s.

where

$$p_t(x) := (2\pi t)^{-d/2} \exp\left(-|x|^2/(2t)\right),$$

$$\lambda_{x,y} := \bigoplus_{\mu=1}^{d} (\lambda(\cdot - x) + \lambda(\cdot - y))(x_\mu - y_\mu).$$

Then[77,136]

$$e^{-tH_0} = \text{s-}\lim_{n\to\infty} \left(Q_{t/2^n}\right)^{2^n}.$$

Let $E_s := J_s J_s^*$. Then E_s is a projection of $L^2(Q_0)$. Define

$$Q_{[a,b]} := \mathfrak{L}\{F \in L^2(Q_0) | F \in \text{Ran} E_s, s \in [a,b]\}.$$

Let $\Sigma_{[a,b]}$ be the smallest σ-field generated by $Q_{[a,b]}$. Suppose that Ψ is measurable with respect to $\Sigma_{[a,b]}$ and Φ with respect to $\Sigma_{[c,d]}$, where $a \le b \le c \le d$. Then, for $b \le s \le c$, it is proven

$$(\Psi, E_s \Phi) = (\Psi, \Phi).$$

This property is called the Markov property[135] of E_s. Using this Markov property, the intertwining property (5.10), i.e.

$$J_s e^{-tK} = e^{-t\widetilde{K}} J_s,$$

where

$$\widetilde{K} := d\Gamma(h(-i\nabla) \otimes 1),$$

and the facts that

$$e^{-|S-T|H_{\rm f}} = J_S^* J_T, \quad S, T \in \mathbb{R},$$

$$J_s e^{\phi(f)} J_s^* = (J_s J_s^*) e^{\phi_0(j_s f)} (J_s J_s^*)$$

as an operator, we have

$$(5.12) = \lim_{n\to\infty} \left(J_0 F_0, \left(e^{-b_1 \widetilde{K}}\right) e^{-ie\phi_0(t_0,t_1)} (J_{t_1} F_1) \left(e^{-b_2 \widetilde{K}}\right) e^{-ie\phi_0(t_1,t_2)} \cdots (J_{t_{m-1}} F_{m-1}) \left(e^{-b_m \widetilde{K}}\right) e^{-ie\phi_0(t_{m-1},t_m)} J_{t_m} F_m\right), \quad (5.13)$$

where

$$\phi_0(t_a, t_b) := \phi_0\left(\bigoplus_{\mu=1}^{d} \int_{t_a}^{t_b} j_s \lambda(\cdot - X_s) dB_\mu(s)\right).$$

Using also the Markov property of $\Xi_t \Xi_t^*$ and the facts that

$$e^{-|S-T|\widetilde{K}} = \Xi_S^* \Xi_T, \quad S, T \in \mathbb{R},$$

$$\Xi_t e^{\phi_0(f)} \Xi_t^* = (\Xi_t \Xi_t^*) e^{\phi_1(\xi_t f)} (\Xi_t \Xi_t^*),$$

The Fock representation	The Schrödinger representation
$\mathcal{H}_{\rm rad} = L^2(\mathbb{R}^d) \otimes \mathcal{F}_{\rm rad}$	$\mathcal{H} = L^2(\mathbb{R}^d) \otimes L^2(Q)$
$\Omega = \{1, 0, 0, \ldots\}$	1
$A_\mu(\hat{\lambda}, x)$	$\mathcal{A}_\mu = \phi(\oplus_{\nu=1}^d \delta_{\mu\nu}\lambda(\cdot - x))$
$d\Gamma_{\rm rad}(h)$	$d\Gamma(h(-i\nabla))$
$H_{\rm rad} = d\Gamma_{\rm rad}(\omega)$	$H_{\rm f} = d\Gamma(\omega(-i\nabla))$
$N_{\rm rad} = d\Gamma(1)$	$N = d\Gamma(1)$
$H_{\rm PF} = \frac{1}{2}(p - eA)^2 + V + H_{\rm rad}$	$H = \frac{1}{2}(p - e\mathcal{A})^2 + V + H_{\rm f}$
$e^{-tH_{\rm rad}}$	$e^{-tH_{\rm f}} = J_0^* J_t$
$e^{-tH_{\rm PF}} G$	$\mathbb{E}_x^Q \left(e^{-\int_0^t V(X_s)ds} \mathcal{J}_t G(X_t)\right)$

Fig. 17. The Fock representation and the Schrödinger representation

we get (5.10). For $V \neq 0$ one can get (5.10) by the Trotter–Kato product formula

$$e^{-tH} = \text{s-}\lim_{n\to\infty} (e^{-(t/n)V} e^{-(t/n)H_0} e^{-(t/n)H_{\rm f}})^n.$$

□

From this theorem we have

$$e^{-tH} G(x) = \mathbb{E}_x^Q \left(e^{-\int_0^t V(X_s)ds} \mathcal{J}_t G(X_t)\right) \tag{5.14}$$

for almost every $x \in \mathbb{R}^d$.

Remark 5.4: In the proof of Theorem 5.3 we use condition $\omega^2 \hat{\lambda} \in L^2(\mathbb{R}^d)$ to show (5.13). This condition is, however, removed in Section 6.

6. Self-Adjointness for Arbitrary $e \in \mathbb{R}$

Klein and Landau[110] and Segal[132] apply functional integrals to prove the self-adjointness or essential self-adjointness of quantum field models. By using a functional integral representation of e^{-tH}, we also prove the self-adjointness of H for arbitrary values of coupling constants.

Before going to discussion we give an outline of a proof.[83,87]

(1) Using the right hand side of (5.11), we construct a *symmetric* semigroup $S(t)$ for arbitrary e. Let us denote the self-adjoint generator of $S(t)$ by $\widehat{H}$, i.e.

$$S(t) = e^{-t\widehat{H}}.$$

(2) We prove that $\widehat{H}$ is a self-adjoint extension of $H\lceil_{\mathcal{D}}$ with some domain $\mathcal{D}$, i.e.

$$\widehat{H} \supset H\lceil_{\mathcal{D}}.$$

(3) We show that $e^{-t\widehat{H}}$ leaves $\mathcal{D}$, i.e.

$$e^{-t\widehat{H}}\mathcal{D} \subset \mathcal{D}.$$

Then $\widehat{H}$ is essentially self-adjoint on $\mathcal{D}$.

(4) We finally prove that $\widehat{H}$ is closed on $\mathcal{D}$, which implies that $\widehat{H}$ is self-adjoint on $\mathcal{D}$, and hence H is.

6.1. *Self-adjoint extensions*

We redefine $Q_s : \mathcal{H} \to \mathcal{H}$ for *arbitrary* $e \in \mathbb{R}$ by

$$Q_0 F(x) := F(x),$$
$$Q_s F(x) := \int_{\mathbb{R}^d} p_s(|x-y|) e^{-i(e/2)\phi(\lambda_{x,y})} F(y) dy, \quad s > 0.$$

for almost every $x \in \mathbb{R}^d$. Define

$$S(t) := \text{s-}\lim_{n\to\infty} \left(Q_{t/2^n}\right)^{2^n}.$$

Suppose

$$\hat{\lambda}, \omega\hat{\lambda}, \omega^2\hat{\lambda} \in L^2(\mathbb{R}^d). \tag{6.1}$$

Thus by a direct calculation we see that $S(t)$ exists and

$$(F, S(t)G) = \int_W \left(F(X_0), e^{-ie\phi(\mathcal{Z}(t))} G(X_t)\right)_{L^2(Q)} dX, \tag{6.2}$$

where

$$\mathcal{Z}(t) := \bigoplus_{\mu=1}^{d} \int_0^t \lambda(\cdot - X_s) dB_\mu(s).$$

From the definition of $S(t)$ we can immediately check that

(1) $|(F, S(t)G)| \le \|F\|\|G\|$,
(2) $(F, S(t)G) = (S(t)F, G)$,
(3) $(F, S(t)S(s)G) = (F, S(s+t)G)$,
(4) $\lim_{t\to\infty}(F, S(t)G) = (F, S(0)G) = (F, G)$.

Thus $S(t)$ is a strongly continuous symmetric one-parameter semigroup in $t \ge 0$. Define the generator θ of $S(t)$ by $\theta F := s-\lim_{t\to 0} t^{-1}(1-S(t))F$ and $D(\theta) := \{F \in \mathcal{H}_{\mathrm{rad}} | {}^\exists s - \lim_{t\to 0} t^{-1}(1-S(t))F\}$. Then the generator θ of $S(t)$ is symmetric and closed. By the Hille–Yoshida theorem the resolvent set of θ contains interval $(-\infty, E]$ with some E. Then $\sigma(\theta) \subset \mathbb{R}$, i.e. θ is self-adjoint with

$$S(t) = e^{-t\theta}.$$

We want to remove the assumption $\omega^2 \hat{\lambda} \in L^2(\mathbb{R}^d)$ in (6.1).

Lemma 6.1: *Suppose* $\hat{\lambda}$, $\omega\hat{\lambda} \in L^2(\mathbb{R}^d)$. *Then there exists a self-adjoint operator* $\widehat{H}_0$ *such that*

$$(F, e^{-t\widehat{H}_0} G) = \int_{\mathcal{W}} \left(F(X_0), e^{-ie\phi(\mathcal{Z}(t))} G(X_t)\right)_{L^2(Q)} dX.$$

Remark 6.2: We do not need assumption $\sqrt{\omega}\hat{\lambda} \in L^2(\mathbb{R}^d)$ in Lemma 6.1.

Proof: Let $\hat{\lambda}_n$ be such that $\hat{\lambda}_n \to \hat{\lambda}, \omega\hat{\lambda}_n \to \omega\hat{\lambda}$ as $n \to \infty$ with $\hat{\lambda}_n, \omega\hat{\lambda}_n, \omega^2\hat{\lambda}_n \in L^2(\mathbb{R}^d)$. We define $\mathcal{Z}_n(t)$ by $\mathcal{Z}(t)$ with $\hat{\lambda}$ replaced by $\hat{\lambda}_n$. Then there exists a self-adjoint operator $\theta(n)$ such that

$$(F, e^{-t\theta(n)} G) = \int_{\mathcal{W}} \left(F(X_0), e^{-ie\phi(\mathcal{Z}_n(t))} G(X_t)\right)_{L^2(Q)} dX. \tag{6.3}$$

We see that

$$\lim_{n\to\infty} \int_{\mathcal{W}} \left(F(X_0), e^{-ie\phi(\mathcal{Z}_n(t))} G(X_t)\right)_{L^2(Q)} dX \\ = \int_{\mathcal{W}} \left(F(X_0), e^{-ie\phi(\mathcal{Z}(t))} G(X_t)\right)_{L^2(Q)} dX. \tag{6.4}$$

The right hand side of (6.4) defines a contractive operator G_t such as

$$\int_{\mathcal{W}} \left(F(X_0), e^{-ie\phi(\mathcal{Z}(t))} G(X_t)\right)_{L^2(Q)} dX = (F, G_t G). \tag{6.5}$$

Then it is shown that $e^{-t\theta(n)}$ strongly converses to G_t as $n \to \infty$. Hence

$$G_t G_s = \mathrm{s-}\lim_{n\to\infty} e^{-t\theta(n)} e^{-s\theta(n)} = \mathrm{s-}\lim_{n\to\infty} e^{-(t+s)\theta(n)} = G_{t+s}$$

and

$$G_t^* = \left(\mathrm{s-}\lim_{n\to\infty} e^{-t\theta(n)}\right)^* = G_t.$$

Furthermore,

$$\mathrm{s-}\lim_{t\to 0} G_t = 1$$

follows from (6.5). Hence $\{G_t\}_{t\geq 0}$ is a strongly continuous symmetric one-parameter semigroup in $t \geq 0$. Thus there exists a self-adjoint operator $\widehat{H}_0$ such that

$$G_t = e^{-t\widehat{H}_0}.$$

Hence the lemma follows. □

Lemma 6.3: *Suppose $\hat{\lambda}, \omega\hat{\lambda} \in L^2(\mathbb{R}^d)$. Then, for all $e \in \mathbb{R}$,*

$$H_0 \lceil_{D(\Delta)\cap D(N)} \subset \widehat{H}_0.$$

Proof: Take $F \in C_0^\infty(\mathbb{R}^d) \widehat{\otimes} L_0^2(Q)$ and $G \in \mathcal{H}$. For the case of $\hat{\lambda}, \omega\hat{\lambda}, \omega^2\hat{\lambda} \in L^2(\mathbb{R}^d)$, we have[77]

$$\left(G, \frac{1}{t}\left(e^{-t\widehat{H}_0} - 1\right) F\right)_{\mathcal{H}} = -\int_0^1 \left(e^{-ts\widehat{H}_0} G, H_0 F\right)_{\mathcal{H}} ds. \tag{6.6}$$

By a limiting argument, it is seen that (6.6) also holds for the case of $\hat{\lambda}, \omega\hat{\lambda} \in L^2(\mathbb{R}^d)$ but $\omega^2\hat{\lambda} \notin L^2(\mathbb{R}^d)$. Since

$$\|H_0 F\| \leq C\left(\|\Delta F\| + \|NF\| + \|F\|\right)$$

with some constant C, we can extend (6.6) to $F \in D(\Delta) \cap D(N)$. Take $G \in D(\widehat{H}_0)$. We have

$$-\left(\widehat{H}_0 G, F\right) = \lim_{t\to\infty} \left(G, \frac{1}{t}\left(e^{-t\widehat{H}_0} - 1\right) F\right)$$
$$= -\int_0^1 ds\,(G, H_0 F) = -(G, H_0 F).$$

Then $F \in D(\widehat{H}_0)$ and $\widehat{H}_0 F = H_0 F$. Hence the lemma follows. □

Lemma 6.4: *Suppose $\hat{\lambda}/\sqrt{\omega}$, $\hat{\lambda}$, $\omega\hat{\lambda} \in L^2(\mathbb{R}^d)$. Then for all $e \in \mathbb{R}$*

$$H_0 \lceil_{D(\Delta)\cap D(H_f)} \subset \widehat{H}_0. \tag{6.7}$$

Proof: Since $\hat{\lambda}/\sqrt{\omega} \in L^2(\mathbb{R}^d)$, it holds that

$$\|H_0 F\| \le C'(\|\Delta F\| + \|H_{\rm f} F\| + \|F\|)$$

with some constant C'. Then (6.6) can be extended to $F \in D(\Delta) \cap D(H_{\rm f})$. The lemma follows in the similar way as Lemma 6.3. □

We define $\widehat{H}$ by a quadratic form sum such as

$$\widehat{H} := \widehat{H}_0 \dot{+} H_{\rm f}.$$

Let $V = 0$. We see that, for the case of $\hat{\lambda}, \omega\hat{\lambda} \in L^2(\mathbb{R}^d)$,

$$H\lceil_{D(\Delta)\cap D(N)\cap D(H_{\rm f})} \subset \widehat{H}, \tag{6.8}$$

and, for the case of $\hat{\lambda}/\sqrt{\omega}$, $\hat{\lambda}$, $\omega\hat{\lambda} \in L^2(\mathbb{R}^d)$,

$$H\lceil_{D(\Delta)\cap D(H_{\rm f})} \subset \widehat{H}. \tag{6.9}$$

Original operators	Self-adjoint extensions
$H_0 = \frac{1}{2}(p - e\mathcal{A})^2$	$\widehat{H}_0$
$H_0 + H_{\rm f}$	$\widehat{H} = \widehat{H}_0 \dot{+} H_{\rm f}$
$H = H_0 + H_{\rm f} + V$	$\widehat{H} \dot{+} V_+ \dot{-} V_-$

Fig. 18. Self-adjoint extensions

6.2. *Invariant domains*

We present a general lemma.

Lemma 6.5: *Let $\mathcal{K}$ be a Hilbert space.*
(1) *Let T be a densely defined closed operator in $\mathcal{K}$ and D a core of T. Suppose that for $g \in \mathcal{K}$,*

$$|(Tf, g)| \le R_g\|f\| \tag{6.10}$$

for all $f \in D$ with some constant R_g. Then $g \in D(T^*)$ with $\|T^*g\| \le R_g$.
(2) *Let S be a densely defined closed operator and U a densely defined linear operator in $\mathcal{K}$. Let D_1 be a core of S and D_2 the domain of S. Suppose that*

$$|(Uf, Sg)| \le R\|f\|\|g\| \tag{6.11}$$

for all $f \in D_1$ and $g \in D_2$ with some constant R. Then U^*S has the unique bounded operator extension $\overline{U^*S}$ and $\|\overline{U^*S}\| \le R$. In particular, if S is bounded, then U^*S is bounded.

Proof: Let $B(f) := (Tf, g)$ for $f \in D$. Then, by (6.10), B can be extended to $\mathcal{K}$. We denote its extension by $\widetilde{B}$. Thus by the Riesz theorem there exists $\phi \in \mathcal{K}$ such that $\widetilde{B}(f) = (f, \phi)$ and $\|\phi\| \leq R_g$. In particular, for $f \in D(T)$, $\widetilde{B}(f) = B(f) = (Tf, g) = (f, \phi)$. Since D is a core, (1) follows. Let $C(f, g) := (Uf, Sg)$ for $f \in D_1$ and $g \in D_2$. By (6.11), C can be extended to $\mathcal{K} \times \mathcal{K}$. We denote its extension by $\widetilde{C}$. Thus by the Riesz theorem there exists a bounded operator K such that $\widetilde{C}(f, g) = (f, Kg)$ and $\|K\| \leq R$. For $f \in D_1$ and $g \in D_2$, $\widetilde{C}(f, g) = C(f, g) = (Uf, Sg) = (f, Kg)$. Since D_1 is a core of U, $Sg \in D(U^*)$ and $U^*Sg = Kg$. Hence (2) follows. □

A Burkholder inequality[104] is useful for estimating stochastic integrals.

Lemma 6.6: *Suppose $\omega^{k/2}\hat{\lambda} \in L^2(\mathbb{R}^d)$. Then for each $x \in \mathbb{R}^d$*

$$\mathbb{E}_x\left(\left\|\left(\widehat{\omega}^{k/2} \otimes 1\right) \int_0^t j_s \lambda(\cdot - X_s) dB_\mu(s)\right\|_{L^2(\mathbb{R}^{d+1})}^{2m}\right) \leq \frac{(2m)!}{2^m} t^m \|\omega^{k/2}\hat{\lambda}\|_{L^2(\mathbb{R}^d)}^{2m}.$$

Proof: See Ref. 83. □

Lemma 6.7: *Suppose $\hat{\lambda}/\sqrt{\omega}$, $\hat{\lambda}$, $\omega\hat{\lambda} \in L^2(\mathbb{R}^d)$. Then*

$$\left(F, e^{-t\widehat{H}} G\right) = \int_{\mathcal{W}} (F_0, \mathfrak{J}_t G_t)_{L^2(Q)} \, dX. \tag{6.12}$$

Here we set $F_0 := F(X_0)$ and $G_t := G(X_t)$.

Proof: Since

$$e^{-t\widehat{H}} = \text{s-}\lim_{n\to\infty} \left(e^{-(t/n)\widehat{H}_0} e^{-(t/n)H_{\rm f}}\right)^n,$$

it is proven in the similar way as Theorem 5.3. □

Lemma 6.8: *Suppose $\hat{\lambda}/\sqrt{\omega}$, $\hat{\lambda}$, $\omega\hat{\lambda} \in L^2(\mathbb{R}^d)$ and $G \in D(H_{\rm f})$. Then*

$$e^{-t\widehat{H}} G \in D(H_{\rm f})$$

and

$$\|H_{\rm f} e^{-t\widehat{H}} G\| \leq C(1 + \sqrt{t} + t)\|(H_{\rm f} + 1)G\|, \tag{6.13}$$

$$\|[H_{\rm f}, e^{-t\widehat{H}}]G\| \leq C(\sqrt{t} + t)\|(H_{\rm f} + 1)G\| \tag{6.14}$$

with some constant C.

Proof: For the case of $\hat{\lambda}/\sqrt{\omega}$, $\hat{\lambda}$, $\omega\hat{\lambda} \in L^2(\mathbb{R}^d)$, it is proven that $\mathfrak{J}_t$ maps $D(H_{\rm f})$ onto itself. Then we have

$$\begin{aligned}(H_{\rm f}F, e^{-t\widehat{H}}G) &= \int_{\mathcal{W}} (H_{\rm f}F_0, \mathfrak{J}_t G_t)_{L^2(Q)} dX \\ &= \int_{\mathcal{W}} (F_0, [H_{\rm f}, \mathfrak{J}_t]G_t)_{L^2(Q)} dX + (F, e^{-t\widehat{H}} H_{\rm f} G).\end{aligned}$$

Then

$$\int_{\mathcal{W}} (F_0, [H_{\rm f}, \mathfrak{J}_t]G_t)_{L^2(Q)} dX = \int_{\mathcal{W}} \left(F_0, J_0^* e^{-ie\phi(\mathcal{K}(t))} S(X) J_t G_t\right)_{L^2(Q)} dX,$$

where

$$S(X) := d\Gamma(\widehat{\omega}\otimes 1) - e\pi_0([\widehat{\omega}\otimes 1]\mathcal{K}(t)) + \frac{e^2}{2} q_0([\widehat{\omega}\otimes 1]\mathcal{K}(t), \mathcal{K}(t)).$$

Note that

$$\begin{aligned}&\|\pi_0([\widehat{\omega}\otimes 1]\mathcal{K}(t)) J_t \Psi\| \\ &\le \frac{d-1}{\sqrt{2}}\left(2\|[\widehat{\omega}\otimes 1]^{1/2}\mathcal{K}(t)\| + \|[\widehat{\omega}\otimes 1]\mathcal{K}(t)\|\right)\|(H_{\rm f}+1)^{1/2}\Psi\|,\end{aligned} \tag{6.15}$$

and

$$|q_0([\widehat{\omega}\otimes 1]\mathcal{K}(t), \mathcal{K}(t))| \le \frac{1}{2}\|[\widehat{\omega}\otimes 1]^{1/2}\mathcal{K}(t)\|^2. \tag{6.16}$$

We have

$$\begin{aligned}|(H_{\rm f}F, e^{-t\widehat{H}}G)| &\le \int_{\mathcal{W}} \|F\| \|[H_{\rm f}, \mathfrak{J}_t]G_t\| dX + \|F\| \|H_{\rm f} G\| \\ &\le \|F\| \left\{\int_{\mathbb{R}^d} (\mathbb{E}_x \|[H_{\rm f}, \mathfrak{J}_t]G_t\|)^2 dx\right\}^{1/2} + \|F\| \|H_{\rm f} G\|.\end{aligned}$$

Using (6.15), (6.16) and Lemma 6.6, it is seen that

$$(\mathbb{E}_x \|[H_{\rm f}, \mathfrak{J}_t]G_t\|)^2 \le \mathbb{E}_x\left(\|[H_{\rm f}, \mathfrak{J}_t]G_t\|^2\right) \le C'(t+t^2)\|(H_{\rm f}+1)G_t\|^2. \tag{6.17}$$

Thus

$$|(H_{\rm f}F, e^{-t\widehat{H}}G)| \le C'(1+\sqrt{t}+t)\|F\| \|(H_{\rm f}+1)G\|.$$

Hence $e^{-t\widehat{H}}G \in D(H_{\rm f})$ and (6.13) holds by Lemma 6.5. Moreover, since

$$(F, [H_{\rm f}, e^{-t\widehat{H}}]G) = \int_{\mathcal{W}} (F_0, [H_{\rm f}, \mathfrak{J}_t]G_t)_{L^2(Q)} dX,$$

equation (6.14) follows from (6.17). □

Let $z \gg 1$. By (6.13) and

$$(\widehat{H}+z)^{-k}\Psi = \Gamma(k)^{-1}\int_0^\infty e^{-t\widehat{H}}e^{-tz}t^{-1+k}\Psi dt,$$

we see that for $\Psi \in D(H_{\rm f})$

$$(\widehat{H}+z)^{-k}\Psi \in D(H_{\rm f}), \tag{6.18}$$

where $\Gamma(k)$ denotes the Gamma function.

Lemma 6.9: *Suppose* $\hat{\lambda}/\sqrt{\omega}$, $\hat{\lambda}$, $\omega\hat{\lambda} \in L^2(\mathbb{R}^d)$. *Then* $H_{\rm f}(\widehat{H}+z)^{-1}$ *is a bounded operator.*

Proof: Let $\Psi \in D(H_{\rm f})$. Then, by (6.18),

$$\begin{aligned}&H_{\rm f}(\widehat{H}+z)^{-1}\Psi\\&= (\widehat{H}+z)^{-1/2}H_{\rm f}(\widehat{H}+z)^{-1/2}\Psi + [(\widehat{H}+z)^{-1/2}, H_{\rm f}](\widehat{H}+z)^{-1/2}\Psi.\end{aligned}$$

It is checked that $(\widehat{H}+z)^{-1/2}H_{\rm f}(\widehat{H}+z)^{-1/2}$ is bounded. On the other hand,

$$\begin{aligned}&[(\widehat{H}+z)^{-1/2}, H_{\rm f}](\widehat{H}+z)^{-1/2}\Psi\\&= \Gamma(k)^{-1}\int_0^\infty \frac{1}{\sqrt{t}}[e^{-t\widehat{H}}, H_{\rm f}]e^{-tz}(\widehat{H}+z)^{-1/2}\Psi dt.\end{aligned}$$

Using (6.14), we check that

$$\|[(\widehat{H}+z)^{-1/2}, H_{\rm f}](\widehat{H}+z)^{-1/2}\Psi\| \le C'\|\Psi\|$$

with some constant C'. Hence

$$\|H_{\rm f}(\widehat{H}+z)^{-1}\Psi\| \le C''\|\Psi\|, \quad \Psi \in D(H_{\rm f}).$$

Since $D(H_{\rm f})$ is dense, the lemma follows. □

Next we discuss N. It is easily proven that $\mathfrak{J}_t$ maps $D(N^m)$ onto itself for arbitrary m. We see that

$$N^m\mathfrak{J}_t = J_0^*e^{-ie\phi(\mathcal{K}(t))}\left(N_0 - e\pi_0(\mathcal{K}(t)) + \frac{e^2}{2}q_0(\mathcal{K}(t),\mathcal{K}(t))\right)^m J_t$$

holds on $D(N^m)$. Set

$$C^\infty(T) := \bigcap_{n=1}^\infty D(T^n).$$

Lemma 6.10: *Suppose* $\hat{\lambda}$, $\omega\hat{\lambda} \in L^2(\mathbb{R}^d)$. *Then both* $e^{-t\widehat{H}}$ *and* $e^{-t\widehat{H}_0}$ *leave* $C^\infty(N)$ *invariant.*

Proof: Let $F, G \in C^\infty(N)$. Then

$$(N^m F, e^{-t\widehat{H}} G) = \int_{\mathcal{W}} (F_0, N^m \mathfrak{J}_t G_t)_{L^2(Q)} dX.$$

We have

$$\|N^m \mathfrak{J}_t G_t\| \le C_1(X) \|(N+1)^m G_t\|.$$

Here $C_1(X)$ is such that

$$\mathbb{E}_x \left(C_1(X)^2 \right) \le P(t)$$

with some polynomial $P(t)$ of degree m. Hence

$$|(N^m F, e^{-t\widehat{H}} G)| \le P(t) \|F\| \|(N+1)^m G\|.$$

Thus from Lemma 6.5 it follows that

$$e^{-t\widehat{H}} C^\infty(N) \subset C^\infty(N).$$

Similarly we can prove that

$$e^{-t\widehat{H}_0} C^\infty(N) \subset C^\infty(N).$$

We complete the proof. □

Lemma 6.11: *Suppose $\hat{\lambda}, \omega\hat{\lambda} \in L^2(\mathbb{R}^d)$, $z > 0$ and $t > 0$. Then both*

$$\Delta(\widehat{H} + z)^{-1} \text{ and } \Delta^k e^{-t\widehat{H}}, \quad k \ge 0$$

are bounded. In particular, $e^{-t\widehat{H}}$ leaves $C^\infty(\Delta)$ invariant.

Proof: We show an outline of a proof. See Ref. 87 for details. We fix $k \ge 0$. From Lemma 6.1 it follows that

$$|(F, e^{-t\widehat{H}} G)| \le (|F|, e^{-t(-(1/2)\Delta + H_f)} |G|). \tag{6.19}$$

Let $F \in C_0^\infty(\mathbb{R}^d) \widehat{\otimes} [L_0^2(Q) \cap C^\infty(H_f)]$. Let F_R denote the real-part of F and F_I its imaginary-part, i.e. $F = F_R + iF_I$. Define subsets of $\mathbb{R}^d \times Q$ by

$$\begin{aligned}
Q_R^+ &:= \{(x, \phi) \in \mathbb{R}^d \times Q | (\Delta^k F_R)(x, \phi) > 0\}, \\
Q_R^- &:= \{(x, \phi) \in \mathbb{R}^d \times Q | (\Delta^k F_R)(x, \phi) < 0\}, \\
Q_I^+ &:= \{(x, \phi) \in \mathbb{R}^d \times Q | (\Delta^k F_I)(x, \phi) > 0\}, \\
Q_I^- &:= \{(x, \phi) \in \mathbb{R}^d \times Q | (\Delta^k F_I)(x, \phi) < 0\}.
\end{aligned}$$

Define also

$$F_1 := \begin{cases} F_{\mathrm{R}}(x,\phi) & (x,\phi) \in Q_{\mathrm{R}}^+, \\ 0 & (x,\phi) \notin Q_{\mathrm{R}}^+, \end{cases}$$

$$F_2 := \begin{cases} -F_{\mathrm{R}}(x,\phi) & (x,\phi) \in Q_{\mathrm{R}}^-, \\ 0 & (x,\phi) \notin Q_{\mathrm{R}}^-, \end{cases}$$

$$F_3 := \begin{cases} F_{\mathrm{I}}(x,\phi) & (x,\phi) \in Q_{\mathrm{I}}^+, \\ 0 & (x,\phi) \notin Q_{\mathrm{I}}^+, \end{cases}$$

$$F_4 := \begin{cases} -F_{\mathrm{I}}(x,\phi) & (x,\phi) \in Q_{\mathrm{I}}^-, \\ 0 & (x,\phi) \notin Q_{\mathrm{I}}^-. \end{cases}$$

Note that $F_i \in D(\Delta^k)$ and $\Delta^k F_i > 0$, $i = 1,2,3,4$. Since

$$|\Delta^k F_{\mathrm{R}}| = \Delta^k F_1 + \Delta^k F_2,$$
$$|\Delta^k F_{\mathrm{I}}| = \Delta^k F_3 + \Delta^k F_4,$$

we have

$$|(\Delta^k F, e^{-t\widehat{H}} G)| \le \sum_{i=1}^{4} (\Delta^k F_i, e^{-t(-(1/2)\Delta + H_{\mathrm{f}})} |G|) \le \sum_{i=1}^{4} \|F_i\| \|G\|.$$

Since $\sum_{i=1}^4 \|F_i\| \le 2\|F\|$,

$$|(\Delta^k F, e^{-t(-(1/2)\Delta + H_{\mathrm{f}})} G)| \le 2\|F\|\|G\|.$$

Hence the boundedness of $\Delta^k e^{-t\widehat{H}}$ follows. By means of (6.19) we have

$$|(F, (\widehat{H}+z)^{-1} G)| \le (|F|, (-(1/2)\Delta + H_{\mathrm{f}} + z)^{-1} |G|).$$

Thus, in the similar way as above, the boundedness of $\Delta(\widehat{H}+z)^{-1}$ follows. □

6.3. *Self-adjointness*

By Lemmas 6.11 and 6.9, we obtain the following lemma.

Lemma 6.12: *Suppose* $\hat{\lambda}/\sqrt{\omega}$, $\hat{\lambda}$, $\omega\hat{\lambda} \in L^2(\mathbb{R}^d)$. *Then*

$$(-(1/2)\Delta + H_{\mathrm{f}})(\widehat{H} + z)^{-1}$$

is a bounded operator.

Theorem 6.13: Suppose that $\hat{\lambda}/\sqrt{\omega}$, $\hat{\lambda}$, $\omega\hat{\lambda} \in L^2(\mathbb{R}^d)$, and V is relatively bounded with respect to Δ with a relative bound < 1. Then H is self-adjoint on $D(\Delta) \cap D(H_{\mathrm{f}})$ and bounded from below. Moreover it is essentially self-adjoint on any core of $(-1/2)\Delta + H_{\mathrm{f}}$.

Proof: Since $e^{-t\widehat{H}}$ leaves $D(\Delta) \cap D(H_{\rm f})$ invariant, $\widehat{H}$ is essentially self-adjoint on $D(\Delta) \cap D(H_{\rm f})$. The boundedness of $(-(1/2)\Delta + H_{\rm f})(\widehat{H} + z)^{-1}$ implies that $\widehat{H}$ is closed on $D(\Delta) \cap D(H_{\rm f})$. Thus $\widehat{H}$ is self-adjoint on $D(\Delta) \cap D(H_{\rm f})$. By (6.19), we see that V is also relatively bounded with respect to $\widehat{H}$ with a relative bound < 1 (Ref. 77). Hence $\widehat{H} + V$ is self-adjoint on $D(\Delta) \cap D(H_{\rm f})$, which yields that H is self-adjoint on $D(\Delta) \cap D(H_{\rm f})$ by the fact that

$$\widehat{H} + V \supset H\lceil_{D(\Delta)\cap D(H_{\rm f})}$$

which follows from (6.7). □

Theorem 6.14: Suppose that $\hat{\lambda}$, $\omega\hat{\lambda} \in L^2(\mathbb{R}^d)$ and V is relatively bounded with respect to Δ with a relative bound < 1. Then

(1) H is essentially self-adjoint on $C^\infty(\Delta) \cap D(H_{\rm f}) \cap C^\infty(N)$ and bounded from below,
(2) H_0 is essentially self-adjoint on $C^\infty(\Delta) \cap C^\infty(N)$ and bounded from below.

Proof: Let $V = 0$. Since $e^{-t\widehat{H}}$ leaves $D(\Delta) \cap D(H_{\rm f}) \cap C^\infty(N)$ invariant, $\widehat{H}$ is essentially self-adjoint on $C^\infty(\Delta) \cap D(H_{\rm f}) \cap C^\infty(N)$. Since V is relatively bounded with respect to $\widehat{H}$ with a relative bound < 1, $H = \widehat{H} + V$ is essentially self-adjoint on $C^\infty(\Delta) \cap D(H_{\rm f}) \cap C^\infty(N)$. A proof for H_0 is identical to (1). □

Corollary 6.15: *Let V be relatively bounded with respect to Δ with a relative bound < 1.*

(1) *Suppose $\hat{\lambda}/\sqrt{\omega}, \hat{\lambda}, \omega\hat{\lambda} \in L^2(\mathbb{R}^d)$. Then*

- **(i)** $H_{\rm PF}$ *is self-adjoint on* $D(\Delta) \cap D(H_{\rm rad})$,
- **(ii)** $H_{\rm PF}^\sigma$ *is self-adjoint on*

$$\mathbb{C}^{2^{[d/2]}} \otimes (D(\Delta) \cap D(H_{\rm rad}))\,.$$

(2) *Suppose $\hat{\lambda}, \omega\hat{\lambda} \in L^2(\mathbb{R}^d)$. Then*

- **(i)** $H_{\rm PF}$ *is essentially self-adjoint on*

$$C^\infty(\Delta) \cap D(H_{\rm rad}) \cap C^\infty(N_{\rm rad}),$$

- **(ii)** $H_{\rm PF}^\sigma$ *is essentially self-adjoint on*

$$\mathbb{C}^{2^{[d/2]}} \otimes (C^\infty(\Delta) \cap D(H_{\rm rad}) \cap C^\infty(N_{\rm rad}))\,.$$

Proof: Note that $H_{\rm spin}$ is infinitesimally small with respect to $-(1/2)\Delta + H_{\rm f}$. The corollary follows from Theorems 6.13, 6.14, (5.1) and (5.2). □

Corollary 6.16: *Suppose that V is relatively bounded with respect to Δ with a relative bound < 1 and $\hat\lambda/\sqrt{\omega}, \hat\lambda, \omega\hat\lambda \in L^2(\mathbb{R}^d)$. Then for arbitrary $e \in \mathbb{R}$,*

$$\|\Delta\Psi\| + \|H_{\rm f}\Psi\| \le C(e)(\|H\Psi\| + \|\Psi\|) \tag{6.20}$$

follows, where $C(e)$ is a positive constant.

Proof: It follows from Theorem 6.13 and the closed graph theorem. □

	Self-adjointness	Domain
H	Self-adjoint	$D(\Delta) \cap D(H_{\rm f})$
H_0	Essentially self-adjoint	$C^\infty(\Delta) \cap C^\infty(N)$

Fig. 19. The case of $\hat\lambda/\sqrt{\omega}, \hat\lambda, \omega\hat\lambda \in L^2(\mathbb{R}^d)$

	Self-adjointness	Domain
H	Essentially self-adjoint	$C^\infty(\Delta) \cap D(H_{\rm f}) \cap C^\infty(N)$
H_0	Essentially self-adjoint	$C^\infty(\Delta) \cap C^\infty(N)$

Fig. 20. The case of $\hat\lambda/\sqrt{\omega} \notin L^2(\mathbb{R}^d)$ and $\hat\lambda, \omega\hat\lambda \in L^2(\mathbb{R}^d)$

6.4. *Functional integral representation II*

We define a set, $V_{\rm Kato}$, of external potentials V as follows.

Definition 6.17: We say $V \in V_{\rm Kato}$ if $V = V_+ - V_-$ satisfies

(i) $V_\pm \ge 0$,
(ii) $V_+ \in L^1_{\rm loc}(\mathbb{R}^d)$,
(iii) V_- is infinitesimally small with respect to Δ in the sense of form.

For $V \in V_{\mathrm{Kato}}$, and $\hat{\lambda}$ such that $\hat{\lambda}$, $\omega\hat{\lambda} \in L^2(\mathbb{R}^d)$, we redefine the Pauli–Fierz Hamiltonian by

$$H := \widehat{H} \dot{+} V_+ \dot{-} V_-. \tag{6.21}$$

Remark 6.18: In terms of the relative bound (6.20) or the diamagnetic inequality (6.19), V_- is infinitesimally small with respect to $\widehat{H}$ in the sense of form.

Theorem 6.19: [Functional integral representation II] Suppose $\hat{\lambda}$, $\omega\hat{\lambda} \in L^2(\mathbb{R}^d)$. Then, for all $e \in \mathbb{R}$, the functional integral representation in Theorem 5.3 holds for (6.21). In particular,

$$(F, e^{-tH}G) = \int_{\mathcal{W}} (F_0, \mathfrak{J}_t G_t)_{L^2(Q)} dX$$

and

$$(f \otimes 1, e^{-tH} f \otimes 1) = \int_{\mathcal{W}} \overline{f(X_0)} f(X_t) e^{-(e^2/2) q_0(\mathcal{K}(t))} dX.$$

Proof: It is a simple limiting argument. □

We immediately see a diamagnetic inequality.[1]

Corollary 6.20: *Suppose $\hat{\lambda}$, $\omega\hat{\lambda} \in L^2(\mathbb{R}^d)$ and H is in (6.21). Then*

$$\left|(F, e^{-tH}G)\right| \le \left(|F|, e^{-t(H_{\mathrm{p}}+H_{\mathrm{f}})}|G|\right).$$

In particular

$$\inf \sigma(H_{\mathrm{p}}) \le \inf \sigma(H).$$

Proof: Since J_t is positivity preserving (see Section 7), we have $|J_t G| = J_t |G|$. Thus the corollary follows directly from Theorem 6.19. □

7. Multiplicity

This section is due to Hiroshima.[81,86,91] See also Gross.[62,63] Let (M, m) be a σ-finite measure space.

Definition 7.1: We say that $\Psi \in L^2(M, dm)$ is positive if $\Psi \ge 0$ ($\Psi \not\equiv 0$) for a.e. m. We say that bounded operator A of $L^2(M, dm)$ is *positivity preserving* (simply we say PP), if

$$(\Psi, A\Phi)_{L^2(M,dm)} \ge 0$$

[1]This kind of inequality is studied in e.g. Kato[106,107] and Kato and Masuda.[108]

for all positive Ψ, Φ, moreover, *positivity improving* (simply we say PI), if

$$(\Psi, A\Phi)_{L^2(M,dm)} > 0$$

for all positive Ψ, Φ.

Let K be a nonnegative self-adjoint operator acting in $L^2(M, dm)$. It is well known that if e^{-tK} is PI, then the ground state of K is strictly positive and then unique up to multiple phase factors. It is established[60,135] that $\Gamma(T)$ is PP if $\|T\| \leq 1$ and that $\Gamma(T)$ is PI if $\|T\| < 1$. In particular J_t is PP. The following propositions are well known.

Proposition 7.2: *$e^{-tH_{\rm f}}$ is PI for all $t > 0$ in $L^2(Q)$.*

Proof: See Refs. 47, 48 and 134. □

Proposition 7.3: *Suppose $V \in V_{\rm Kato}$. Then $e^{-tH_{\rm p}}$ is PI for all $t > 0$ in $L^2(\mathbb{R}^d)$.*

Proof: See Ref. 136. □

Proposition 7.4: *Suppose $V \in V_{\rm Kato}$. Then $e^{-t(H_{\rm p}\otimes 1 + 1 \otimes H_{\rm f})}$ is PI for all $t > 0$ in $\mathcal{H}$.*

Proof: See Refs. 26 and 81. □

Proposition 7.4 does not directly follow from Propositions 7.2 and 7.3. It is seen that $e^{-t(H_{\rm p}\otimes 1 + 1\otimes H_{\rm f})} = e^{-t(H_{\rm p}\otimes 1)} e^{-t(1\otimes H_{\rm f})}$. Both $e^{-t(H_{\rm p}\otimes 1)}$ and $e^{-t(1\otimes H_{\rm f})}$, however, are *not* PI, but PP in $\mathcal{H}$. By Proposition 7.4, $H_{\rm p} + H_{\rm f}$ has the strictly positive unique ground state $\phi_{\rm p} \otimes 1$, where $\phi_{\rm p}$ is the ground state of $H_{\rm p}$.

7.1. *Positivity improving semigroups and uniqueness of a ground state*

We define

$$U := \exp\left(i\frac{\pi}{2}N\right).$$

Note that

$$J_t e^{iaN} = e^{iaN_0} J_t, \quad a \in \mathbb{R}.$$

Thus we have

$$(F, U^{-1}e^{-tH}UG) = \int_{\mathcal{W}} \left(F(X_0), J_0^* e^{-ie\pi_0(\mathcal{K}(t))} J_t G(X_t)\right)_{L^2(Q)} dX.$$

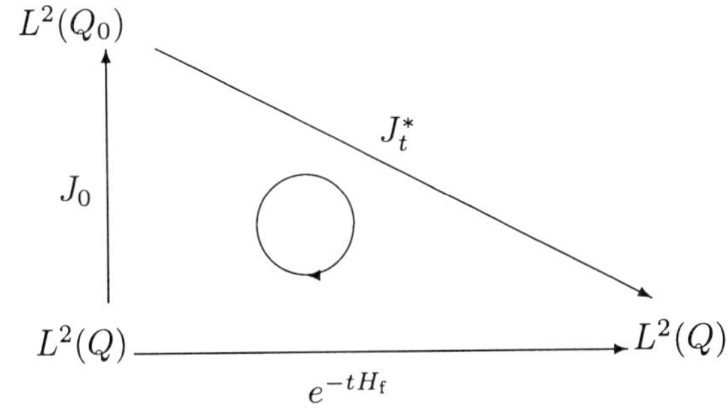

Fig. 21. Proposition 7.2

The purpose of this subsection is to prove that $U^{-1}e^{-tH}U$ is PI for all $t > 0$. We define the subspace of coherent states in $L^2(Q)$ by

$$L^2_{\mathrm{C}}(Q) := \{F(\phi(f_1), \ldots, \phi(f_n)) | \\ F \in \mathcal{S}(\mathbb{R}^n), f_j \in \oplus^d L^2(\mathbb{R}^d), j = 1, \ldots, n, n \in \mathbb{N}\},$$

where $\mathcal{S}(\mathbb{R}^n)$ denotes the set of Schwartz test functions on $\mathbb{R}^n$.

Lemma 7.5: *Let $F \in L^2(Q)$ be positive. Then there exists a positive sequence $F_n \in L^2_{\mathrm{C}}(Q)$ such that* $\mathrm{s}\text{-}\lim_{n\to\infty} F_n = F$.

Proof: See Refs. 96, 60 and 135. □

We learned the following lemma from J. Fröhlich in a private communication.

Lemma 7.6: *Let $h \in \oplus^d L^2(\mathbb{R}^d)$. Then $e^{i\pi(h)}$ is PP in $L^2(Q)$.*

Proof: Let

$$F := \int_{\mathbb{R}^N} f(t) e^{i\sum_{j=1}^N t_j \phi(f_j)} dt, \quad G := \int_{\mathbb{R}^M} g(t) e^{i\sum_{j=1}^M t_j \phi(g_j)} dt$$

with f and g the Fourier transform of positive Schwartz test functions. By the Weyl relations

$$e^{i\pi(h)} e^{i\phi(g)} = e^{i2q(h,g)} e^{i\phi(g)} e^{i\pi(h)}, \tag{7.1}$$

$$e^{i\pi(h)} 1 = e^{-q(h)} e^{-\phi(h)} 1, \tag{7.2}$$

we have

$$\left(F, e^{i\pi(h)} G\right) \\ = \int dt \int ds \overline{f(t)} g(s) \left(e^{i\sum_{j=1}^N t_j \phi(f_j)} 1, e^{i\pi(h)} e^{i\sum_{j=1}^M s_j \phi(g_j)} 1\right) \geq 0. \tag{7.3}$$

From Lemma 7.5, $(F, e^{i\pi(h)}G) \geq 0$ follows for arbitrary positive F, G in $L^2(Q)$. □

Lemma 7.7: *Let $f \in \oplus^d L^2(\mathbb{R}^{d+1})$. Then we have*

$$J_0^* e^{i\pi_0(f)} J_t = e^{-(q_0(f)+q([j_0^*]f))} \overline{J_0^* e^{-\phi_0(f)} J_t e^{i\pi([j_0^*]f)} e^{\phi([j_t^*]f)} \lceil L^2_{\mathrm{C}}(Q)},$$

where $\overline{A}$ denotes the closure of A.

Proof: Note that

$$q_0([j_t]f, g) = q(f, [j_t^*]g).$$

Suppose that $G \in L^2_{\mathrm{C}}(Q)$ is of the form

$$G = G(\phi(f_1), \ldots, \phi(f_n)) = \int_{\mathbb{R}^n} g(t) e^{i\phi(\sum_{j=1}^n t_j f_j)} dt, \quad g \in \mathcal{S}(\mathbb{R}^n).$$

Then by (7.1) and (7.2), we have

$$\begin{aligned}
& e^{i\pi_0(f)} J_t G 1 \\
&= e^{i\pi_0(f)} G(\phi_0(j_t f_1), \ldots, \phi_0(j_t f_n)) 1 \\
&= G(\phi_0(j_t f_1) + 2q_0([j_t]f_1, f), \ldots, \phi_0(j_t f_n) + 2q_0([j_t]f_n, f)) e^{i\pi_0(f)} 1 \\
&= e^{-q_0(f)} e^{-\phi_0(f)} G(\phi_0(j_t f_1) + 2q(f_1, [j_t^*]f), \ldots, \phi_0(j_t f_n) + 2q(f_n, [j_t^*]f)) 1 \\
&= e^{-q_0(f)} e^{-\phi_0(f)} J_t e^{i\pi([j_0^*]f)} G e^{-i\pi([j_t^*]f)} 1 \\
&= e^{-(q_0(f)+q([j_t^*]f))} e^{-\phi_0(f)} J_t e^{i\pi([j_0^*]f)} e^{\phi([j_0^*]f)} G 1.
\end{aligned}$$

Since $L^2_{\mathrm{C}}(Q)$ is dense, the lemma follows. □

Remark 7.8: It can be regarded that $e^{i\pi_0(f)}$ is a shift operator in $L^2(Q_0)$.

Let $f \in \oplus^d L^2(\mathbb{R}^{d+1})$. Define the bounded operator Q_M on $L^2(Q)$ by

$$Q_M := J_0^* \left(e^{-\phi_0(f)}\right)_M J_t,$$

where

$$\left(e^{-\phi_0(f)}\right)_M := \begin{cases} e^{-\phi_0(f)} & e^{-\phi_0(f)} < M, \\ M & e^{-\phi_0(f)} \geq M. \end{cases}$$

Lemma 7.9: *Q_M is PI.*

Proof: Let θ_1, θ_2 be positive. It is easily proven that

$$(\theta_1, Q_M \theta_2) \geq 0.$$

Hence it is enough to prove that

$$(\theta_1, Q_M\theta_2) \neq 0. \tag{7.4}$$

Assume that $(\theta_1, Q_M\theta_2) = 0$. Since J_t and J_0 are PP, we have

$$\left\{\mathrm{supp}\left(e^{-\phi_0(f)}\right)_M J_t\theta_2\right\} \bigcap \{\mathrm{supp} J_0\theta_1\} = \emptyset.$$

Moreover $\left(e^{-\phi_0(f)}\right)_M \neq 0$ a.e., since $\int_{Q_0} |\phi_0(f)|^2 d\nu_0(\phi_0) < \infty$. Hence

$$(\mathrm{supp} J_t\theta_2) \cap (\mathrm{supp} J_0\theta_1) = \emptyset,$$

which deduces that

$$0 = (J_0\theta_1, J_t\theta_2) = \left(\theta_1, e^{-tH_\mathrm{f}}\theta_2\right). \tag{7.5}$$

Since e^{-tH_f} is PI by Proposition 7.2, we have

$$\left(\theta_1, e^{-tH_\mathrm{f}}\theta_2\right) > 0.$$

Thus we have the contradiction with (7.5). Thus (7.4) follows. □

Lemma 7.10: *Let $f \in \oplus^d L^2(\mathbb{R}^{d+1})$. Then $J_0^* e^{i\pi_0(f)} J_t$ is PI.*

Proof: Let

$$P_M := e^{-(q_0(f)+q([j_0^*]f))} J_0^* \left(e^{-\phi_0(f)}\right)_M J_t e^{i\pi([j_0^*]f)} \left(e^{\phi([j_t^*]f)}\right)_M.$$

Note that P_M is PI by Lemmas 7.6 and 7.9. For positive $F \in L^2_\mathrm{C}(Q)$,

$$\begin{aligned} P_M F &\leq e^{-(q_0(f)+q([j_0^*]f))} J_0^* e^{-\phi_0(f)} J_t e^{i\pi([j_0^*]f)} e^{\phi([j_t^*]f)} F \\ &= J_0^* e^{i\pi_0(f)} J_t F. \end{aligned}$$

Thus, by a limiting argument, for arbitrary positive $F \in L^2(Q)$, we have

$$P_M F \leq J_0^* e^{i\pi_0(f)} J_t F.$$

Since $P_M F > 0$, the lemma holds. □

Theorem 7.11: Suppose $V \in V_\mathrm{Kato}$. Then $U^{-1}e^{-tH}U$ is PI for all $t > 0$.

Proof: Suppose that both $F = F(x,\phi)$ and $G = G(x,\phi)$ are positive in $\mathcal{H}$. Define

$$\begin{aligned} D_F &:= \{x \in \mathbb{R}^d | F(x,\cdot) \not\equiv 0\}, \\ D_G &:= \{x \in \mathbb{R}^d | G(x,\cdot) \not\equiv 0\}, \\ D_{FG} &:= \{X_\cdot \in \mathcal{W} | X_t \in D_F, X_0 \in D_G\}. \end{aligned}$$

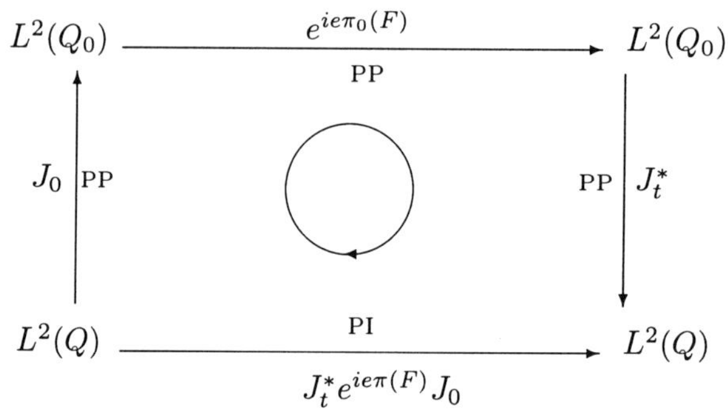

Fig. 22. Lemma 7.10

It is checked that

$$\int_{D_{FG}} dX = \int_{D_G} dx \int_{C([0,\infty);\mathbb{R}^d)} 1_{\{X_t \in D_F\}} dP$$
$$= \int_{D_G} dx \int_{D_F} p_t(x-y)dy > 0.$$

Thus D_{FG} has a positive measure. Since $J_0^* e^{-ie\pi_0(\mathcal{K}(t))} J_t$ is PI on $L^2(Q)$, and $F(X_0,\cdot) \not\equiv 0$ and $G(X_t,\cdot) \not\equiv 0$ on D_{FG}, we have

$$(F, U^{-1}e^{-tH}UG)$$
$$\geq \int_{D_{FG}} dX e^{-\int_0^t V(X_s)ds} \left(F(X_0), J_0^* e^{-ie\pi_0(\mathcal{K}(t))} J_t G(X_t)\right) > 0.$$

We get the desired result. □

Corollary 7.12: *Suppose $V \in V_{\mathrm{Kato}}$. We assume that H has a ground state Ψ_{g}. Then it is unique and $U\Psi_{\mathrm{g}}$ is strictly positive. In particular, for all positive $f \in L^2(\mathbb{R}^d)$,*

$$(f \otimes 1, \Psi_{\mathrm{g}}) > 0. \tag{7.6}$$

Proof: The positiveness of $U\Psi_{\mathrm{g}}$ follows from Theorem 7.11. (7.6) follows from the fact $U(f \otimes 1) = f \otimes 1$. □

7.2. *Degenerate ground states with singular potentials*

In this subsection we give some simple examples of V for which H has degenerate ground states. Let D_j, $j = 1, \ldots, J$, be open sets such that

$$\bigcup_{j=1}^{J} \overline{D_j} = \mathbb{R}^d, \quad \bigcap_{j=1}^{J} D_j = \emptyset,$$

and the Lebesgue measure of the boundary $S := \partial\left(\bigcup_{j=1}^{J} D_j\right)$ is zero.

Assumption 6: We suppose the following (1)–(6) in this subsection.

(1) $V_+ \in L^1_{\text{loc}}(\mathbb{R}^d \setminus S)$,
(2) $D(\Delta) \cap D(V_+)$ is dense in $L^2(\mathbb{R}^d)$,
(3) V_- is infinitesimally small with respect to Δ in the sense of form,
(4) if $X_0 \in D_i$ and $X_t \in D_j$, $i \neq j$, then

$$\int_0^t V_+(X_s)ds = 0, \tag{7.7}$$

(5) operator $H_{\mathrm{p}_j} := H_{\mathrm{p}} \lceil_{L^2(D_j)}$, $j = 1, \ldots, J$, is essentially self-adjoint on $C_0^\infty(D_j)$, and on each $L^2(D_j)$,

$$\|\Delta f\| \leq a\|H_{\mathrm{p}_j} f\| + b\|f\|, \quad j = 1, \ldots, J,$$

for $f \in C_0^\infty(D_j)$ with some positive constants a and b,
(6) the spectral gap of H_{p_j} is positive, i.e.

$$E_{\mathrm{p}_j} := \inf \sigma(H_{\mathrm{p}_j}) \in \sigma_{\text{disc}}(H_{\mathrm{p}_j}), \quad j = 1, \ldots, J.$$

Lemma 7.13: *Let P_j be the projection of $L^2(\mathbb{R}^d)$ to $L^2(D_j)$. Then*

$$e^{-tH} P_j = P_j e^{-tH}, \quad t \geq 0.$$

Proof: Let $F, G \in C_0^\infty(D_j) \widehat{\otimes} L_0^2(Q)$. We can extend functional integral representations in Theorem 5.3 to external potentials satisfying (1) and (3). We see that, by (7.7),

$$\begin{aligned}
&(F, e^{-tH} P_j G)_{\mathcal{H}} \\
&= \sum_{j=1,2,3} \int_{\mathcal{W}_j} e^{-\int_0^t V(X_s)ds} \left(F(X_0), \mathfrak{J}_t (P_j G)(X_t)\right)_{L^2(Q)} dX \\
&= \sum_{j=1,2,3} \int_{\mathcal{W}_j} e^{-\int_0^t V(X_s)ds} \left((P_j F)(X_0), \mathfrak{J}_t G(X_t)\right)_{L^2(Q)} dX \\
&= (P_j F, e^{-tH} G)_{\mathcal{H}},
\end{aligned}$$

where $\mathcal{W}_j$ is the set of paths such that $X_s \in D_j$ for $0 \le s \le t$. Thus the lemma follows. □

Lemma 7.14: *Suppose $|e| \ll 1$. Then H_j is reduced by $L^2(D_j) \otimes L^2(Q)$, and*

$$H_j := H\lceil_{L^2(D_j)\otimes L^2(Q)}$$

is essentially self-adjoint on $C_0^\infty(D_j)\,\widehat{\otimes}\,[L_0^2(Q) \cap D(H_{\mathrm{f}})]$. Moreover the ground state of H_j exists and it is unique.

Proof: By Lemma 7.13, H_j is reduced by $L^2(D_j) \otimes L^2(Q)$. Since H_{p_j} is essentially self-adjoint on $C_0^\infty(D_j)$, we see that H_j is essentially self-adjoint on $C_0^\infty(D_j)\,\widehat{\otimes}\,[L_0^2(Q) \cap D(H_{\mathrm{f}})]$ by the assumption (5) and the Kato–Rellich theorem. In the similar way as the proofs of Theorems 4.14 and 7.11, one can prove the existence and uniqueness of the ground state of H_j. □

Let mult(a) denote the multiplicity of point spectrum a of H.

Theorem 7.15: Suppose $|e| \ll 1$. Set

$$E_j := \inf \sigma(H_j), \quad j = 1, \ldots, J.$$

Then E_j is an eigenvalue of H and

$$\mathrm{mult}(E_j) \ge \#\{E_k | E_k = E_j, k = 1, \ldots, J\}.$$

Moreover

$$\lim_{e\to 0} E_j = \inf \sigma(H_{\mathrm{p}_j}).$$

Proof: Let $\mathcal{H}_j := L^2(D_j) \otimes L^2(Q)$ and Ψ_j be the unique ground state of H_j. Since $H \cong \bigoplus_{j=1}^J H_j$ under the identification $\mathcal{H} \cong \bigoplus_{j=1}^J \mathcal{H}_j$, $\bigoplus_{j=1}^J \delta_{ij} \Psi_j$ is an eigenvector associated with eigenvalue E_j. Thus the theorem follows. □

Corollary 7.16: *Suppose $|e| \ll 1$. Set*

$$E := \min_k E_k = \inf \sigma(H),$$

$$\overline{H} := H - E - \sum_{j=1}^J (E_j - E) 1_{D_j}.$$

Then $\overline{H}$ has a J-fold degenerate ground state.

Proof: It follows from Theorem 7.15. □

A typical example is as follows: let $d = 3$, $J = 3$, and

$$D_1 := \{x \in \mathbb{R}^3 | x_1 > 0, x_2 > 0, x_3 > 0\},$$
$$D_2 := \{x \in \mathbb{R}^3 | x_1 < 0, x_2 < 0, x_3 < 0\},$$
$$D_3 := \mathbb{R}^3 \setminus \overline{D_1 \cup D_2},$$
$$D := \bigcup_{j=1}^{3} D_j.$$

Define

$$V_\nu(x) := \frac{\nu}{|x - \partial D|^3} + |x|^2 + m1_{D_1} + n1_{D_2},$$

where ν, m and n are positive constants, and ∂D denotes the boundary of D. For all $\nu > 0$, we see that $-(1/2)\Delta + V_\nu\lceil_{L^2(D_j)}$ is essentially self-adjoint on[m] $C_0^\infty(D_j)$ and satisfies (1)–(6). $H(\nu)$ is defined by H with the external potential V_ν. Then $H(\nu)$ has embedded eigenvalues (Figs. 23 and 24). Moreover it follows that

$$\begin{aligned}\lim_{\nu\to 0}\left(F, e^{-tH(\nu)}G\right) &= \sum_{j=1,2,3}\int_{\mathcal{W}_j} e^{-\int_0^t V_0(X_s)ds}(F_0, \mathfrak{J}_t G_t)_{L^2(Q)}dX \\ &\neq \int_{\mathcal{W}} e^{-\int_0^t V_0(X_s)ds}(F_0, \mathfrak{J}_t G_t)_{L^2(Q)}dX \\ &= \left(F, e^{-tH(0)}G\right).\end{aligned}$$

Hence

$$\text{s}-\lim_{\nu\to 0} e^{-tH(\nu)} \neq e^{-tH(0)}.$$

This phenomena carries an interesting consequence that once turned on an effect of a singular potential cannot be completely turned off. See Ezawa and Klauder.[45,46]

7.3. *Twofold degenerate ground states with spin*

In this subsection we take the Fock representation. Suppose $d = 3$ and $\hat{\lambda}$ is *rotation invariant*, i.e. for an arbitrary rotation $R \in \mathrm{SO}(3)$,

$$\hat{\lambda}(Rk) = \hat{\lambda}(k).$$

---

[m] See Kalf and Walter.[103]

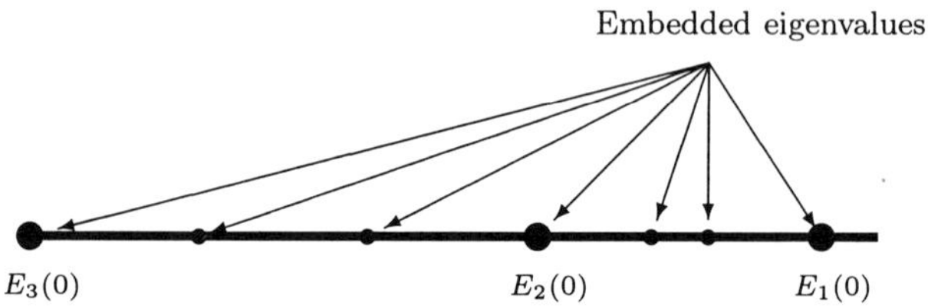

Fig. 23. The spectrum of $H(\nu)$ for $e = 0$ and for $1 \ll n < m$

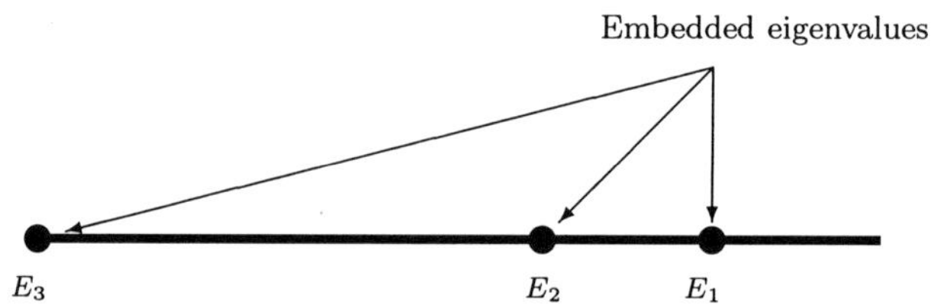

Fig. 24. The spectrum of $H(\nu)$ for $|e| \ll 1$ and for $1 \ll n < m$

7.3.1. *Fixed total momentum*

Let us define the field momentum by

$$P_{\mathrm{f}\mu} := \int k_\mu a^{*r}(k) a^r(k) dk, \quad \mu = 1, 2, 3.$$

Then the total momentum is given by

$$P_\mu := p_\mu + P_{\mathrm{f}\mu}, \quad \mu = 1, 2, 3.$$

Let $V = 0$. Then H_{PF}^σ turns out to be translation invariant, i.e.

$$[H_{\mathrm{PF}}^\sigma, P_\mu] = 0, \quad \mu = 1, 2, 3.$$

In particular H_{PF}^σ has no ground states. We decompose H_{PF}^σ as

$$\mathcal{H}_\sigma = \int_{\mathbb{R}^3}^\oplus \mathbb{C}^2 \otimes \mathcal{F}_{\mathrm{rad}} dp,$$

$$H_{\mathrm{PF}}^\sigma = \int_{\mathbb{R}^3}^\oplus H_{\mathrm{PF}}^\sigma(p) dp,$$

Hilbert space	$\mathbb{C}^2 \otimes \mathfrak{F}_{\text{rad}}$
$A_\mu(0)$	$\frac{1}{\sqrt{2}} \int e^r_\mu(k) \left\{ a^{*r}(k)\hat{\lambda}(-k) + a^r(k)\hat{\lambda}(k) \right\} dk$
$B_\mu(0)$	$\frac{-i}{\sqrt{2}} \int (k \times e^r(k))_\mu \left\{ a^{*r}(k)\hat{\lambda}(-k) - a^r(k)\hat{\lambda}(k) \right\} dk$
P_{f}	$\int k a^{*r}(k) a^r(k) dk$
$H^\sigma_{\text{PF}}(p)$	$\frac{1}{2}(p - P_{\text{f}} - eA(0))^2 + H_{\text{rad}} - \frac{e}{2}\sigma B(0)$

Fig. 25. The Pauli–Fierz Hamiltonian with fixed total momentum $p \in \mathbb{R}^3$

where

$$H^\sigma_{\text{PF}}(p) := \frac{1}{2}(p - P_{\text{f}} - eA(0))^2 + H_{\text{rad}} - \frac{e}{2}\sigma B(0), \quad p \in \mathbb{R}^3,$$

$$A_\mu(0) := \frac{1}{\sqrt{2}} \int e^r_\mu(k)\hat{\lambda}(k) \left\{ a^{*r}(k) + a^r(k) \right\} dk,$$

$$B_\mu(0) := \frac{-i}{\sqrt{2}} \int (k \times e^r(k))_\mu \hat{\lambda}(k) \left\{ a^{*r}(k) - a^r(k) \right\} dk.$$

$H^\sigma_{\text{PF}}(p)$ is called the Pauli–Fierz Hamiltonian with fixed total momentum $p \in \mathbb{R}^3$.

In this subsection we assume that a radiation field has a positive mass m,

$$\omega(k) := \sqrt{|k|^2 + m^2}.$$

Suppose $e = 0$. Then, for $|p| < \sqrt{2}m$,

$$\frac{1}{2}(p - P_{\text{f}})^2 + H_{\text{rad}}$$

has a twofold degenerate ground state spanned by

$$\phi_+ := \begin{pmatrix} 1 \\ 0 \end{pmatrix} \otimes \Omega \quad \text{and} \quad \phi_- := \begin{pmatrix} 0 \\ 1 \end{pmatrix} \otimes \Omega$$

associated with eigenvalue $|p|^2/2$. We set

$$\Sigma := \inf \sigma_{\text{ess}} \left(\frac{1}{2}(p - P_{\text{f}})^2 + H_{\text{rad}} \right) = \inf_{k \in \mathbb{R}^d} \left(\frac{1}{2}|p - k|^2 + \omega(k) \right).$$

A trivial lower bound of Σ is $\Sigma \geq m$. Then in the case where $m \geq (1/2)|p|^2$, i.e. $|p| < \sqrt{2m}$, the positive spectral gap of $\frac{1}{2}(p - P_{\rm f})^2 + H_{\rm rad}$ exists. (See Fig. 26.)

Fig. 26. The spectrum of $\frac{1}{2}(p - P_{\rm f})^2 + H_{\rm rad}$ for $|p| < \sqrt{2m}$

The ground state energy of $H_{\rm PF}^{\sigma}(p)$ is denoted by $E(p)$ and the bottom of the essential spectrum by $E_{\rm c}(p)$. It has been proven in Fröhlich[53] that

$$\inf_{k\in\mathbb{R}^3} \{E(p-k) + \omega(k)\} \leq E_{\rm c}(p),$$

and in Spohn[138] that $\inf_{k\in\mathbb{R}^3} \{E(p-k) + \omega(k)\}$ is an accumulate point in $\sigma(H_{\rm PF}^{\sigma}(p))$. Then it has been established that

$$E_{\rm c}(p) = \inf_{k\in\mathbb{R}^3} \{E(p-k) + \omega(k)\}.$$

It is natural to set

$$\Delta(p) := E_{\rm c}(p) - E(p) = \inf_{k\in\mathbb{R}^3} \{E(p-k) + \omega(k) - E(p)\}.$$

It is also seen that

$$\{(p,e) \in \mathbb{R}^3 \times \mathbb{R} | \Delta(p) > 0\} \ni (0,0),$$

and $E(p)$ is continuous in e and p. Then it follows that

$$\{(p,e) \in \mathbb{R}^3 \times \mathbb{R} | \Delta(p) > 0, (p,e)\} \setminus \{(0,0)\} \neq \emptyset.$$

Remark 7.17: If in addition one would impose an implicit but rather natural condition

$$E(0) \leq E(p), \tag{7.8}$$

one concludes that $\Delta(p) > 0$ for $|p| < p_*$ with some p_* for arbitrary e. See Fröhlich.[52] Unfortunately (7.8) is not yet established for $H_{\rm PF}^{\sigma}(p)$.

By the pull-through formula, for an arbitrary ground state Ψ_g, we have[91]

$$(\Psi_g, N_{\rm rad}\Psi_g) \leq 2e^2 \int_{\mathbb{R}^3} \frac{(|k|^2/4) + 6E(p)}{(E(p-k) + \omega(k) - E(p))^2} |\hat{\lambda}(k)|^2 dk. \tag{7.9}$$

We set

$$\theta := 2 \int_{\mathbb{R}^3} \frac{(|k|^2/4) + 6E(p)}{(E(p-k) + \omega(k) - E(p))^2} |\hat{\lambda}(k)|^2 dk.$$

Let P_e, $e \in \mathbb{R}$, be the projection onto the subspace spanned by ground states of $H^\sigma_{\rm PF}(p)$.

Lemma 7.18: *Suppose* $\Delta(p) > 0$ *and* $|e| < 1/\sqrt{3\theta}$. *Then* $\mathrm{Tr}P_e \leq 2$.

Proof: By (7.9) we can see that

$$\mathrm{Tr}P_e - \mathrm{Tr}(P_e P_0) = \mathrm{Tr}P_e(1 - P_0) \leq \mathrm{Tr}(P_e N_{\rm rad}) \leq e^2\theta \mathrm{Tr}P_e,$$

and $\mathrm{Tr}(P_e P_0) \leq 2$. Then we have

$$\mathrm{Tr}P_e \leq \frac{2}{1 - e^2\theta} < 3.$$

The lemma follows. □

Lemma 7.19: *Let* $\tau \in \mathbb{C}^2$. *Then there exist* $e_0 > 0$ *and* $a(t) > 0$ *independent of* x *such that for* $|e| < e_0$

$$(\tau \otimes \Omega, e^{-tH^\sigma_{\rm PF}(p)} \tau \otimes \Omega) = a(t)(\tau, \tau)_{\mathbb{C}^2}. \tag{7.10}$$

Proof: We show an outline of a proof. See Hiroshima and Spohn[91] for details. Let

$$\overline{H^\sigma_{\rm PF}(p)} := H^\sigma_{\rm PF}(p) - E(p).$$

Take $|e|$ sufficiently small, say $|e| < e_0$. Then it is seen that

$$(\tau \otimes \Omega, e^{-t\overline{H^\sigma_{\rm PF}(p)}} \tau \otimes \Omega) = \lim_{n\to\infty} \left(\tau \otimes \Omega, \left(1 + \frac{t}{n}\overline{H^\sigma_{\rm PF}(p)}\right)^{-n} \tau \otimes \Omega \right).$$

By a direct calculation, we derive that

$$\left(1 + \frac{t}{n}\overline{H^\sigma_{\rm PF}(p)}\right)^{-n} = a_n + i\sigma \cdot b_n.$$

Here a_n and $b_{n\mu}$ ($\mu = 1, 2, 3$) are unbounded operators so that

$$(\Omega, a_n\Omega) \in \mathbb{R}, \quad (\Omega, b_{n\mu}\Omega) \in \mathbb{R}. \tag{7.11}$$

Hence we have

$$(\tau\otimes\Omega, e^{-t\overline{H^\sigma_{\rm PF}(p)}}\tau\otimes\Omega) = \lim_{n\to\infty}\left\{(\tau,\tau)(\Omega,a_n\Omega)+i(\tau,\sigma_\mu\tau)(\Omega,b_{n\mu}\Omega)\right\}.$$

Since the left-hand side above and $(\tau,\sigma_\mu\tau)$ are *real*, we conclude that

$$(\tau,\sigma_\mu\tau)(\Omega,b_{n\mu}\Omega)=0.$$

Thus the lemma follows with $a(t):=\lim_{n\to\infty}(\Omega,a_n\Omega)$. □

Lemma 7.20: *Suppose* $\Delta(p)>0$ *and* $|e|<1/\sqrt{\theta}$. *Then*

$$(\Psi_{\rm g}, P_0\Psi_{\rm g})\neq 0.$$

Proof: Since $P_\Omega+N_{\rm rad}\geq 1$, we have by (7.9)

$$(\Psi_{\rm g},P_0\Psi_{\rm g})\geq\|\Psi_{\rm g}\|^2-\|(1\otimes N_{\rm rad}^{1/2})\Psi_{\rm g}\|^2\geq 1-e^2\theta>0.$$

Thus the lemma follows. □

Lemma 7.21: *Suppose* $|e|<e_0$ *and* $|e|<1/\sqrt{\theta}$. *Then there exists* $a>0$ *such that*

$$P_0P_eP_0=aP_0.$$

Proof: Note that

$$P_e={\rm s}-\lim_{t\to\infty}e^{-t\overline{H^\sigma_{\rm PF}(p)}}.$$

Then, by Lemma 7.19, for *all* $\tau\in\mathbb{C}$,

$$(\tau\otimes\Omega,P_e\tau\otimes\Omega)=\lim_{t\to\infty}(\tau\otimes\Omega,e^{-t\overline{H^\sigma_{\rm PF}(p)}}\tau\otimes\Omega)=\lim_{t\to\infty}a(t)(\tau,\tau).$$

Set

$$P_0\Psi_{\rm g}=\begin{pmatrix}(\phi_+,\Psi_{\rm g})\\(\phi_-,\Psi_{\rm g})\end{pmatrix}\otimes\Omega:=\tau'\otimes\Omega.$$

By Lemma 7.20,

$$(\Psi_{\rm g},\tau'\otimes\Omega)\neq 0$$

follows. Hence

$$(\tau'\otimes\Omega,P_e\tau'\otimes\Omega)\neq 0. \tag{7.12}$$

This implies that $a:=\lim_{t\to\infty}a(t)$ is positive. Thus the lemma follows. □

Lemma 7.22: *Suppose* $|e|<e_0$ *and* $|e|<1/\sqrt{\theta}$. *Then* ${\rm Tr}P_e\geq 2$.

Proof: Assume that $\mathrm{Tr}P_e = 1$. Then

$$P_0 = |\phi_+\rangle\langle\phi_+| + |\phi_-\rangle\langle\phi_-|,$$
$$P_e = |\Psi_{\mathrm{g}}\rangle\langle\Psi_{\mathrm{g}}|.$$

Lemma 7.21 leads to

$$\begin{aligned}P_0P_eP_0 &= (|\phi_+\rangle\langle\phi_+| + |\phi_-\rangle\langle\phi_-|)|\Psi_{\mathrm{g}}\rangle\langle\Psi_{\mathrm{g}}|(|\phi_+\rangle\langle\phi_+| + |\phi_-\rangle\langle\phi_-|)\\ &= |(\phi_+,\Psi_{\mathrm{g}})|^2|\phi_+\rangle\langle\phi_+| + |(\phi_-,\Psi_{\mathrm{g}})|^2|\phi_-\rangle\langle\phi_-|\\ &\quad +(\phi_+,\Psi_{\mathrm{g}})(\Psi_{\mathrm{g}},\phi_-)|\phi_+\rangle\langle\phi_-| + (\phi_-,\Psi_{\mathrm{g}})(\Psi_{\mathrm{g}},\phi_+)|\phi_-\rangle\langle\phi_+|\\ &= a(|\phi_+\rangle\langle\phi_+| + |\phi_-\rangle\langle\phi_-|).\end{aligned}$$

It follows that

$$(\phi_+,\Psi_{\mathrm{g}})(\Psi_{\mathrm{g}},\phi_-) = 0.$$

Let us assume $(\Psi_{\mathrm{g}},\phi_-) = 0$. Then

$$|(\phi_+,\Psi_{\mathrm{g}})|^2|\phi_+\rangle\langle\phi_+| = a(|\phi_+\rangle\langle\phi_+| + |\phi_-\rangle\langle\phi_-|).$$

This contradicts $a \neq 0$. Thus the lemma follows. □

We define

$$e^* := \sup\left\{|e| \,\middle|\, |e| < 1/\sqrt{3\theta}, |e| < e_0\right\}.$$

Theorem 7.23: Suppose $\Delta(p) > 0$ and $|e| < e^*$. Then $\mathrm{Tr}P_e = 2$.

Proof: It follows from Lemmas 7.18 and 7.22. □

We will see that the Hamiltonian $H_{\mathrm{PF}}^{\sigma}(p)$ has

"rotation+helicity+$SU(2)$" –symmetry.

We define the field angular momentum relative to the origin by

$$J_{\mathrm{f}} := -i\int (k\wedge\nabla_k)a^{*r}(k)a^r(k)dk$$

and the helicity by

$$S_{\mathrm{f}} := i\int (k/|k|)\left\{a^{*2}(k)a^1(k) - a^{*1}(k)a^2(k)\right\}dk.$$

Let us define the total angular momentum J by

$$J := J_{\mathrm{f}} + S_{\mathrm{f}} + \frac{1}{2}\sigma.$$

Then a direct calculation shows that for arbitrary $\varphi\in\mathbb{R}$

$$e^{i\varphi\hat{p}\cdot J}H_{\mathrm{PF}}^{\sigma}(p)e^{-i\varphi\hat{p}\cdot J} = H_{\mathrm{PF}}^{\sigma}(p),$$

where $\hat{p} := p/|p|$. Since

$$\sigma(\hat{p} \cdot J) = \mathbb{Z}_{1/2} := \{\pm 1/2, \pm 3/2, \pm 5/2, \dots\},$$

$H^{\sigma}_{\mathrm{PF}}(p)$ is decomposable as

$$\mathbb{C}^2 \otimes \mathcal{F}_{\mathrm{rad}} = \bigoplus_{z \in \mathbb{Z}_{1/2}} \mathcal{H}(z),$$

$$H^{\sigma}_{\mathrm{PF}}(p) = \bigoplus_{z \in \mathbb{Z}_{1/2}} H^{\sigma}_{\mathrm{PF}}(p, z).$$

Theorem 7.24: Suppose $\Delta(p) > 0$ and $|e| < e^*$. Then $H^{\sigma}_{\mathrm{PF}}(p)$ has two orthogonal ground states, $\Psi_{\mathrm{g}_\pm}$, such that

$$\Psi_{\mathrm{g}_\pm} \in \mathcal{H}(\pm 1/2).$$

Proof: Without loss of generality we may assume that $p = (0, 0, 1)$. Then

$$\phi_\pm \in \mathcal{H}(\pm 1/2).$$

Let $\Psi_{\mathrm{g}_\pm}$ be normalized ground states of $H^{\sigma}_{\mathrm{PF}}(p)$ such that $\Psi_{\mathrm{g}_+} \in \mathcal{H}(z)$ and $\Psi_{\mathrm{g}_-} \in \mathcal{H}(z')$ with some $z, z' \in \mathbb{Z}_{1/2}$. Since $P_0 P_e P_0 = a P_0$ we see that

$$(\phi_\pm, P_e \phi_\pm) = (\phi_\pm, P_0 P_e P_0 \phi_\pm) = a > 0. \tag{7.13}$$

Let $Q_\pm$ be the projection onto $\mathcal{H}(\pm 1/2)$. Then from (7.13) it follows that

$$Q_+ P_e \phi_+ \neq 0,$$
$$Q_- P_e \phi_- \neq 0.$$

The alternative

$$Q_+ \Psi_{\mathrm{g}_+} \neq 0 \text{ or } Q_+ \Psi_{\mathrm{g}_-} \neq 0$$

holds, and the alternative

$$Q_- \Psi_{\mathrm{g}_+} \neq 0 \text{ or } Q_- \Psi_{\mathrm{g}_-} \neq 0$$

holds. Suppose $Q_+ \Psi_{\mathrm{g}_+} \neq 0$. Then $\Psi_{\mathrm{g}_+} \in \mathcal{H}(+1/2)$ and $\Psi_{\mathrm{g}_-} \in \mathcal{H}(-1/2)$. Thus the theorem follows. □

In the spinless case $H^{\sigma}_{\mathrm{PF}}(p)$ simplifies to

$$H_{\mathrm{PF}}(p) := \frac{1}{2}(p - P_{\mathrm{f}} - eA(0))^2 + H_{\mathrm{rad}}.$$

To show the uniqueness of the ground state only the pull-through argument seems to be available.

Theorem 7.25: Suppose $\Delta(p) > 0$ and

$$e^2 \leq \frac{1}{2}\left\{\int_{\mathbb{R}^3} \frac{E(p)}{(E(p-k)+\omega(k)-E(p))^2}|\hat{\lambda}(k)|^2 dk\right\}^{-1}.$$

Then the ground state of $H_{\text{PF}}(p)$ is unique.

Proof: See Hiroshima and Spohn.[91] □

7.3.2. *External potentials*

Next we consider the case of $V \not\equiv 0$, and we revive

$$\omega(k) = |k|.$$

We shall prove that H_{PF}^σ has a twofold degenerate ground state. We assume Assumption V_n (Assumption 4) and

$$V(x) = V(-x). \tag{7.14}$$

For arbitrary ground state Ψ_{g}', in the same way as in Section 4.3 we can prove that

$$(\Psi_{\text{g}}', P_0 \Psi_{\text{g}}') \geq (1 - e^2 c')\|\Psi_{\text{g}}'\|^2$$

with some constant c'. Hence in the similar way as Lemma 7.18 we can prove

$$\text{Tr} P_e \leq 2 \tag{7.15}$$

for $|e| < 1/\sqrt{3c'}$. The realness argument of (7.11) requires the extra condition (7.14) of V. Let

$$T := (F \otimes 1) e^{ix \otimes P_{\text{f}}},$$

where F denotes the Fourier transformation of $L^2(\mathbb{R}^3)$. Then T maps $D(-\Delta) \cap D(H_{\text{rad}})$ onto $D(|x|^2) \cap D(H_{\text{rad}})$, with

$$T H_{\text{PF}}^\sigma T^{-1} = \frac{1}{2}\{\sigma(x - P_{\text{f}} - eA(0))\}^2 + FVF^{-1} + H_{\text{rad}}.$$

By (7.14), we see that FVF^{-1} is a reality preserving operator. Let f be the ground state of $-(1/2)\Delta + V$. Then it can be proven that there exists $e_* > 0$ such that for $|e| < e_*$,

$$(\tau \otimes f \otimes \Omega, e^{-t(T H_{\text{PF}}^\sigma T^{-1})} \tau \otimes f \otimes \Omega) = a(t)(\tau \otimes f, \tau \otimes f)$$

for all $\tau \in \mathbb{C}^2$ with some $a(t) > 0$. Moreover, at least one ground state Ψ_g satisfies $(\Psi_g, P_0\Psi_g) > 0$. Hence,

$$(\tau' \otimes f \otimes \Omega, P_e\tau' \otimes f \otimes \Omega) \neq 0$$

for some $\tau' \in \mathbb{C}^2$, which implies that there exists

$$\lim_{t\to\infty} a(t) := a > 0$$

such that

$$P_0 P_e P_0 = aP_0.$$

Then we can prove the following lemma:

Lemma 7.26: *Let* $|e| < e_*$. *Then* $\mathrm{Tr}P_e \geq 2$.

Theorem 7.27: Let $|e| < \min\{e_*, 1/\sqrt{3c'}\}$. Then $\mathrm{Tr}P_e = 2$.

Proof: It follows from (7.15) and Lemma 7.26. □

8. Ground State Energy, Charge Density and Effective Mass

8.1. *Ground state energy*

By using the overlap $(f \otimes 1, \Psi_g) \neq 0$ for $f \geq 0$ we can see qualitative properties of the ground state energy of H.

Theorem 8.1: Suppose $V \in V_{\mathrm{Kato}}$. We assume that H has a ground state. Then, for all $f \geq 0$,

$$E(e^2) := \inf\sigma(H) = -\lim_{t\to\infty}\frac{1}{t}\log\int_{\mathcal{W}} f(X_0)f(X_t)e^{-(e^2/2)q_0(\mathcal{K}(t))}dX. \quad (8.1)$$

In particular $E(e^2)$ is a continuous, monotonously increasing, concave function in e^2.[n]

Proof: By Corollary 7.12, $(f \otimes 1, \Psi_g) \neq 0$ holds. From this it follows that

$$E(e^2) = -\lim_{t\to\infty}\frac{1}{t}\log\left(f \otimes 1, e^{-tH}f \otimes 1\right).$$

[n]Lieb and Loss[113,114] obtained a bound of the ground state energy of H_{PF}. Hirokawa[69,70] studied the ground state energy of a spin-boson model.

By Theorem 5.3, (8.1) follows. By the Hölder inequality we see that $E(e^2)$ is concave, which implies that $E(e^2)$ is continuous in $e^2 > 0$. Since H converges to H_{dec} as $e \to 0$ uniformly in the sense of resolvent, $\lim_{e^2 \to 0} E(e^2) = E(0)$. Hence $E(e^2)$ is continuous in $e^2 \geq 0$. It is known that the concave and continuous function $E(e^2)$ can be represented as

$$E(e^2) = E(0) + \int_0^{e^2} \phi(t)dt$$

with some increasing function $\phi(t)$. By the diamagnetic inequality,

$$E(0) \leq E(e^2),$$

we have $\phi(t) \geq 0$. Thus $E(e^2)$ is monotonously increasing. □

8.2. *Charge density in ground states*

In this subsection we show that $\|\Psi_{\mathrm{g}}(x)\|_{L^2(Q)}$ decays exponentially in $x \in \mathbb{R}^d$ pointwise. Bach, Fröhlich and Sigal,[26] and Griesemer, Lieb and Loss[61] show that $\|\Psi_{\mathrm{g}}(x)\|_{L^2(Q)}$ decays exponentially in a different way but not pointwise.

We suppose Assumption V_n (Assumption 4). Then recall that

$$e^{-tH_{\mathrm{P}}} : L^2(\mathbb{R}^d) \to L^\infty(\mathbb{R}^d)$$

with

$$\|e^{-tH_{\mathrm{P}}} f\|_\infty \leq c_{\mathrm{p}} \|f\|_{L^2(\mathbb{R}^d)}, \tag{8.2}$$

where c_{p} is a constant.

Lemma 8.2: *There exists a constant* c_{b} *such that*

$$\sup_{x \in \mathbb{R}^d} \|\Psi_{\mathrm{g}}(x)\|_{L^2(Q)} < c_{\mathrm{b}} \|\Psi_{\mathrm{g}}\|_{\mathcal{H}}. \tag{8.3}$$

Proof: Since $\Psi_{\mathrm{g}} = e^{tE} e^{-tH} \Psi_{\mathrm{g}}$, we have for almost every $x \in \mathbb{R}^d$,

$$\Psi_{\mathrm{g}}(x) = e^{tE} \mathbb{E}_x^Q \left(e^{-\int_0^t V(X_s)ds} \mathfrak{J}_t \Psi_{\mathrm{g}}(X_t) \right). \tag{8.4}$$

Hence for almost every $x \in \mathbb{R}^d$,

$$\begin{aligned} \|\Psi_{\mathrm{g}}(x)\|_{L^2(Q)} &\leq e^{tE} \mathbb{E}_x \left(e^{-\int_0^t V(X_s)ds} \|\Psi_{\mathrm{g}}(X_t)\|_{L^2(Q)} \right) \\ &= e^{tE} \left(e^{-tH_{\mathrm{P}}} \|\Psi_{\mathrm{g}}(\cdot)\|_{L^2(Q)} \right)(x). \end{aligned}$$

By (8.2) it follows that

$$\sup_{x\in\mathbb{R}^d} \left|\left(e^{-tH_{\mathrm{p}}}\|\Psi_{\mathrm{g}}(\cdot)\|_{L^2(Q)}\right)(x)\right| < c_{\mathrm{p}}\|\Psi_{\mathrm{g}}\|_{\mathcal{H}}.$$

Take

$$c_{\mathrm{b}} := e^{tE}c_{\mathrm{p}}.$$

Then the proof is complete. □

The following lemma is known as Carmona's estimate.

Lemma 8.3:

(1) *Let $V \in V(n)$, $n \geq 1$, and $t(x) := |x|^{1-n}$. Then there exist D and δ such that*

$$e^{t(x)E}\mathbb{E}_x\left(e^{-\int_0^{t(x)} V(X_s)ds}\right) \leq De^{-\delta|x|^{n+1}}. \tag{8.5}$$

(2) *Let $V \in V(0)$. Then there exist D, δ, and β such that for $t(x) := \beta|x|$*

$$e^{t(x)E}\mathbb{E}_x\left(e^{-\int_0^{t(x)} V(X_s)ds}\right) \leq De^{-\delta|x|}. \tag{8.6}$$

Proof: See Carmona.[37] □

Theorem 8.4: We have, for almost every $x \in \mathbb{R}^d$,

$$\|\Psi_{\mathrm{g}}(x)\|_{L^2(Q)} \leq c_{\mathrm{b}}\|\Psi_{\mathrm{g}}\|_{\mathcal{H}}D\exp\left(-\delta|x|^{n+1}\right). \tag{8.7}$$

Here, D and δ is in Lemma 8.3.

Proof: By Lemma 8.2 and (8.4) we have for almost every $x \in \mathbb{R}^d$,

$$\|\Psi_{\mathrm{g}}(x)\|_{L^2(Q)} \leq c_{\mathrm{b}}\|\Psi_{\mathrm{g}}\|_{\mathcal{H}}e^{tE}\mathbb{E}_x\left(e^{-\int_0^t V(X_s)ds}\right).$$

Here t is an arbitrary positive number. The lemma follows from (8.5) and (8.6). □

From Theorem 8.4, it follows that

$$\||x|^k\Psi_{\mathrm{g}}\|_{\mathcal{H}} < \infty$$

for all $k \in \mathbb{N}$. The next corollary presents a more strong statement.

Corollary 8.5: *For all $k \geq 0$, there exists a constant c_k such that*

$$\||x|^k\Psi_{\mathrm{g}}\|_{\mathcal{H}} \leq c_k\|\Psi_{\mathrm{g}}\|_{\mathcal{H}}.$$

Proof: By Theorem 8.4, we obtain

$$\begin{aligned}\||x|^k\Psi_{\rm g}\|^2_{\mathcal H} &= \int \left(|x|^k\|\Psi_{\rm g}(x)\|_{L^2(Q)}\right)^2 dx \\ &\le \left(c_{\rm b}^2 D^2 \int |x|^{2k} e^{-2\delta|x|^{n+1}} dx\right) \times \|\Psi_{\rm g}\|^2_{\mathcal H}.\end{aligned}$$

Set $c_k := c_{\rm b} D \left(\int |x|^{2k} e^{-\delta|x|^{n+1}} dx\right)^{1/2}$. Then the corollary follows. □

Remark 8.6: Let $E_m = \inf\sigma(H^m_{\rm PF})$, where $H^m_{\rm PF}$ is defined by $H_{\rm PF}$ with ω replaced by $\omega + m$. We can rewrite c_k as

$$c_k = e^{tE_m} c_{\rm p} D \left(\int |x|^{2k} e^{-\delta|x|^{n+1}} dx\right)^{1/2}, \quad t > 0,$$

and $c_{\rm p}$ and D are independent of m. Since $\lim_{m\to 0} E_m = \inf\sigma(H_{\rm PF})$, we can see that there exists c'_k independent of m such that $c_k \le c'_k$. Then we obtain the bound

$$\||x|^k \Psi_{\rm g}^{(m)}\| \le c'_k \|\Psi_{\rm g}^{(m)}\|.$$

Remark 8.7: Using the identity

$$N^{k/2}\Psi_{\rm g} = e^{-tH} e^{tE} N^{k/2}\Psi_{\rm g} + e^{tE}[N^{k/2}, e^{-tH}]\Psi_{\rm g}$$

and functional integral representations of e^{-tH}, we can also prove that $\|N^{k/2}\Psi_{\rm g}(x)\|_{L^2(Q)}$ decays exponentially for almost every $x \in \mathbb{R}^d$. See Hiroshima[88] for details.

8.3. *Effective mass and mass renormalization*

Let us revive the mass M of an electron for $H^\sigma_{\rm PF}$ and we assume that $d = 3$ and $V = 0$. Moreover let

$$\hat\lambda(k) = \begin{cases} 0 & |k| < \kappa, \\ 1/\sqrt{(2\pi)^3\omega(k)} & \kappa \le |k| \le \Lambda, \\ 0 & |k| > \Lambda \end{cases}$$

with some positive constants Λ and κ. As was seen in Section 7.3, we can decompose $H^\sigma_{\rm PF}$ as

$$H^\sigma_{\rm PF} = \int_{\mathbb{R}^3}^{\oplus} H^\sigma_{\rm PF}(p) dp.$$

Since $H^\sigma_{\rm PF}(p)$ is an analytic family[130] of type (A) for $(p,e^2) \in \mathbb{R}^3 \times \mathbb{R}$ near $(0,0)$, $E(p) := \inf \sigma(H^\sigma_{\rm PF}(p))$ is analytic for $(p,e^2) \in \mathbb{R}^3 \times \mathbb{R}$ near $(0,0)$. Then we can define the effective mass $M_{\rm eff}$ of $H^\sigma_{\rm PF}$ by

$$E(p) - E(0) = \frac{1}{2M_{\rm eff}}|p|^2 + O(|p|^3),$$

i.e. it follows that

$$M_{\rm eff} = M_{\rm eff}(M,\Lambda,\kappa,e^2) := \left(\frac{1}{3}\Delta_p E(p)\lceil_{p=0}\right)^{-1}.$$

The ultimate goal of the mass renormalization of $H^\sigma_{\rm PF}$ is to find constants $\beta = \beta(e^2) < 0$ and $M_{\rm ph} > 0$ such that

$$\lim_{\Lambda\to\infty} M_{\rm eff}(M\Lambda^\beta, \Lambda, \kappa\Lambda^\beta, e^2) = M_{\rm ph}.$$

Namely the scaled mass $M\Lambda^\beta$ goes to zero and the effective mass goes to a physical mass $M_{\rm ph}$ as $\Lambda \to \infty$.

Remark 8.8: In the dipole approximation, by Theorem 4.17, the effective mass is

$$M_{\rm eff} = M + \frac{e^2}{4\pi}\frac{4}{3\pi}(\Lambda - \kappa).$$

Hence the mass renormalization is not available.

It is established that $M/M_{\rm eff}$ is a function of Λ/m and κ/M. Assume that

$$\lim_{\Lambda\to\infty} \frac{M}{M_{\rm eff}}\left(\frac{\Lambda}{M}\right)^\gamma = M_*$$

with some $0 < \gamma < 1$. Then the choice

$$\beta = -\frac{\gamma}{1-\gamma}$$

yields that

$$\lim_{\Lambda\to\infty} M_{\rm eff}(M\Lambda^\beta, \Lambda, \kappa\Lambda^\beta, e^2) = M^{1-\gamma} M_*.$$

For simplicity we consider the spinless case. Let

$$\frac{M}{M_{\rm eff}} = \sum_{n=0}^\infty \frac{a_n(\Lambda/M, \kappa/M)}{n!}(e^2/4\pi)^n.$$

The formula

$$\frac{M}{M_{\rm eff}} = 1 - \frac{2}{3}\frac{((P_{\rm f} + eA(0))_\mu \Psi_{\rm g}(0), (H_{\rm PF}(0) - E(0))^{-1}(P_{\rm f} + eA(0))_\mu \Psi_{\rm g}(0))}{(\Psi_{\rm g}(0), \Psi_{\rm g}(0))}$$

is derived from a perturbation theory, where

$$A_\mu(0) := \frac{1}{\sqrt{2}} \int e^r_\mu(k)\hat{\lambda}(k) \{a^{*r}(k) + a^r(k)\} \, dk.$$

Using this, we can exactly show that

$$\begin{aligned} a_0(\Lambda/M, \kappa/M) &= 1, \\ a_1(\Lambda/M, \kappa/M) &= -\frac{8}{3\pi} \log\left(\frac{(\Lambda/M)+2}{(\kappa/M)+2}\right). \end{aligned}$$

Hence it may be written as, for sufficiently small e^2,

$$M_{\text{eff}} \approx \frac{M}{1 - \dfrac{e^2}{4\pi}\dfrac{8}{3\pi}\log\left(\dfrac{(\Lambda/M)+2}{(\kappa/M)+2}\right)} \quad .$$

The conventional expectation to the next order is that the asymptotics of the absolute value of coefficient $a_2(\Lambda/M, \kappa/M)$ would be $[\log(\Lambda/M)]^2$ as $\Lambda \to \infty$. The following theorem has been proven, however, by Hiroshima and Spohn.[92]

Theorem 8.9: We have (1) $a_2(\Lambda/M, \kappa/\Lambda) < 0$ for sufficiently large Λ and (2) $a_2(\Lambda/M, \kappa/\Lambda)/\sqrt{\Lambda/m}$ converges as $\Lambda \to \infty$ such as

$$c_1 < \lim_{\Lambda\to\infty} \frac{|a_2(\Lambda/M, \kappa/M)|}{\sqrt{\Lambda/M}} < c_2,$$

with some positive constants c_1 and c_2.

From this theorem we conjecture $\gamma = 1/2$.

Conjecture 1: We conjecture that $\beta = -1$, i.e. $\gamma = 1/2$, for sufficiently small e but $e \neq 0$.

In the case including spin we can also prove that

$$a_1(\Lambda/M, \kappa/M) \sim \log[\Lambda/M]$$

as $\Lambda \to \infty$. As far as we know, however, the asymptotics of $a_2(\Lambda/M, \kappa/M)$ as $\Lambda \to \infty$ is unknown. See also Hainzl and Seiringer,[67] and Lieb and Loss.[114]

9. Gibbs Measures

In this section we discuss properties of a ground state by means of infinite volume Gibbs measures. Works related to this section are Refs. 33–35, 82, 115–117, 123 and 140.

Throughout this section we assume that $V \in V_{\rm Kato}$ (Definition 6.17) and $\inf V > -\infty$.

9.1. *Infinite volume Gibbs measures*

Let $f_j \in L^\infty(\mathbb{R}^d)$, $j = 1, 2, \ldots, m$, and $f \in L^2(\mathbb{R}^d)$ be a positive function. Then, for $-T \le t_1 \le \cdots \le t_m \le T$, we see that from Theorem 5.3:

$$\frac{\left(f \otimes 1, e^{-(T+t_1)H} f_1 e^{-(t_2-t_1)H} \cdots e^{-(t_m-t_{m-1})H} f_m e^{-(T-t_m)H} f \otimes 1\right)}{(f \otimes 1, e^{-2TH} f \otimes 1)}$$
$$= \int_{\mathrm{C}([0,\infty);\mathbb{R}^d)\times\mathbb{R}^d} f_1(X_{T+t_1}) \cdots f_m(X_{T+t_m}) dW_{2T}^f. \tag{9.1}$$

Here dW_{2T}^f is a probability measure on $\mathrm{C}([0,\infty);\mathbb{R}^d) \times \mathbb{R}^d$ defined by

$$dW_{2T}^f := W_{2T}^f dP \otimes dx,$$

where

$$W_{2T}^f = W_{2T}^f(X) := \frac{1}{Z_{2T}} f(X_0) f(X_{2T}) e^{-\int_0^{2T} V(X_s) ds} e^{-(e^2/2) q_0(\mathcal{K}(2T))}$$

and

$$Z_{2T} := (f \otimes 1, e^{-2TH} f \otimes 1)$$

is regarded as the normalizing constant such as

$$\int_{\mathrm{C}([0,\infty);\mathbb{R}^d)\times\mathbb{R}^d} dW_{2T}^f = 1.$$

Formally we write $q_0(\mathcal{K}(2T))$ as

$$q_0(\mathcal{K}(2T)) = \frac{1}{2} \int_0^{2T} dB_\mu(s) \int_0^{2T} dB_\nu(s')$$
$$\times \int_{\mathbb{R}^d} d_{\mu\nu}(k) e^{-|s-s'|\omega(k)} e^{ik(X_s - X_{s'})} |\hat{\lambda}(k)|^2 dk.$$

Define the time-shift θ_T by

$$\theta_T : \mathrm{C}([0,\infty);\mathbb{R}^d) \to \mathrm{C}([-T,\infty);\mathbb{R}^d),$$
$$(\theta_T q)(s) := q(T+s), \quad -T \le s,$$

and the projection Π_T by

$$\Pi_T : \mathrm{C}([-T,\infty);\mathbb{R}^d) \to \mathrm{C}(\mathbb{R};\mathbb{R}^d),$$
$$(\Pi_T q)(s) := \begin{cases} q(s) & s \geq -T, \\ q(-T) & s < -T. \end{cases}$$

Set

$$D_T := \Pi_T \circ \theta_T : \mathrm{C}([0,\infty);\mathbb{R}^d) \to \mathrm{C}(\mathbb{R};\mathbb{R}^d).$$

We define the stochastic process $Y := (Y_t)_{t \geq -T}$ on $\mathrm{C}(\mathbb{R};\mathbb{R}^d) \times \mathbb{R}^d$ by

$$Y_t = Y_t(q,x) := X_{t+T}(D_T^{-1}(q), x), \quad -T \leq t. \tag{9.2}$$

Note that

$$Y_t(q,x) = B_{t+T}(D_T^{-1}(q)) + x = q(t) + x.$$

The measure $d\mu_T$ on $\mathrm{C}(\mathbb{R};\mathbb{R}^d)$ is defined by

$$d\mu_T := W_{2T}^f(Y) d(P \circ D_T^{-1}).$$

Hence we have

$$\int_{\mathrm{C}([0,\infty);\mathbb{R}^d)\times\mathbb{R}^d} f_1(X_{T+t_1})\cdots f_m(X_{T+t_m}) dW_{2T}^f$$
$$= \int_{\mathrm{C}(\mathbb{R};\mathbb{R}^d)\times\mathbb{R}^d} f_1(Y_{t_1})\cdots f_m(Y_{t_m}) d\mu_T dx. \tag{9.3}$$

Lemma 9.1: *We see that, for $-T \leq t_1 \leq \ldots \leq t_m \leq T$,*

$$\frac{(f\otimes 1, e^{-(T+t_1)H} f_1 e^{-(t_2-t_1)H} \ldots e^{-(t_m-t_{m-1})H} f_m e^{-(T-t_m)H} f\otimes 1)}{(f\otimes 1, e^{-2TH} f \otimes 1)}$$
$$= \int_{\mathrm{C}(\mathbb{R};\mathbb{R}^d)\times\mathbb{R}^d} f_1(Y_{t_1})\cdots f_m(Y_{t_m}) d\mu_T dx.$$

Proof: It follows from (9.1) and (9.3). □

9.2. *Kolmogorov extension theorem*

Let Ψ_{g} be the normalized ground state of H. By the Kolmogorov extension theorem there exists a unique probability measure μ_{K} on $(\mathbb{R}^d)^{\mathbb{Q}}$ such that for $f_j \in L^\infty(\mathbb{R}^d)$, $j = 1, 2, \ldots, m$,

$$(\Psi_{\mathrm{g}}, f_1 e^{-(t_2-t_1)\overline{H}} f_2 \cdots e^{-(t_m-t_{m-1})\overline{H}} f_m \Psi_{\mathrm{g}})$$
$$= \int_{(\mathbb{R}^d)^{\mathbb{Q}}} f_1(Z_{t_1})\cdots f_m(Z_{t_m}) d\mu_{\mathrm{K}}, \tag{9.4}$$

where $\overline{H} := H - E$, $-\infty < t_1 \leq \cdots \leq t_m < \infty$ with $t_j \in \mathbb{Q}$, $j = 1, 2, \ldots, m$, and

$$Z := (Z_t)_{t\in\mathbb{Q}} = (Z_{\mu,t})_{1\leq\mu\leq d, t\in\mathbb{Q}}$$

is the stochastic process on $(\mathbb{R}^d)^{\mathbb{Q}}$ given by

$$Z_s(q) := q(s), \quad s \in \mathbb{Q}.$$

Lemma 9.2: *Let C_n be the constant such that*

$$\int_{C([0,\infty);\mathbb{R}^d)} |B(t) - B(s)|^{2n} dP = C_n |t - s|^n, \quad n \geq 0.$$

Then we have[o]

$$\int_{(\mathbb{R}^d)^{\mathbb{Q}}} |Z_t - Z_s|^{2n} d\mu_{\mathrm{K}} \leq |t - s|^n \sqrt{C_{2n}} e^{|t-s|(E - \inf V)}, \quad t, s \in \mathbb{Q}.$$

Proof: Suppose $t \geq s$. Let Z_s^a and X_s^a be truncated stochastic processes defined by

$$Z_{\nu,s}^a := \begin{cases} Z_{\nu,s} & |Z_{\nu,s}| \leq a \\ -a & Z_{\nu,s} < -a \\ a & Z_{\nu,s} > a \end{cases}, \quad X_{\nu,s}^a := \begin{cases} X_{\nu,s} & |X_{\nu,s}| \leq a \\ -a & X_{\nu,s} < -a \\ a & X_{\nu,s} > a \end{cases}.$$

Moreover we define

$$h_\nu^a(x) := \begin{cases} x_\nu & |x_\nu| \leq a \\ -a & x_\nu < -a \\ a & x_\nu > a \end{cases}.$$

[o]Note that $E - \inf V \geq E - \inf \sigma(H_{\mathrm{p}}) \geq 0$.

Since $|Z_s^a - Z_t^a|$ is bounded, we have

$$\begin{aligned}
&\int_{(\mathbb{R}^d)^{\mathbb{Q}}} |Z_s^a - Z_t^a|^{2n} d\mu_{\mathrm{K}} \\
&= \sum_{k=0}^{2n} \binom{2n}{k} (-1)^k \int_{(\mathbb{R}^d)^{\mathbb{Q}}} \left(Z_{\nu,s}^a\right)^k \left(Z_{\nu,t}^a\right)^{2n-k} d\mu_{\mathrm{K}} \\
&= \sum_{k=0}^{2n} \binom{2n}{k} (-1)^k (\Psi_{\mathrm{g}}, (h_\nu^a)^k e^{-(t-s)\overline{H}} (h_\nu^a)^{2n-k} \Psi_{\mathrm{g}})_{\mathcal{H}} \\
&= \sum_{k=0}^{2n} \binom{2n}{k} (-1)^k \int_{\mathcal{W}} (X_{\nu,0}^a)^k (X_{\nu,t-s}^a)^{2n-k} e^{-\int_0^{t-s}(V(X_{s'})-E)ds'} \\
&\qquad\qquad \times (\Psi_{\mathrm{g}}(X_0), \mathcal{J}_{t-s}\Psi_{\mathrm{g}}(X_{t-s}))_{L^2(Q)}\, dX \\
&\le \int_{\mathcal{W}} |B(0) - B(t-s)|^{2n} e^{|t-s|(E-\inf V)} \\
&\qquad\qquad \times \|\Psi_{\mathrm{g}}(X_0)\|_{L^2(Q)} \|\Psi_{\mathrm{g}}(X_{t-s})\|_{L^2(Q)}\, dX \\
&\le \left(\int_{\mathrm{C}([0,\infty);\mathbb{R}^d)} |B(0) - B(t-s)|^{4n} dP\right)^{1/2} e^{|t-s|(E-\inf V)} \|\Psi_{\mathrm{g}}\|_{\mathcal{H}}^2 \\
&\le \sqrt{C_{2n}} |t-s|^n e^{|t-s|(E-\inf V)}.
\end{aligned}$$

Since $|Z_{\nu,t}^a - Z_{\nu,s}^a| \uparrow |Z_{\nu,t} - Z_{\nu,s}|$ as $a \uparrow \infty$, the lemma follows by the Lebesgue monotone convergence theorem. □

By Lemma 9.2 we see that there exists a continuous version $\tilde{Z} = (\tilde{Z}_t)_{t\in\mathbb{R}}$, of $(Z_t)_{t\in\mathbb{Q}}$. Then continuous version $\tilde{Z}$ induces the probability measure $d\mu$ on $\mathrm{C}(\mathbb{R};\mathbb{R}^d)$, i.e.

$$d\mu := d(\mu_{\mathrm{K}} \circ \tilde{Z}^{-1}). \tag{9.5}$$

Lemma 9.3: *We see that, for $-\infty < t_1 \le \ldots \le t_m < \infty$,*

$$\begin{aligned}
&(\Psi_{\mathrm{g}}, f_1 e^{-(t_2-t_1)\overline{H}} f_2 \cdots e^{-(t_m-t_{m-1})\overline{H}} f_m \Psi_{\mathrm{g}}) \\
&= \int_{\mathrm{C}(\mathbb{R};\mathbb{R}^d)} f_1(q(t_1)) \cdots f_m(q(t_m)) d\mu.
\end{aligned}$$

Proof: This follows from (9.4) and (9.5). □

Theorem 9.4: We assume that H has a ground state. Let $f_j \in L^\infty(\mathbb{R}^d)$, $j = 1, \ldots, m$. We see that, for $-\infty < t_1 \le \cdots \le t_m < \infty$,

$$\begin{aligned}
&\lim_{T\to\infty} \int_{\mathrm{C}(\mathbb{R};\mathbb{R}^d)\times\mathbb{R}^d} f_1(Y_{t_1}) \cdots f_m(Y_{t_m}) d\mu_T dx \\
&= \int_{\mathrm{C}(\mathbb{R};\mathbb{R}^d)} f_1(q(t_1)) \cdots f_m(q(t_m)) d\mu.
\end{aligned}$$

Proof: We note that since $(f\otimes 1, \Psi_g)>0$ and $\|\Psi_g\|=1$,

$$\text{s-}\lim_{T\to\infty}\frac{e^{-(T+a)H}(f\otimes 1)}{\|e^{-TH}(f\otimes 1)\|}=e^{-aE}\frac{(f\otimes 1,\Psi_g)}{|(f\otimes 1,\Psi_g)|}\frac{\Psi_g}{\|\Psi_g\|}=e^{-aE}\Psi_g,\quad a\in\mathbb{R}.$$

Hence it is obtained by Lemmas 9.1 and 9.3 that

$$\begin{aligned}
&\lim_{T\to\infty}\int_{C(\mathbb{R};\mathbb{R}^d)\times\mathbb{R}^d} f_1(Y_{t_1})\cdots f_m(Y_{t_m})d\mu_T\\
&=\lim_{T\to\infty}\frac{\left(f\otimes 1, e^{-(T+t_1)H}f_1e^{-(t_2-t_1)H}\cdots e^{-(t_m-t_{m-1})H}f_me^{-(T-t_m)H}f\otimes 1\right)}{(f\otimes 1, e^{-2TH}f\otimes 1)}\\
&=\lim_{T\to\infty}\frac{\left(f\otimes 1, e^{-(T+t_1)\overline{H}}f_1e^{-(t_2-t_1)\overline{H}}\cdots e^{-(t_m-t_{m-1})\overline{H}}f_me^{-(T-t_m)\overline{H}}f\otimes 1\right)}{(f\otimes 1, e^{-2T\overline{H}}f\otimes 1)}\\
&=(\Psi_g, f_1e^{-(t_2-t_1)\overline{H}}f_2\cdots e^{-(t_m-t_{m-1})\overline{H}}f_m\Psi_g)e^{(t_m-t_1)\inf\sigma(\overline{H})}\\
&=(\Psi_g, f_1e^{-(t_2-t_1)\overline{H}}f_2\cdots e^{-(t_m-t_{m-1})\overline{H}}f_m\Psi_g)\\
&=\int_{C(\mathbb{R};\mathbb{R}^d)} f_1(q(t_1))\cdots f_m(q(t_m))d\mu.
\end{aligned}$$

This yields the theorem. □

Remark 9.5: Theorem 9.4 holds for all $e\in\mathbb{R}$.

9.3. *Expectation values*

In this subsection we assume that H has a normalized ground state Ψ_g. Let

$$\langle T\rangle := (\Psi_g, T\Psi_g)_{\mathcal{H}}.$$

Corollary 9.6: *We have*

$$\lim_{|t-s|\to\infty}\int_{C(\mathbb{R};\mathbb{R}^d)} q(t)q(s)d\mu=\sum_{\mu=1}^d\langle x_\mu\rangle^2.$$

Proof: By a limiting argument we have

$$\int_{C(\mathbb{R};\mathbb{R}^d)} q(t)q(s)d\mu=\sum_{\mu=1}^d\langle x_\mu e^{-|t-s|\overline{H}}x_\mu\rangle.$$

Thus the corollary follows. □

Corollary 9.7: *Suppose that $V\in V(n)$ for some $n\geq 0$ (see Definition 4.5). Then, for sufficiently small $\delta>0$, we have*

$$\int_{C(\mathbb{R};\mathbb{R}^d)} e^{\delta|q(t)|^{n+1}}d\mu=\langle e^{\delta|x|^{n+1}}\rangle<\infty. \tag{9.6}$$

Proof: We have

$$\langle e^{\delta|x|^{n+1}}\lceil_m\rangle = \int_{C(\mathbb{R};\mathbb{R}^d)} e^{\delta|q(t)|^{n+1}}\lceil_m d\mu,$$

where $f(x)\lceil_m := f(x)$ if $f(x) \le m$, otherwise $f(x)\lceil_m = m$. Since

$$\|\Psi_g(\cdot)\|\, e^{\delta|\cdot|^{n+1}} \in L^2(\mathbb{R}^d),$$

(9.6) follows from the Lebesgue monotone convergence theorem. □

Lemma 9.8: *Let $f \in L^2(\mathbb{R}^d)$ be positive and $K = d\Gamma(h(-i\nabla))$ with some positive function h. Then*

$$\frac{(f\otimes 1, e^{-TH}e^{-\beta K}e^{-TH} f\otimes 1)}{(f\otimes 1, e^{-2TH} f\otimes 1)} = \int_{C([0,\infty);\mathbb{R}^d)\times\mathbb{R}^d} e^{(e^2/2)F(X,\beta)}dW_{2T}^f \tag{9.7}$$

$$= \int_{C(\mathbb{R};\mathbb{R}^d)\times\mathbb{R}^d} e^{(e^2/2)F(Y,\beta)}d\mu_T dx, \tag{9.8}$$

where

$$\begin{aligned} F(X,\beta) := \; & 2q_0\left(\bigoplus_{\mu=1}^d \int_0^T j_s\lambda(\,\cdot - X_s)dB_\mu(s),\right. \\ & \left.\qquad\qquad \bigoplus_{\mu=1}^d \int_T^{2T} j_{s'}\lambda(\,\cdot - X_{s'})dB_\mu(s')\right) \\ & -q_1\left(\bigoplus_{\mu=1}^d \int_0^T \xi_0(h)j_s\lambda(\,\cdot - X_s)dB_\mu(s),\right. \\ & \left.\qquad\qquad \bigoplus_{\mu=1}^d \int_T^{2T} \xi_\beta(h)j_{s'}\lambda(\,\cdot - X_{s'})dB_\mu(s')\right). \end{aligned}$$

and

$$F(Y,\beta):=2q_0\left(\bigoplus_{\mu=1}^d\int_{-T}^0 j_s\lambda(\cdot-Y_s)dB_\mu^T(s),\right.$$
$$\left.\bigoplus_{\mu=1}^d\int_0^T j_{s'}\lambda(\cdot-Y_{s'})dB_\mu^T(s')\right)$$
$$-q_1\left(\bigoplus_{\mu=1}^d\int_{-T}^0 \xi_0(h)j_s\lambda(\cdot-Y_s)dB_\mu^T(s),\right.$$
$$\left.\bigoplus_{\mu=1}^d\int_0^T \xi_\beta(h)j_{s'}\lambda(\cdot-Y_{s'})dB_\mu^T(s')\right)$$

with $B^T(t):=B(t+T)$ *for* $-T\le t$.

Proof: By Theorems 5.3 and 6.19, we have

$$\begin{aligned}&\text{L.H.S. (9.8)}\\&=\frac{1}{Z_{2T}}\int_{\mathcal{W}}e^{-\int_0^{2T}V(X_s)ds}f(X_0)f(X_{2T})\left(1,e^{-ie\phi_1(W)}1\right)_{L^2(Q_1)}dX\\&=\frac{1}{Z_{2T}}\int_{\mathcal{W}}e^{-\int_0^{2T}V(X_s)ds}f(X_0)f(X_{2T})e^{-(e^2/2)q_1(W)}dX,\end{aligned}$$

where

$$W:=\bigoplus_{\mu=1}^d\left(\int_0^T\xi_0(h)j_s\lambda(\cdot-X_s)dB_\mu(s)+\int_T^{2T}\xi_\beta(h)j_s\lambda(\cdot-X_s)dB_\mu(s)\right).$$

Since

$$q_1(W)=q_0(\mathcal{K}(2T))-F(X,\beta), \tag{9.9}$$

we get (9.7). (9.8) follows from the definition of $d\mu_T$. □

Remark 9.9: Note that $F(Y,\beta)$ is independent of $x\in\mathbb{R}^d$. Formally we write $F(Y,\beta)$ as[p]

$$F(Y,\beta)=\int_{-T}^0 dB_\mu^T(s)\int_0^T dB_\nu^T(s')\times\int_{\mathbb{R}^d}\left(1-e^{-\beta h(k)}\right)d_{\mu\nu}(k)e^{-|s-s'|\omega(k)}e^{ik(Y_s-Y_{s'})}|\hat\lambda(k)|^2dk.$$

[p]This kind of expressions appears in Refs. 51, 65, 81 and 139.

Let $F_T(Y)$ be defined by

$$F_T(Y) := q_0 \left(\bigoplus_{\mu=1}^{d} \int_{-T}^{0} j_s \lambda(\,\cdot - Y_s) dB_\mu^T(s), \right.$$

$$\left. \bigoplus_{\mu=1}^{d} \int_{0}^{T} j_{s'} \lambda(\,\cdot - Y_{s'}) dB_\mu^T(s') \right)$$

$$= \int_{-T}^{0} dB_\mu^T(s) \int_0^T dB_\nu^T(s') \times \int_{\mathbb{R}^d} d_{\mu\nu} e^{-|s-s'|\omega(k)} e^{ik(Y_s - Y_{s'})} |\hat{\lambda}(k)|^2 dk.$$

Note again that $F_T(Y)$ is independent of $x \in \mathbb{R}^d$.

Lemma 9.10: *We have*

$$\left(\Psi_{\mathrm{g}}, e^{-\beta N} \Psi_{\mathrm{g}}\right) = \lim_{T\to\infty} \int_{\mathrm{C}(\mathbb{R};\mathbb{R}^d)} e^{(e^2/2)(1-e^{-\beta})F_T(Y)} d\widetilde{\mu}_T,$$

where $d\widetilde{\mu}_T$ is the marginal probability measure on $\mathrm{C}(\mathbb{R};\mathbb{R}^d)$ given by

$$d\widetilde{\mu}_T := \int_{\mathbb{R}^d} d\mu_T dx = \left(\int_{\mathbb{R}^d} W_{2T}^f(Y) dx \right) d(P \circ D_T^{-1}).$$

Proof: Take $h(k) = 1$ in Lemma 9.8. By means of (9.8) we have

$$\begin{aligned}
\left(\Psi_{\mathrm{g}}, e^{-\beta N} \Psi_{\mathrm{g}}\right) &= \lim_{T\to\infty} \frac{(f \otimes 1, e^{-TH} e^{-\beta N} e^{-TH} f \otimes 1)}{(f \otimes 1, e^{-2TH} f \otimes 1)} \\
&= \lim_{T\to\infty} \int_{\mathrm{C}(\mathbb{R};\mathbb{R}^d)\times\mathbb{R}^d} e^{(e^2/2)(1-e^{-\beta})F_T(Y)} d\mu_T dx \\
&= \lim_{T\to\infty} \int_{\mathrm{C}(\mathbb{R};\mathbb{R}^d)} e^{(e^2/2)(1-e^{-\beta})F_T(Y)} d\widetilde{\mu}_T.
\end{aligned}$$

Hence the lemma follows. □

Our conjecture is as follows:

Conjecture 2: We conjecture that there exists a function F_∞ on $\mathrm{C}(\mathbb{R};\mathbb{R}^d)$ such that

$$\langle e^{-\beta N} \rangle = \int_{\mathrm{C}(\mathbb{R};\mathbb{R}^d)} e^{(e^2/2)(1-e^{-\beta})F_\infty(q)} d\mu. \tag{9.10}$$

This conjecture is justified for the Nelson model stated in the next section. See (10.2).

10. The Nelson Model

10.1. *The Nelson Hamiltonian*

The Nelson model is introduced by Nelson,[121] which describes an interaction between nuclei and mesons. The Nelson Hamiltonian $H_{\rm N}$ is defined as a symmetric operator acting in the Hilbert space

$$\mathcal{H}_{\rm N} := L^2(\mathbb{R}^d) \otimes \mathcal{F}.$$

We identify $\mathcal{H}_{\rm N}$ with the set of $\mathcal{F}$-valued L^2-functions on $\mathbb{R}^d$, i.e.

$$\mathcal{H}_{\rm N} \cong \int_{\mathbb{R}^d}^{\oplus} \mathcal{F} dx.$$

Hamiltonian $H_{\rm N}$ is defined by

$$H_{\rm N} := H_{\rm p} + H_{\rm b} + g\phi(\hat{\lambda}),$$

where g is a coupling constant,

$$H_{\rm p} := -\frac{1}{2M}\Delta + V,$$
$$H_{\rm b} := \int \omega_\nu(k) a^*(k) a(k) dk$$

with

$$\omega_\nu(k) := \sqrt{|k|^2 + \nu^2}, \quad \nu \geq 0,$$

and

$$\phi(\hat{\lambda}) := \int_{\mathbb{R}^d}^{\oplus} \phi(\hat{\lambda}, x) dx$$

with a scalar quantum field:

$$\phi(\hat{\lambda}, x) := \frac{1}{\sqrt{2}} \int \left\{ a^*(k)\hat{\lambda}(-k)e^{-ikx} + a(k)\hat{\lambda}(k)e^{ikx} \right\} dk.$$

Suppose $\hat{\lambda}, \hat{\lambda}/\sqrt{\omega} \in L^2(\mathbb{R}^d)$ and $\hat{\lambda}(-k) = \overline{\hat{\lambda}(k)}$. Since it is seen that (see Section 4)

$$\|\phi(\hat{\lambda})\Psi\| \leq (1/\sqrt{2})(\|\hat{\lambda}\| + 2\|\hat{\lambda}/\sqrt{\omega}\|)\|(H_{\rm b} + 1)^{1/2}\Psi\|,$$

$H_{\rm N}$ is self-adjoint on $D(\Delta) \cap D(H_{\rm b})$ and bounded from below by the Kato–Rellich theorem. Nelson[121] proved the following proposition.

Proposition 10.1: *Suppose* $\nu > 0$ *and* $d = 3$. *Let*

$$\hat{\lambda}(k) := \begin{cases} 0 & |k| < \kappa \\ 1/\sqrt{(2\pi)^3} & \kappa \leq |k| \leq \Lambda \\ 0 & |k| > \Lambda \end{cases}$$

with some positive constants κ and Λ, and

$$E_{\rm ren}(\Lambda) := -g^2 \int_{\mathbb{R}^3} \frac{|\hat{\lambda}(k)|^2}{\omega_\nu(k) + |k|^2/(2M)} dk.$$

Then there exists a self-adjoint operator $H_{\mathrm{N}\infty}$ such that

$$\lim_{\Lambda\to\infty} e^{-it(H_{\mathrm{N}} - E_{\rm ren}(\Lambda))} = e^{-itH_{\mathrm{N}\infty}}.$$

Here the limit is taken in the operator norm.

Remark 10.2: Suppose $\nu > 0$. Then $\inf\sigma(H_{\mathrm{N}})$ is analytic in g^2 near $g^2 = 0$. Let

$$\inf\sigma(H_{\mathrm{N}}) = \sum_{n=0}^{\infty} a_n(\Lambda) g^{2n}.$$

Then it is seen by a perturbation theory that

$$a_0(\Lambda) = \inf\sigma(H_{\mathrm{p}}), \quad a_1(\Lambda)g^2 = E_{\rm ren}(\Lambda).$$

It is expected that each $a_j(\Lambda)$, $j \geq 2$, converges as $\Lambda \to \infty$.

10.2. *Functional integral representations for the Nelson model*

For simplicity let us set

$$M = 1.$$

It is similar, rather easier than the Pauli–Fierz model, to see a functional integral representation of a matrix element for the Nelson model. Since H_{N} has the linear interaction $g\phi(\hat{\lambda})$, there does *not* appear stochastic integrals such as H_{PF} in a functional integral representation. We will quickly see a functional integral representation of the Nelson model.

Let $(Q^{\mathrm{s}}, \nu^{\mathrm{s}})$ be a probability space endowed with a Gaussian random process $\phi^{\mathrm{s}}(f)$ indexed by real $f \in L^2(\mathbb{R}^d)$ with mean zero and the covariance

$$\int \phi^{\mathrm{s}}(f)\phi^{\mathrm{s}}(g) d\nu^{\mathrm{s}}(\phi^{\mathrm{s}}) = \frac{1}{2}(f, g).$$

In addition to this, let $(Q_i^{\mathrm{s}}, \nu_i^{\mathrm{s}})$, $i = 0, 1$, be probability spaces and $\phi_i^{\mathrm{s}}(f)$, $f \in L^2(\mathbb{R}^{d+1+i})$, the Gaussian random process with mean zero and the covariance

$$\int \phi_i^{\mathrm{s}}(f)\phi_i^{\mathrm{s}}(g) d\nu_i^{\mathrm{s}}(\phi^{\mathrm{s}}) = \frac{1}{2}(f, g)_{L^2(\mathbb{R}^{d+1+i})}.$$

$\mathcal{H}_{\rm N}$	$L^2(\mathbb{R}^d)\otimes\mathcal{F}$
$H_{\rm p}$	$-\dfrac{1}{2M}\Delta+V$
$H_{\rm b}$	$\displaystyle\int\omega_\nu(k)a^*(k)a(k)dk$
$H_{\rm dec}$	$H_{\rm p}+H_{\rm b}$
CCR	$[a(k),a^*(k')]=\delta(k-k')$
$\phi(\hat\lambda)$	$\displaystyle\frac{1}{\sqrt{2}}\int\left\{a^*(k)\hat\lambda(-k)e^{-ikx}+a(k)\hat\lambda(k)e^{ikx}\right\}dk$
$\pi(\hat\lambda)$	$\displaystyle\frac{i}{\sqrt{2}}\int\left\{a^*(k)\hat\lambda(-k)e^{-ikx}-a(k)\hat\lambda(k)e^{ikx}\right\}dk$
CCR	$\displaystyle[\phi(\hat\lambda),\pi(\hat\rho)]=i\int_{\mathbb{R}^d}\hat\lambda(-k)\hat\rho(k)dk$
$H_{\rm N}$	$-\dfrac{1}{2M}\Delta+V+g\phi(\hat\lambda)+H_{\rm b}$

Fig. 27. The Nelson model

For a contraction operator T of $L^2(\mathbb{R}^d)$, $\Gamma^{\rm s}(T)$ is defined by

$$\Gamma^{\rm s}(T)1:=1,$$
$$\Gamma^{\rm s}(T)\ :\phi^{\rm s}(f_1)\cdots\phi^{\rm s}(f_n):\,=\,:\phi^{\rm s}(Tf_1)\cdots\phi^{\rm s}(Tf_n):\ .$$

The infinitesimal generator of $\Gamma^{\rm s}(e^{ith})$ with self-adjoint operator h is denoted by $d\Gamma^{\rm s}(h)$, i.e.

$$\Gamma^{\rm s}(e^{ith})=e^{itd\Gamma^{\rm s}(h)}.$$

Let us define $H_{\rm f}^{\rm s}$ by

$$H_{\rm f}^{\rm s}:=d\Gamma^{\rm s}(\omega_\nu(-i\nabla)).$$

Hence we can get a probabilistic description of $H_{\rm N}$ as follows:

$$\mathcal{H}_{\rm N}\cong L^2(\mathbb{R}^d)\otimes L^2(Q^{\rm s})$$

and

$$H_{\rm N} \cong \widetilde{H}_{\rm N} := H_{\rm p} + g\phi_\lambda^{\rm s} + H_{\rm f}^{\rm s},$$

where, under the identification

$$L^2(\mathbb{R}^d) \otimes L^2(Q^{\rm s}) \cong \int_{\mathbb{R}^d}^{\oplus} L^2(Q^{\rm s})dx,$$

$\phi_\lambda^{\rm s}$ is given by

$$\phi_\lambda^{\rm s} := \int_{\mathbb{R}^d}^{\oplus} \phi^{\rm s}(\lambda(\cdot - x))dx.$$

To derive a functional integral representation we define J_t, $t \in \mathbb{R}$, by

$$\begin{aligned} &J_t^{\rm s} : L^2(Q^{\rm s}) \to L^2(Q_0^{\rm s}), \\ &J_t^{\rm s} 1 := 1, \\ &J_t^{\rm s} \ {:}\phi^{\rm s}(f_1)\cdots\phi^{\rm s}(f_n){:} = {:}\phi^{\rm s}(j_t f_1)\cdots\phi^{\rm s}(j_t f_n){:}\,, \end{aligned}$$

and Ξ_t, $t \in \mathbb{R}$, by

$$\begin{aligned} &\Xi_t^{\rm s} : L^2(Q_0^{\rm s}) \to L^2(Q_1^{\rm s}), \\ &\Xi_t^{\rm s} 1 = 1, \\ &\Xi_t^{\rm s} \ {:}\phi^{\rm s}(f_1)\cdots\phi^{\rm s}(f_n){:} = {:}\phi^{\rm s}(\xi_t f_1)\cdots\phi^{\rm s}(\xi_t f_n){:}\,, \end{aligned}$$

where j_t and ξ_t are defined in (5.3) and (5.6), respectively. Let

$$L_0^2(Q^{\rm s}) := \mathfrak{L}\{\,{:}\phi^{\rm s}(f_1)\cdots\phi^{\rm s}(f_n){:}\,, 1 | f_j \in \oplus^d L^2(\mathbb{R}^d), j = 1, \ldots, n, n \in \mathbb{N}\}.$$

Theorem 10.3: Suppose that $\hat{\lambda}/\sqrt{\omega}$, $\hat{\lambda}$, $\omega\hat{\lambda} \in L^2(\mathbb{R}^d)$. Let h be a nonnegative multiplication operator of $L^2(\mathbb{R}^d)$. Assume that $F_0, F_m \in L^2(\mathbb{R}^d) \otimes L^2(Q^{\rm s})$, and $F_1, \cdots, F_{m-1} \in L^\infty(\mathbb{R}^d) \,\widehat{\otimes}\, L_0^2(Q^{\rm s})$. Set

$$K := d\Gamma^{\rm s}(h(-i\nabla))$$

and

$$\widehat{F}_j := \Xi_{\tau_j}^{\rm s} J_{t_j}^{\rm s} F_j.$$

Then, for $0 := t_0 \le t_1 \le \cdots \le t_m$, and $0 \le \tau_1 \le \tau_2 \le \cdots \le \tau_m$, it holds that

$$\begin{aligned} &\Big(F_0, e^{-\tau_1 K} e^{-t_1 \widetilde{H}_{\rm N}} F_1 e^{-(\tau_2-\tau_1)K} e^{-(t_2-t_1)\widetilde{H}_{\rm N}} F_2 \cdots \\ &\qquad\qquad \cdots F_{m-1} e^{-(\tau_m-\tau_{m-1})K} e^{-(t_m-t_{m-1})\widetilde{H}_{\rm N}} F_m\Big) \\ &= \int_{\mathcal{W}\times Q_1^{\rm s}} e^{-g\phi_1^{\rm s}(\mathcal{L}^{\rm s}(t_0,t_1,\ldots,t_m))} \widehat{F}_0(X_0)\widehat{F}_{t_1}(X_{t_1})\cdots\widehat{F}_{t_m}(X_{t_m}) d\nu_1^{\rm s}(\phi_1^{\rm s})dX, \end{aligned}$$

where

$$\mathcal{L}^{\mathrm{s}}(t_0, t_1, \ldots, t_m) := \sum_{j=1}^{m} \int_{t_{j-1}}^{t_j} \xi_{T_j} j_s \lambda(\cdot - X_s) ds.$$

In particular

$$\left(F, e^{-t\widetilde{H}_{\mathrm{N}}} G\right) = \int_{\mathrm{W}} e^{-\int_0^t V(X_s)ds} \left(F(X_0), J_0^{\mathrm{s}*} e^{-g\phi_0^{\mathrm{s}}(\mathcal{K}^{\mathrm{s}}(t))} J_t^{\mathrm{s}} G(X_t)\right)_{L^2(Q^{\mathrm{s}})} dX,$$

where

$$\mathcal{K}^{\mathrm{s}}(t) := \int_0^t j_s \lambda(\cdot - X_s) ds.$$

Proof: Since the interaction $g\phi_\lambda^{\mathrm{s}}$ is linear, the theorem directly follows from the fact

$$e^{-|t-s|H_{\mathrm{f}}^{\mathrm{s}}} = J_t^{\mathrm{s}*} J_s^{\mathrm{s}}$$

and the Trotter–Kato product formula

$$e^{-t\widetilde{H}_{\mathrm{N}}} = \mathrm{s-}\lim_{n\to\infty} \left\{ e^{(t/n)\Delta/2} e^{-(t/n)V} e^{-(t/n)g\phi_\lambda^{\mathrm{s}}} e^{-(t/n)H_{\mathrm{f}}^{\mathrm{s}}} \right\}^n.$$

□

From Theorem 10.3 the diamagnetic inequality[74] for H_{N} is derived:

$$|(F, e^{-tH_{\mathrm{N}}} G)_{\mathcal{H}_{\mathrm{N}}}| \le (\|F\|_{\mathcal{F}}, e^{-t(H_{\mathrm{p}} - g^2 \|\hat{\lambda}/\sqrt{\omega}\|^2)} \|G\|_{\mathcal{F}})_{L^2(\mathbb{R}^d)}.$$

From this it follows that

$$\inf \sigma(H_{\mathrm{p}}) - g^2 \|\hat{\lambda}/\sqrt{\omega}\|^2 \le \inf \sigma(H_{\mathrm{N}}).$$

10.3. *Superexponential decay and infrared singularities*

In this subsection we suppose the following assumption.

Assumption 7: We assume that

(1) $\omega_{\nu=0}(k) = \omega(k) = |k|$,
(2) $\hat{\lambda}/\sqrt{\omega}, \hat{\lambda}, \omega\hat{\lambda} \in L^2(\mathbb{R}^d)$ and $\hat{\lambda}/\omega \in L^2(\mathbb{R}^d)$,
(3) H_{p} has a positive spectral gap.

The following proposition can be proven in the similar manner as that of $H_{\rm PF}$.

Proposition 10.4: *There exists a constant $e_* > 0$ such that for $|e| < e_*$, $\widetilde{H}_{\rm N}$ has the ground state $\Psi_{\rm g}$ and it is unique. Moreover*

$$(\Psi_{\rm g}, f \otimes 1) > 0 \tag{10.1}$$

follows for all positive but non zero $f \in L^2(\mathbb{R}^d)$.

Let

$$W_2(Y) := \left\| \int_{-T}^{T} j_s \lambda(\cdot - Y_s) ds \right\|^2_{L^2(\mathbb{R}^{d+1})},$$

where $Y = (Y_t)_{t \geq -T}$ is the stochastic process on $\mathrm{C}(\mathbb{R}; \mathbb{R}^d) \times \mathbb{R}^d$ defined by (9.2). Let us fix a positive but non zero $f \in L^2(\mathbb{R}^d)$. We define the measure $d\mu_T^{\rm s}$ on $\mathrm{C}(\mathbb{R}; \mathbb{R}^d)$ by

$$d\mu_T^{\rm s} := \frac{1}{Z_{2T}^{\rm s}} f(Y_{-T}) f(Y_T) e^{-\int_{-T}^{T} V(Y_s) ds} e^{(g^2/4) W_2(Y)} d(P \circ D_T^{-1}),$$

where $Z_{2T}^{\rm s} := (f \otimes 1, e^{-2TH_{\rm N}} f \otimes 1)$.

Lemma 10.5: *We see that, for $-T \leq t_1 \leq \ldots \leq t_m \leq T$,*

$$\frac{(f \otimes 1, e^{-(T+t_1)H_{\rm N}} f_1 e^{-(t_2-t_1)H_{\rm N}} \cdots e^{-(t_m - t_{m-1})H_{\rm N}} f_m e^{-(T-t_m)H_{\rm N}} f \otimes 1)}{(f \otimes 1, e^{-2TH_{\rm N}} f \otimes 1)}$$
$$= \int_{\mathrm{C}(\mathbb{R};\mathbb{R}^d) \times \mathbb{R}^d} f_1(Y_{t_1}) \cdots f_m(Y_{t_m}) d\mu_T^{\rm s} dx.$$

Proof: The proof is the same as that of Lemma 9.1. □

Lemma 10.6: *Let $\Psi_{\rm g}$ be a normalized ground state of $\widetilde{H}_{\rm N}$. Then there exists a probability measure $d\mu^{\rm s}$ on $\mathrm{C}(\mathbb{R}; \mathbb{R}^d)$ such that, for $-\infty < t_1 \leq \cdots \leq t_m < \infty$, we have*

$$(\Psi_{\rm g}, f_1 e^{-(t_2-t_1)\overline{H}_{\rm N}} f_2 \cdots e^{-(t_m-t_{m-1})\overline{H}_{\rm N}} f_m \Psi_{\rm g})$$
$$= \int_{\mathrm{C}(\mathbb{R};\mathbb{R}^d)} f_1(q(t_1)) \cdots f_m(q(t_m)) d\mu^{\rm s},$$

where $\overline{H}_{\rm N} := \widetilde{H}_{\rm N} - \inf \sigma(\widetilde{H}_{\rm N})$.

Proof: The proof is the same as that of Lemma 9.3. i.e. measure $d\mu^{\rm s}$ is constructed through the Kolmogorov extension theorem and a continuous version argument such as (9.5). □

Theorem 10.7: We assume that $H_{\rm N}$ has a ground state. Let $f_j \in L^\infty(\mathbb{R}^d)$, $j = 1, \ldots, m$. Then we see that, for $-\infty < t_1 \leq \cdots \leq t_m < \infty$,

$$\lim_{T\to\infty} \int_{{\rm C}(\mathbb{R};\mathbb{R}^d)\times\mathbb{R}^d} f_1(Y_{t_1})\cdots f_m(Y_{t_m}) d\mu_T^{\rm s} dx$$
$$= \int_{{\rm C}(\mathbb{R};\mathbb{R}^d)} f_1(q(t_1))\cdots f_m(q(t_m)) d\mu^{\rm s}.$$

Proof: Using Lemmas 10.5 and 10.6, in the similar way as Theorem 9.4 we prove the theorem. □

Let $d\tilde{\mu}_T^{\rm s}$ be the marginal probability measure on ${\rm C}(\mathbb{R};\mathbb{R}^d)$ defined by

$$d\tilde{\mu}_T^{\rm s} := \int_{\mathbb{R}^d} d\mu_T^{\rm s} dx.$$

Let $N^{\rm s}$ be the number operator in $L^2(Q^{\rm s})$, i.e.

$$N^{\rm s} := d\Gamma^{\rm s}(1).$$

Recall that $\langle \mathcal{O}\rangle = (\Psi_{\rm g}, \mathcal{O}\Psi_{\rm g})$.

Lemma 10.8: *It follows that*

$$\langle e^{-\beta N^{\rm s}}\rangle = \lim_{T\to\infty} \int_{{\rm C}(\mathbb{R};\mathbb{R}^d)} e^{-(g^2/2)(1-e^{-\beta})F_T^{\rm s}(Y)} d\tilde{\mu}_T^{\rm s},$$

where

$$F_T^{\rm s}(Y) := \int_{-T}^0 ds \int_0^T dt \int_{\mathbb{R}^d} e^{-|s-t|\omega(k)} e^{ik(Y_s - Y_t)} |\hat{\lambda}(k)|^2 dk.$$

Proof: Using (10.1) and Theorem 10.3 with $h(k) = 1$, we can see that

$$(\Psi_{\rm g}, e^{-\beta N^{\rm s}}\Psi_{\rm g})$$
$$= \lim_{T\to\infty} \frac{1}{Z_{2T}^{\rm s}} (e^{-T\widetilde{H}_{\rm N}} f\otimes 1, e^{-\beta N^{\rm s}} e^{-T\widetilde{H}_{\rm N}} f\otimes 1)$$
$$= \lim_{T\to\infty} \frac{1}{Z_{2T}^{\rm s}} \int_{{\rm C}(\mathbb{R};\mathbb{R}^d)\times\mathbb{R}^d} f(Y_{-T}) f(Y_T)$$
$$\times e^{-\int_{-T}^T V(Y_s)ds} e^{(g^2/4)W_1(Y)} d(P\circ D_T^{-1}) dx,$$

where

$$W_1(Y) := \left\| \int_{-T}^{0} \xi_0(1) j_s \lambda(\cdot - Y_s) ds + \int_0^T \xi_\beta(1) j_s \lambda(\cdot - Y_s) ds \right\|^2_{L^2(\mathbb{R}^{d+2})}$$

Let

$$K_{st} := \int_{\mathbb{R}^d} e^{-|s-t|\omega(k)} e^{ik(Y_s - Y_t)} |\hat{\lambda}(k)|^2 dk.$$

Since

$$\xi_\beta^*(1)\xi_0(1) = e^{-\beta},$$

we have

$$\begin{aligned} W_1(Y) &= \left\| \int_{-T}^{0} \xi_0(1) j_s \lambda(\cdot - Y_s) ds + \int_0^T \xi_\beta(1) j_s \lambda(\cdot - Y_s) ds \right\|^2_{L^2(\mathbb{R}^{d+2})} \\ &= \int_{-T}^{0} ds \int_{-T}^{0} dt\, K_{st} + 2e^{-\beta} \int_{-T}^{0} ds \int_0^{-T} dt\, K_{st} + \int_0^T ds \int_0^T dt\, K_{st} \\ &= \int_{-T}^{T} ds \int_{-T}^{T} dt\, K_{st} + 2(e^{-\beta} - 1) \int_{-T}^{0} ds \int_0^T dt\, K_{st} \\ &= \left\| \int_{-T}^{T} j_s \lambda(\cdot - X_s) ds \right\|^2_{L^2(\mathbb{R}^{d+1})} + 2(e^{-\beta} - 1) \int_{-T}^{0} ds \int_0^T dt\, K_{st} \\ &= W_2(Y) + 2(e^{-\beta} - 1) F_T^{\mathrm{s}}(Y). \end{aligned}$$

Then the lemma follows. □

Remark 10.9: Note that $F_T^{\mathrm{s}}(Y)$ is independent of $x \in \mathbb{R}^d$ and

$$|F_T^{\mathrm{s}}(Y)| \leq \|\hat{\lambda}/\omega\|^2$$

follows uniformly in path and T.

Remark 10.10: Let $\hat{\lambda}$ be a rotation invariant function. Then we can directly calculate $F_T^{\mathrm{s}}(Y)$, i.e.

$$\begin{aligned} F_T^{\mathrm{s}}(Y) = \sqrt{(2\pi)^d} \int_{-T}^{0} ds \int_0^T dt \int_0^\infty |\hat{\lambda}(r)|^2 e^{-|t-s|\omega(r)} \\ \times r^{(d-1)/2} \sqrt{r|Y_s - Y_t|} \frac{J_{(d-2)/2}(r|Y_t - Y_s|)}{|Y_t - Y_s|^{(d-1)/2}} dr, \end{aligned}$$

where J_ν is the Bessel function defined by

$$J_\nu(x) = \left(\frac{x}{2}\right)^\nu \sum_{n=0}^\infty \frac{(-1)^n}{n!\Gamma(n+\nu+1)} \left(\frac{x}{2}\right)^{2n}.$$

Since

$$J_{1/2}(x) = \sqrt{\frac{2}{\pi}} \frac{\sin x}{\sqrt{x}},$$

in particular in the case of $d = 3$ and $\hat{\lambda}(k) = \hat{\rho}(k)/\sqrt{(2\pi)^3\omega(k)}$ with a rotation invariant function $\hat{\rho}$, we have

$$F_T^{\mathrm{s}}(Y) = \frac{1}{2\pi^2} \int_{-T}^0 ds \int_0^T dt \int_0^\infty |\hat{\rho}(r)|^2 e^{-|t-s|\omega(r)} \frac{r}{\omega(r)} \frac{\sin(r|Y_t - Y_s|)}{|Y_t - Y_s|} dr.$$

The following lemma has been established by Betz, Hiroshima, Lőrinczi, Minlos and Spohn.[34]

Lemma 10.11: *It follows that*

$$\lim_{T\to\infty} \int_{C(\mathbb{R};\mathbb{R}^d)} e^{-(g^2/2)(1-e^{-\beta})F_T^{\mathrm{s}}(Y)} d\mu_T^{\mathrm{s}} = \int_{\mathrm{C}(\mathbb{R};\mathbb{R}^d)} e^{-(g^2/2)(1-e^{-\beta})F_\infty^{\mathrm{s}}(q)} d\mu^{\mathrm{s}},$$

where

$$F_\infty^{\mathrm{s}}(q) := \int_{-\infty}^0 dt \int_0^\infty ds \int_{\mathbb{R}^d} |\hat{\lambda}(k)|^2 e^{-|t-s|\omega(k)} e^{ik(q(s)-q(t))} dk.$$

In particular we have

$$\langle e^{-\beta N^{\mathrm{s}}} \rangle = \int_{\mathrm{C}(\mathbb{R};\mathbb{R}^d)} e^{-(g^2/2)(1-e^{-\beta})F_\infty^{\mathrm{s}}(q)} d\mu^{\mathrm{s}}. \tag{10.2}$$

Theorem 10.12: For *all* $\beta \in \mathbb{C}$, we have $\Psi_{\mathrm{g}} \in D(e^{\beta N^{\mathrm{s}}})$ and it holds that

$$\langle e^{+\beta N^{\mathrm{s}}} \rangle = \int_{\mathrm{C}(\mathbb{R};\mathbb{R}^d)} e^{-(g^2/2)(1-e^{+\beta})F_\infty^{\mathrm{s}}(q)} d\mu^{\mathrm{s}}. \tag{10.3}$$

Proof: Set $g(\beta) :=$ R.H.S. (10.2). It is easily seen that $g(\beta)$ can be analytically continued into $\mathbb{C}$ in β. We denote its analytic continuation by $\tilde{g}$. Let $\mathbb{C}_{\mathrm{R}} := \{z \in \mathbb{C} \mid \Re z \geq 0\}$ and $\mathbb{C}_{\mathrm{L}}$ be its complement. Let $\beta_0 \in \mathbb{C}_{\mathrm{R}}$ be such that $|\Re\beta_0| = \epsilon$ with some $\epsilon > 0$. Fix an arbitrary R such that $R > \epsilon$. We see that

$$\tilde{g}(\beta) = \sum_{n=0}^\infty (\beta - \beta_0)^n b_n(\beta_0) \tag{10.4}$$

for $\beta \in U := \{x \in \mathbb{C} \mid |\beta_0 - x| < R\}$, and (10.4) absolutely converges. Let $\nu(d\rho)$ denote the spectral projection of N^{s} to Ψ_{g}. Then

$$f(\beta) := \langle e^{-\beta N^{\mathrm{s}}} \rangle = \int_0^\infty e^{-\beta\rho} \nu(d\rho)$$

for $\beta \geq 0$. Note that $f(\beta)$ is analytic in the interior of $\mathbb{C}_{\mathrm{R}}$. Then

$$f(\beta) = \sum_{n=0}^\infty (\beta - \beta_0)^n \frac{1}{n!} \int_0^\infty (-\rho)^n e^{-\beta_0\rho} \nu(d\rho)$$

for β so that $|\beta - \beta_0| < \epsilon$. Since $f(\beta) = \tilde{g}(\beta)$ for β such that $|\beta - \beta_0| < \epsilon$, we see that

$$b_n(\beta_0) = \frac{1}{n!} \int_0^\infty (-\rho)^n e^{-\beta_0\rho} \nu(d\rho).$$

Then we have

$$\tilde{g}(\beta) = \sum_{n=0}^\infty (\beta_0 - \beta)^n \frac{1}{n!} \int_0^\infty \rho^n e^{-\beta_0\rho} \nu(d\rho) \tag{10.5}$$

for $\beta \in U$. In particular R.H.S. (10.5) absolutely converges for $\beta \in U$, and

$$U \cap \mathbb{C}_{\mathrm{L}} \neq \emptyset.$$

In particular, for $\beta \in \mathbb{R} \cap U \cap \mathbb{C}_{\mathrm{L}}$, i.e. for

$$-(R - \epsilon) < \beta < R, \tag{10.6}$$

by Fatou's Lemma we have for any $M > 0$,

$$\begin{aligned}
\int_0^M e^{-\beta\rho} \nu(d\rho) &\leq \sum_{n=0}^\infty |\beta_0 - \beta|^n \frac{1}{n!} \int_0^M \rho^n e^{-\beta_0\rho} \nu(d\rho) \\
&\leq \sum_{n=0}^\infty |\beta_0 - \beta|^n \frac{1}{n!} \int_0^\infty \rho^n e^{-\beta_0\rho} \nu(d\rho) < \infty.
\end{aligned}$$

Thus

$$\int_0^\infty e^{-\beta\rho} \nu(d\rho) < \infty$$

for β with (10.6), which implies that $\Psi_{\mathrm{g}} \in D(e^{-(\beta/2)N^{\mathrm{s}}})$ and (10.2) holds for β with (10.6). Since R is an arbitrary large number, we get (10.2) for all $\beta \in \mathbb{C}$. □

Corollary 10.13: *Let $m \geq 0$. Then*

$$\langle (N^{\mathrm{s}})^m \rangle = \int_{C(\mathbb{R};\mathbb{R}^d)} \left(\frac{g^2}{2} \right)^m F_\infty(q)^m d\mu^{\mathrm{s}}. \tag{10.7}$$

Proof: By Theorem 10.12, we have

$$\langle (N^{\rm s})^m\rangle = \frac{d^m}{d\beta^m}\langle e^{\beta N^{\rm s}}\rangle\lceil_{\beta=0}.$$

Thus the corollary follows from (10.3). □

Corollary 10.14: *There exist positive constants a and b such that*

$$-a+b\|\hat\lambda/\omega\|^2 \le \langle N^{\rm s}\rangle \le \frac{g^2}{2}\|\hat\lambda/\omega\|^2. \tag{10.8}$$

Proof: From the expression (10.7) we have

$$\langle N^{\rm s}\rangle = \frac{g^2}{2}\int_{\mathbb{R}^d} dk|\hat\lambda(k)|^2\int_{-\infty}^0 ds\int_0^\infty dt e^{-|t-s|\omega(k)}\int_{C(\mathbb{R};\mathbb{R}^d)} e^{ik(q(s)-q(t))}d\mu^{\rm s}. \tag{10.9}$$

Since

$$\int_{-\infty}^0 ds\int_0^\infty dt e^{-|t-s|\omega(k)} = \int_{-\infty}^0 ds e^{s\omega(k)}\int_0^\infty dt e^{-t\omega(k)} = \frac{1}{\omega(k)^2},$$

we see that together with (10.9),

$$\langle N^{\rm s}\rangle \le \frac{g^2}{2}\|\hat\lambda/\omega\|^2.$$

Since

$$\int_{C(\mathbb{R};\mathbb{R}^d)} e^{ik(q(s)-q(t))}d\mu^{\rm s} = \left(e^{ikx}\Psi_{\rm g}, e^{-|t-s|\overline{H}_{\rm N}}e^{ikx}\Psi_{\rm g}\right),$$

we get

$$\langle N^{\rm s}\rangle = \frac{g^2}{2}\int_{\mathbb{R}^d}\frac{\hat\lambda(k)^2}{\omega(k)^2}(e^{ikx}\Psi_{\rm g}, \omega(k)^2(\overline{H}_{\rm N}+\omega(k))^{-2}e^{ikx}\Psi_{\rm g})dk.$$

It is easily seen that

$$\lim_{k\to 0}(e^{ikx}\Psi_{\rm g}, \omega(k)^2(\overline{H}_{\rm N}+\omega(k))^{-2}e^{ikx}\Psi_{\rm g}) = 1.$$

Thus, by the continuity in k, we see that for arbitrary $0<\epsilon<1$ there exists δ such that, for $|k|<\delta$,

$$0<1-\epsilon \le (e^{ikx}\Psi_{\rm g}, \omega(k)^2(\overline{H}_{\rm N}+\omega(k))^{-2}e^{ikx}\Psi_{\rm g}).$$

Hence, we have

$$-a+b\|\hat\lambda/\omega\|^2 \le \langle N^{\rm s}\rangle,$$

where

$$a := \frac{g^2}{2}(2-\epsilon)\frac{1}{\delta^2}\|\hat{\lambda}\|^2, \quad b := \frac{g^2}{2}(1-\epsilon).$$

The lemma follows. □

Corollary 10.15: *We have*

$$\lim_{\|\hat{\lambda}/\omega\|\to\infty} \langle N^{\mathrm{s}} \rangle = \infty.$$

Proof: It follows from (10.8). □

A physical folklore is that a ground state with $\langle N^{\mathrm{s}} \rangle = \infty$ does not exist. Corollary 10.15 in a sense connects (i) infrared singularity (10.12), (ii) $\langle N^{\mathrm{s}} \rangle = \infty$, and (iii) the absence of ground states. See subsection 10.5.

10.4. *Boson density*

By using μ^{s} we can explicitly express both a momentum density, $\langle a^*(k)a(k) \rangle$, and a spatial density, $\langle a^*(x)a(x) \rangle$. A strategy being a suitable variant of (10.7) is the following (1)–(3).

(1) For a real positive $h \in C_0^\infty(\mathbb{R}^d)$ we construct a functional integral representation of $\langle e^{-\beta d\Gamma^{\mathrm{s}}(h(-i\nabla))} \rangle$ through

$$\langle e^{-\beta d\Gamma^{\mathrm{s}}(h(-i\nabla))} \rangle = \lim_{t\to\infty} \frac{(e^{-t\widetilde{H}_{\mathrm{N}}} f \otimes 1, e^{-\beta d\Gamma^{\mathrm{s}}(h(-i\nabla))} e^{-t\widetilde{H}_{\mathrm{N}}} f \otimes 1)}{(f \otimes 1, e^{-2t\widetilde{H}_{\mathrm{N}}} f \otimes 1)}.$$

(2) We prove that $\langle e^{-\beta d\Gamma^{\mathrm{s}}(h(-i\nabla))} \rangle$ can be analytically continued in β to $\mathbb{C}$.

(3) We conclude that

$$\langle d\Gamma^{\mathrm{s}}(h(-i\nabla)) \rangle = - \left. \frac{d\langle e^{-\beta d\Gamma^{\mathrm{s}}(h(-i\nabla))} \rangle}{d\beta} \right|_{\beta=0}$$

$$= \frac{g^2}{2} \int_{\mathbb{R}^d} dk h(k) |\hat{\lambda}(k)|^2 \tag{10.10}$$

$$\times \int_{-\infty}^0 dt \int_0^\infty ds e^{-|t-s|\omega(k)} \int_{C(\mathbb{R};\mathbb{R}^d)} e^{ik(q(t)-q(s))} d\mu^{\mathrm{s}}. \tag{10.11}$$

For $h = h_+ - h_- \in C_0^\infty(\mathbb{R}^d)$, $h_\pm \geq 0$, we define

$$d\Gamma^{\mathrm{s}}(h(-i\nabla)) := d\Gamma^{\mathrm{s}}(h_+(-i\nabla)) - d\Gamma^{\mathrm{s}}(h_-(-i\nabla)).$$

Furthermore for $h \in C_0^\infty(\mathbb{R}^d)$, we define

$$d\Gamma^{\mathrm{s}}(h(-i\nabla)) := d\Gamma^{\mathrm{s}}(\Re h(-i\nabla)) + d\Gamma^{\mathrm{s}}(\Im h(-i\nabla)).$$

Then it is obtained that

$$|\langle d\Gamma^{\mathrm{s}}(h(-i\nabla))\rangle| \le \frac{g^2}{2}\|h\|_\infty \|\hat{\lambda}/\omega\|^2.$$

Hence if $\|\hat{\lambda}/\omega\| < \infty$, then the map

$$\vartheta : C_0^\infty(\mathbb{R}^d) \ni h \longmapsto \langle d\Gamma^{\mathrm{s}}(h(-i\nabla))\rangle \in \mathbb{C}$$

defines a distribution. We define $\langle a^*(k)a(k)\rangle$ as the kernels of distribution ϑ.

Theorem 10.16: We have

$$\langle a^*(k)a(k)\rangle = \frac{g^2}{2}|\hat{\lambda}(k)|^2 \int_{-\infty}^0 dt \int_0^\infty ds e^{-|t-s|\omega(k)} \int_{C(\mathbb{R};\mathbb{R}^d)} e^{ik(q(s)-q(t))} d\mu^{\mathrm{s}}.$$

In particular we obtain that

$$|\langle a^*(k)a(k)\rangle| \le \frac{g^2}{2}\frac{|\hat{\lambda}(k)|^2}{\omega(k)^2} \quad \text{a.e.}$$

Proof: It follows from (10.11). □

On the other hand a spatial density $\langle a^*(x)a(x)\rangle$ is defined as the kernel of the distribution

$$\tilde{\vartheta} : C_0^\infty(\mathbb{R}^d) \ni h \longmapsto \langle d\Gamma^{\mathrm{s}}(h)\rangle \in \mathbb{C}.$$

We can also express $\langle a^*(x)a(x)\rangle$ by measure $d\mu^{\mathrm{s}}$. We just write a result:

Theorem 10.17: We have

$$\langle a^*(x)a(x)\rangle = \frac{g^2}{2}\int_{-\infty}^0 dt \int_0^\infty ds \int_{C(\mathbb{R};\mathbb{R}^d)} \overline{\hat{f}(q(s),s,x)}\hat{f}(q(t),t,x) d\mu^{\mathrm{s}},$$

where

$$f(q(s),s,k) := e^{-|s|\omega(k)}\hat{\lambda}(k)e^{ik\cdot q(s)}.$$

In particular we obtain that

$$|\langle a^*(x)a(x)\rangle| \le \frac{g^2}{2}\frac{1}{4e}\|\hat{\lambda}/\omega\|^2.$$

Proof: See Ref. 34 for details. □

	$\|\hat{\lambda}/\omega\| < \infty$	$\|\hat{\lambda}/\omega\| = \infty$
$H_{\rm PF}$	Exist	Exist
$H_{\rm N}$	Exist	Not exist for $d = 3$

Fig. 28. The existence of ground states and infrared singularity

10.5. *Notes*

Gérard[57] and Spohn[142] proved the existence of a ground state of $H_{\rm N}$ for arbitrary $g \in \mathbb{R}$ for suitable external potentials. On the other side Lőrinczi, Minlos and Spohn[116] proved the absence of ground states of the Nelson Hamiltonian with some confined external potentials in three dimensions (Fig. 28) in the case of

$$\int \frac{|\hat{\lambda}(k)|^2}{\omega(k)^2} dk = \infty. \tag{10.12}$$

Recently Hirokawa[71] extends Lőrinczi, Minlos and Spohn[116] for general external potentials by a different method. See also Arai, Hirokawa and Hiroshima,[21] Chen,[39] Minlos[119] and, Minlos and Spohn.[120]

Arai[16] showed that even under (10.12) by taking a non-Fock representation, the Nelson Hamiltonian has a ground state. The infrared regular Nelson model such as Arai[16] is also considered by Lőrinczi, Minlos and Spohn[117] from a path measure point of view. In a non-Fock representation Hirokawa, Hiroshima and Spohn[72] proved the existence of a ground state of the Nelson Hamiltonian without infrared and ultraviolet cutoffs.

Acknowledgments

I am very grateful to S. Albeverio, A. Arai, H. Araki, V. Bach, V. Betz, H. Ezawa, J. Fröhlich, M. Hirokawa, T. Ichinose, K. R. Ito, J. Klauder, E. Lieb, J. Lőrinczi, M. Loss, R. A. Minlos, T. Nakamura, H. Siedentop, H. Spohn, K. Watanabe for variable comments and useful discussions. Special thanks go to H. Araki and H. Ezawa for giving a chance to have a lecture in "the summer school of mathematical physics 2000" held in Gakushuin University in September of 2000.

Bibliography

1. S. Albeverio, An introduction to some mathematical aspects of scattering theory in models of quantum fields, *Scattering theory in mathematical physics* ed. by J. A. LaVita and J. P. Marchand (1974), 299–381.

2. Z. Ammari, Asymptotic completeness for a renormalized non-relativistic hamiltonian in quantum field theory: the Nelson model, *Math. Phys., Anal. Geom.* **3** (2000), 217–285.
3. A. Arai, Self-adjointness and spectrum of Hamiltonians in nonrelativistic quantum electrodynamics, *J. Math. Phys.* **22** (1981), 534–537.
4. A. Arai, On a model of a harmonic oscillator coupled to a quantized, massless, scalar field. I, *J. Math. Phys.* **22** (1981), 2539–2548.
5. A. Arai, On a model of a harmonic oscillator coupled to a quantized, massless, scalar field. II, *J. Math. Phys.* **22** (1981), 2549–2552.
6. A. Arai, Rigorous theory of spectra and radiation for a model in quantum electrodynamics, *J. Math. Phys.* **24** (1983), 1896–1910.
7. A. Arai, Note on scattering theory in non-relativistic quantum electrodynamics, *J. Phys. A: Math. Gen.* **16** (1983), 49–70.
8. A. Arai, Spectral analysis of a quantum harmonic oscillator coupled to infinitely many scalar bosons, *J. Math. Anal. Appl.* **140** (1989), 270–288.
9. A. Arai, Perturbation of embedded eigenvalues: A general class of exactly soluble models in Fock spaces, *Hokkaido Math. J.* **19** (1990), 1–34.
10. A. Arai, Long-time behavior of two-point functions of quantum harmonic oscillator interacting with bosons, *J. Math. Phys.* **30** (1989), 1277–1288.
11. A. Arai, An asymptotic analysis and its applications to the nonrelativistic limit of the Pauli–Fierz and a spin-boson model, *J. Math. Phys.* **31** (1990), 2653–2663.
12. A. Arai, Long-time behavior of an electron interacting with a quantized radiation field, *J. Math. Phys.* **32** (1991), 2224–2242.
13. A. Arai, Noninvertible Bogoliubov transformations and instability of embedded eigenvalues, *J. Math. Phys.* **32** (1991), 1838–1846.
14. A. Arai, Essential spectrum of a self-adjoint operator on an abstract Hilbert space of Fock type and applications to quantum field Hamiltonians, *J. Math. Anal. Appl.* **246** (2000), 189–216.
15. A. Arai, A particle-field Hamiltonian in relativistic quantum electrodynamics, *J. Math. Phys.* **41** (2000), 4271–4283.
16. A. Arai, Ground state of the massless Nelson model without infrared cutoff in a non-Fock representation, *Rev. Math. Phys.* **13** (2001), 1075–1094.
17. A. Arai, Mathematical theory of quantum particles interacting with a quantum field, *Non-Commutativity, Infinite-Dimensionality and Probability at the Crossroads*, ed. by N. Obata, T. Matsui and A. Hora, World Scientific, 2002, 1–50.
18. A. Arai and M. Hirokawa, On the existence and uniqueness of ground states of a generalized spin-boson model, *J. Funct. Anal.* **151** (1997), 455–503.
19. A. Arai and M. Hirokawa, Ground states of a general class of quantum field Hamiltonians, *Rev. Math. Phys.* **12** (2000), 1085–1135.
20. A. Arai and M. Hirokawa, Stability of ground states in sectors and its application to the Wigner-Weisskopf model, *Rev. Math. Phys.* **13** (2001), 513–527.
21. A. Arai, M. Hirokawa, and F. Hiroshima, On the absence of eigenvectors of Hamiltonians in a class of massless quantum field models without infrared cutoff, *J. Funct. Anal.* **168** (1999), 470–497.

22. A. Arai and H. Kawano, Enhanced binding in a general class of quantum field models, *Rev. Math. Phys.* **15** (2003), 387–423.
23. H. Araki and E. Woods, Representations of the canonical commutation relations describing a non-relativistic infinite free gas, *J. Math. Phys.* **4** (1963), 637–662.
24. V. Bach, J. Fröhlich, I. M. Sigal, Mathematical theory of non-relativistic matter and radiation, *Lett. Math. Phys.* **34** (1995), 183–201.
25. V. Bach, J. Fröhlich, I. M. Sigal, Renormalization group analysis of spectral problems in quantum field theory, *Adv. Math.* **137** (1998), 205–298.
26. V. Bach, J. Fröhlich, I. M. Sigal, Quantum electrodynamics of confined nonrelativistic particles, *Adv. Math.* **137** (1998), 299–395.
27. V. Bach, J. Fröhlich, I. M. Sigal, Spectral analysis for systems of atoms and molecules coupled to the quantized radiation field, *Commun. Math. Phys.* **207** (1999), 249–290.
28. V. Bach, J. Fröhlich, I. M. Sigal, Return to equilibrium, *J. Math. Phys.* **41** (2000), 3985–4060.
29. V. Bach, J. Fröhlich, I. M. Sigal, and A. Soffer, Positive commutators and the spectrum of Pauli–Fierz Hamiltonian of atoms and molecules, *Commun. Math. Phys.* **207** (1999), 557–587.
30. J. Barbaroux, M. Dimassi and J. C. Guillot, Quantum electrodynamics of relativistic bound states with cutoffs II, mp-arc 02-134, preprint, 2002.
31. E. A. Berezin, *The Method of Second Quantization* (Academic Press, 1966).
32. H. A. Bethe, The electromagnetic shift of energy levels, *Phys. Rev.* **72** (1947), 339–342.
33. V. Betz, Gibbs measures relative to Brownian motion and Nelson's model, a thesis for a doctorate, 2002.
34. V. Betz, F. Hiroshima, J. Lőrinczi, R. A. Minlos and H. Spohn, Gibbs measure associated with particle-field system, *Rev. Math. Phys.*, **14** (2002), 173–198.
35. V. Betz and J. Lőrinczi, A Gibbsian description of $P(\phi)_1$-processes, preprint, 1999,
36. G. Breit, Quantum theory of dispersion, *Rev. Mod. Phys.* **4** (1932), 504–576.
37. R. Carmona, Pointwise bounds for Schrödinger operators, *Commun. Math. Phys.* **62** (1978), 97–106.
38. I. Catto and C. Hainzl, Self-energy of one electron in non-relativistic QED, arXiv:math-ph/0207036, preprint, 2002.
39. T. Chen, Operator-theoretic infrared renormalization and construction of dressed 1-particle states in non-relativistic QED, mp-arc 01-301, preprint, 2001.
40. T. Chen, V. Vougalter and S. Vugalter, The increase of binding energy and enhanced binding in non-relativistic QED, *J. Math. Phys.* **44** (2003), 1961–1970.
41. E. B. Davies, Particle-boson interactions and the weak coupling limit, *J. Math. Phys.* **20** (1979), 345–351.
42. E. B. Davies, Asymptotic analysis of some abstract evolution equations, *J. Funct. Anal.* **25** (1977), 81–101.

43. J. Dereziński and C. Gérard, Asymptotic completeness in quantum field theory. Massive Pauli–Fierz Hamiltonian, *Rev. Math. Phys.* **11** (1999), 383–450.
44. M. Dimassi and J. C. Guillot, The quantum electrodynamics of relativistic bound states with cutoffs I, mp-arc 01-325, preprint, 2001.
45. H. Ezawa, J. Klauder, L. A. Shepp, Vestigial effects of singular potentials in diffusion theory and quantum mechanics, *J. Math. Phys.* **16** (1975), 783–799.
46. H. Ezawa, J. Klauder, L. A. Shepp, On the divergence of certain integrals of the wiener progress, *Ann. Inst. Fourier, Grenoble* **24** (1974), 189–193.
47. W. Faris, Invariant cones and uniqueness of the ground state for Fermion systems, *J. Math. Phys.* **13** (1972), 1285–1290.
48. W. Faris and B. Simon, Degenerate and nondegenerate ground states for Schrödinger operators, *Duke Math. J.* **42** (1975), 559–567.
49. C. Fefferman, J. Fröhlich, G. M. Graf, Stability of ultraviolet-cutoff quantum electrodynamics with non-relativistic matter, *Commun. Math. Phys.* **190** (1997), 309–330.
50. E. Fermi, Quantum theory of radiation, *Rev. Mod. Phys.* **4** (1932), 87–132.
51. R. Feynman, Mathematical formulation of the quantum theory of electromagnetic interaction, *Phys. Rev.* **80** (1950), 440–457.
52. J. Fröhlich, On the infrared problem in a model of scalar electrons and massless, scalar bosons, *Ann. Inst. Henri Poincaré* **19** (1973), 1–103.
53. J. Fröhlich, Existence of dressed one electron states in a class of persistent models, *Fortschritte der Physik* **22** (1974), 159–198.
54. J. Fröhlich and Y. M. Park, Correlation inequalities and thermodynamic limit for classical and quantum continuous systems II. Bose-Einstein and Fermi-Dirac statistics, *J. Stat. Phys.* **23** (1980), 701–753.
55. J. Fröhlich, M. Griesemer and B. Schlein, Asymptotic electromagnetic fields in a mode of quantum-mechanical matter interacting with the quantum radiation field, *Adv. in Math.* **164** (2001), 349–398.
56. J. Fröhlich, M. Griesemer and B. Schlein, Asymptotic completeness for Rayleigh scattering, *Ann. Henri Poincaré* **3** (2002), 107–170.
57. C. Gérard, On the existence of ground states for massless Pauli–Fierz Hamiltonians, *Ann. Henri Poincaré* **1** (2000), 443–459.
58. C. Gérard, On the scattering theory of massless Nelson models, *Rev. Math. Phys.* **14** (2002), 1165–1280.
59. J. Glimm and A. Jaffe, The $\lambda(\phi^4)_2$ quantum field theory without cutoffs. I. *Phys. Rev.* **176** (1968), 1945–1951.
60. J. Glimm and A. Jaffe, *Quantum Physics* (Springer-Verlag, 1987).
61. M. Griesemer, E. Lieb and M. Loss, Ground states in non-relativistic quantum electrodynamics, *Invent. Math.* **145** (2001), 557–595.
62. L. Gross, A noncommutative extension of the Perron-Frobenius theorem, *Bull. Amer. Math. Soc.* **77** (1971), 343–347.
63. L. Gross, Existence and uniqueness of physical ground states, *J. Funct. Anal.* **10** (1972), 52–109.
64. L. Gross, The free euclidean Proca and electromagnetic fields, *Functional*

integral and its application, proceedings of the international conference, the Cumberland Lodge, Windsor Great Park, London in 1974. A. M. Arthurs (editor), (Clarendon Press, Oxford, 1975).

65. Z. Haba, Feynman integral in regularized nonrelativistic quantum electrodynamics, *J. Math. Phys.* **39** (1998), 1766–1787.
66. C. Hainzl, V. Vougalter and S. A. Vugalter, Enhanced binding in nonrelativistic QED, *Commun. Math. Phys.* **233** (2003), 13–26.
67. C. Hainzl and R. Seiringer, Mass renormalization and energy level shift in non-relativistic QED, arXiv:math-ph/0205044, preprint, 2003.
68. T. Hida, *Brownian Motion* (Springer-Verlag, 1980).
69. M. Hirokawa, An expression of the ground state energy of the Spin-Boson model, *J. Funct. Anal.* **162** (1999), 178–218.
70. M. Hirokawa, Remarks on the ground state energy of the Spin-Boson model. An application of the Wigner-Weisskopf model, *Rev. Math. Phys.* **13** (2001), 221–251.
71. M. Hirokawa, Divergence of soft photon number and absence of ground state for Nelson's model, arXiv:math-ph/0211051, preprint, 2002.
72. M. Hirokawa, F. Hiroshima, and H. Spohn, Ground state for point particles interacting through a massless scalar Bose field, arXiv:math-ph/0211050, preprint, 2002, to be published in *Adv. Math.*
73. F. Hiroshima, Scaling limit of a model of quantum electrodynamics, *J. Math. Phys.* **34** (1993), 4478-4518.
74. F. Hiroshima, Diamagnetic inequalities for systems of nonrelativistic particles with a quantized field, *Rev. Math. Phys.* **8** (1996), 185–203.
75. F. Hiroshima, Scaling limit of a model of quantum electrodynamics with many nonrelativistic particles, *Rev. Math. Phys.* **9** (1997), 201-225.
76. F. Hiroshima, Asymptotic behaviors of an interaction Hamiltonian, *Nonlinear Analysis, Theory, Methods and Application* **30** (1997), 4863-4874.
77. F. Hiroshima, Functional integral representation of a model in quantum electrodynamics, *Rev. Math. Phys.* **9** (1997), 489–530.
78. F. Hiroshima, Weak coupling limit and a removal of an ultraviolet cut-off for a Hamiltonian of particles interacting with a massive scalar field, *Infinite Dimensional Analysis, Quantum Probability and Related Topics* **1** (1998), 407-423.
79. F. Hiroshima, Weak coupling limit removing an ultraviolet cut-off for a Hamiltonian of particles interacting with a quantized scalar field, *J. Math. Phys.* **40** (1999), 1215-1236.
80. F. Hiroshima, Ground states of a model in nonrelativistic quantum electrodynamics I, *J. Math. Phys.* **40** (1999), 6209–6222.
81. F. Hiroshima, Ground states of a model in nonrelativistic quantum electrodynamics II, *J. Math. Phys.* **41** (2000), 661–674.
82. F. Hiroshima, Euclidean Gell-Mann-Low formula and double stochastic integrals, *Stochastic Processes, Physics and Geometry: New Interplays II, Conference proceedings, Canadian Math. Soc.* **29** (2000), 285–292.
83. F. Hiroshima, Essential self-adjointness of translation-invariant quantum field models for arbitrary coupling constants, *Commun. Math. Phys.* **211** (2000), 585–613.

84. F. Hiroshima, Ground states and spectrum of quantum electrodynamics of non-relativistic particles, *Trans. Amer. Math. Soc.* **353** (2001), 4497–4528.
85. F. Hiroshima, Observable effects and parametrized scaling limits of a model in nonrelativistic quantum electrodynamics, *J. Math. Phys.*, **43** (2002), 1755–1795.
86. F. Hiroshima, Embedded eigenvalues, localization and asymptotics of quantum field models: a functional integral approach, *J. Phys. A: Math. Gen.* **35** (2002), 351–375.
87. F. Hiroshima, Self-adjointness of the Pauli–Fierz Hamiltonian for arbitrary values of coupling constants, *Ann. Henri Poincaré*, **3** (2002), 171–201.
88. F. Hiroshima, Localization of the number of photons in nonrelativistic QED, *Rev. Math. Phys.* **15** (2003), 271–312.
89. F. Hiroshima, Nonrelativistic QED at large momentum of photons, *Garden of quanta* ed. by A. Tonomura et. al., World Scientific, 2003, 167–196.
90. F. Hiroshima and H. Spohn, Enhanced binding through coupling to a quantum field, *Ann. Henri Poincaré* **2** (2001), 1159–1187.
91. F. Hiroshima and H. Spohn, Ground state degeneracy of the Pauli–Fierz model with spin, *Adv. Theor. Math. Phys.* **5** (2001), 1091–1104.
92. F. Hiroshima and H. Spohn, Mass renormalization in nonrelativistic QED, arXiv:math-ph/0310043, preprint, 2003.
93. R. Høegh-Krohn, Asymptotic fields in some models of quantum field theory I, *J. Math. Phys.* **9** (1968), 2075–2080.
94. R. Høegh-Krohn, Asymptotic fields in some models of quantum field theory II, *J. Math. Phys.* **10** (1969), 639–643.
95. R. Høegh-Krohn, Asymptotic fields in some models of quantum field theory III, *J. Math. Phys.* **11** (1969), 185–189.
96. R. Høegh-Krohn and B. Simon, Hypercontractive semigroup and two dimensional self-coupled Bose fields, *J. Funct. Anal.* **9** (1972), 121–180.
97. M. Hübner and H. Spohn, Radiative decay: nonperturbative approaches, *Rev. Math. Phys.* **7** (1995), 363–387.
98. M. Hübner and H. Spohn, Spectral properties of the spin-boson Hamiltonian, *Ann. Inst. Henri Poincaré* **62** (1995), 289–323.
99. M. Hübner and H. Spohn, The spectrum of the spin-boson Hamiltonian, *Operator Theory Adv. Appl.* **70** (1996), 233–338.
100. V. Jakšić and C. A. Pillet, On a model for quantum friction I. Fermi's golden rule and dynamics at zero temperature, *Ann. Inst. Henri Poincaré* **62** (1995), 47–68.
101. V. Jakšić and C. A. Pillet, On a model for quantum friction II. Fermi's golden rule and dynamics for positive temperature, *Commun. Math. Phys.* **176** (1996), 619–644.
102. V. Jakšić and C. A. Pillet, On a model for quantum friction III. Ergodic properties of the Spin-Boson system, *Commun. Math. Phys.* **178** (1996), 627–651.
103. H. Kalf and J. Walter, Strongly singular potentials and essentially self-adjointness of singular elliptic operators in $C_0^\infty(\mathbb{R}^n \setminus \{0\})$, *J. Funct. Anal.* **10** (1972), 114–130.

104. I. Karatzas and S. E. Shreve, *Brownian Motion and Stochastic Calculus*, Graduate Texts in Mathematics **113** (Springer-Verlag, 1997).
105. T. Kato, *Perturbation Theory for Linear Operators* (Springer-Verlag, 1966).
106. T. Kato, Schrödinger operator with singular potentials, *Israel J. Math.* **13** (1972), 135–148.
107. T. Kato, Remarks on Schrödinger operators with vector potentials, *Integral Eq. Op. Theory* **1** (1978), 103–113.
108. T. Kato and K. Masuda, Trotter's product formula for nonlinear semigroups generated by the subdifferentiables of convex functionals, *J. Math. Soc. Japan* **30** (1978), 169–178.
109. Y. Kato and N. Mugibayashi, Regular perturbation and asymptotic limits of operators in quantum field theory, *Prog. Theoret. Phys.* **30** (1963), 103–133.
110. A. Klein and L. J. Landau, Stochastic process associated with KMS states, *J. Funct. Anal.* **42** (1981), 368–428.
111. W. Lamb and R. Retherford, Fine structure of the hydrogen atom by a microwave method, *Phys. Rev.* **72** (1947), 241–243.
112. E. Lieb, Bounds on the eigenvalues of the Laplace and Schrödinger operators, *Bull. Amer. Math. Soc.* **82** (1976), 751–753.
113. E. Lieb and M. Loss, Self-energy of electrons in non-perturbative QED, preprint, 1999.
114. E. Lieb and M. Loss, A bound on binding energies and mass renormalization in models of quantum electrodynamics, *J. Stat. Phys.* **108** (2002) 1057–1069.
115. J. Lőrinczi and R. A. Minlos, Gibbs measures for Brownian paths under the effect of an external and a small pair potential, *J. Stat. Phys.* **105** (2001), 605–647.
116. J. Lőrinczi, R. A. Minlos and H. Spohn, The infrared behaviour in Nelson's model of a quantum particle coupled to a massless scalar field, *Ann. Henri Poincaré* **3** (2001), 1–28.
117. J. Lőrinczi, R. A. Minlos and H. Spohn, Infrared regular representation of the three dimensional massless Nelson model, *Lett. Math. Phys.* **59** (2002), 189–198.
118. P. Milonni, *The Quantum Vacuum* (Academic Press, 1994).
119. R. A. Minlos, Lower branch of the spectrum of a fermion interacting with a bosonic gas (polaron), *Teoreticheskaya i Matematicheskaya Fizika* **92** (1992), 225–268.
120. R. A. Minlos and H. Spohn, Existence and uniqueness of ground state of system: scalar field interacting with a particle, unpublished note, 1999.
121. E. Nelson, Interaction of nonrelativistic particles with a quantized scalar field, *J. Math. Phys.* **5** (1964), 1190–1197.
122. T. Okamoto and K. Yajima, Complex scaling technique in non-relativistic massive QED, *Ann. Int. Henri. Poincaré* **42** (1985), 311–327.
123. H. Osada and H. Spohn, Gibbs measures relative to Brownian motion, to be published in *Ann. Probab.*
124. W. Pauli and M. Fierz, Zur Theorie der Emission langwelliger Lichtquanten, *Nuovo Cimento* **15** (1938), 167–188; Theory of the emission of long-wave light quanta, *Early Quantum Electrodynamics* (in English), ed. by I. Miller, (Cambridge Univ. Press, 1995), 227–243.

125. S. N. M. Ruijsenaars, On Bogoliubov transformations for systems of relativistic charged particles, *J. Math. Phys.* **18** (1977), 517–526.
126. S. N. M. Ruijsenaars, On Bogoliubov transformations. II. The general case. *Ann. Phys.* **116** (1978), 105–134.
127. M. Reed and B. Simon, *Methods of Modern Mathematical Physics I* (Academic Press, 1980).
128. M. Reed and B. Simon, *Methods of Modern Mathematical Physics II* (Academic Press, 1975).
129. M. Reed and B. Simon, *Methods of Modern Mathematical Physics III* (Academic Press, 1979).
130. M. Reed and B. Simon, *Methods of Modern Mathematical Physics IV* (Academic Press, 1978).
131. S. Schweber, *QED and men who made it: Dyson, Feynman, Schwinger, and Tomonaga* (Princeton Univ. Press, 1976).
132. I. Segal, Construction of non-linear local quantum process: I, *Ann. Math.* **92** (1970), 462–481.
133. D. Shale, Linear symmetries of free boson fields, *Trans. Amer. Math. Soc.* **103** (1962), 149–167.
134. B. Simon, Ergodic semigroups of positivity preserving self-adjoint operators, *J. Funct. Anal.* **12** (1973), 335–339.
135. B. Simon, *The $P(\phi)_2$ Euclidean (Quantum) Field Theory* (Princeton Univ. Press, 1974).
136. B. Simon, *Functional Integral Representation and Quantum Physics* (Academic Press, 1979).
137. B. Simon, Schrödinger semigroups, *Bull. Amer. Math. Soc.* **7** (1982), 447–526; *J. Funct. Anal.* **32** (1979), 97–101.
138. H. Spohn, The polaron at large total momentum, *J. Phys. A: Math. Gen.* **21** (1988), 1199–1211.
139. H. Spohn, Effective mass of the polaron: A functional integral approach, *Ann. Phys.* **175** (1987), 278–318.
140. H. Spohn, Ground state(s) of the spin-boson Hamiltonian, *Commun. Math. Phys.* **123** (1989), 277–304.
141. H. Spohn, Asymptotic completeness for Rayleigh scattering, *J. Math, Phys.* **38** (1997), 2281–2296.
142. H. Spohn, Ground state of quantum particle coupled to a scalar boson field, *Lett. Math. Phys.* **44** (1998), 9–16.
143. H. Spohn, *The Dynamics of Charged Particles and Radiation*, mp-arc 99-316, preprint, 1999.
144. S. Teufel, Effective N-body dynamics for the massless Nelson model and adiabatic decoupling without spectral gap, *Ann. Henri Poincaré* **3** (2002), 939–965.
145. C. C. Tannoudji, J. D. Roc and G. Grynberg, *Photons and Atoms* (Wiley-International Publication, 1997).
146. T. Welton, Some observable effects of the quantum-mechanical fluctuations of the electromagnetic field, *Phys. Rev.* **74** (1948), 1157–1167.

Index